INTRODUCTORY STATISTICS

Introductory Statistics

THOMAS H. WONNACOTT

Associate Professor of Mathematics
University of Western Ontario

RONALD J. WONNACOTT

Professor of Economics
University of Western Ontario

JOHN WILEY & SONS, INC.

New York · London · Sydney · Toronto

10 9 8 7 6 5 4 3 2 1

Library of Congress Catalog Card Number: 69–16041
SBN 471 95965 0
Printed in the United States of America

To Monique and Eloise

Preface

Our objective has been to write a text that would come into the statistics market between the two texts written by Paul G. Hoel (or the two texts written by John E. Freund). We have tried to cover most of the material in their mathematical statistics books, but we have used mathematics only slightly more difficult than that used in their elementary books. Calculus is used only in sections where the argument is difficult to develop without it; although this puts the calculus student at an advantage, we have made a special effort to design these sections so that a student without calculus can also follow.

By requiring a little more mathematics than many other elementary texts we have been able to treat many important topics normally covered only by books in mathematical statistics: for example, the relation of sampling and inference to the theory of probability and random variables. Another objective has been to show the logical relation between topics that have often appeared in texts as separate and isolated chapters: for example, the equivalence of interval estimation and hypothesis testing, of the t test and F test, and of analysis of variance and regression using dummy variables. In every case our motivation has been twofold: to help the student appreciate—indeed enjoy—the underlying logic, and to help him arrive at answers to practical problems.

We have placed high priority on the regression model, not only because regression is widely regarded as the most powerful tool of the practicing statistician, but also because it provides a good focal point for understanding such related techniques as correlation and analysis of variance.

Our original aim was to write an introduction to statistics for economic students, but as our efforts increased, so it seems did our ambitions. Accordingly, this book is now written for students in economics and other social sciences, for business schools, and for service courses in statistics provided by mathematics departments. Some of the topics covered are typically omitted from introductory courses, but are of interest to such a broad audience: for example, multiple comparisons, multiple regression, Bayesian decisions, and game theory.

A statistics text aimed at several audiences—including students with and without calculus—raises major problems of evenness and design. The text itself is kept simple, with the more difficult interpretations and developments reserved for footnotes and starred sections. In all instances these are optional; a special effort has been made to allow the more elementary student to skip these completely without losing continuity. Moreover, some of the finer points are deferred to the instructor's manual. Thus the instructor is allowed, at least to some degree, to tailor the course to his students' background.

Problems are also starred (*) if they are more difficult, or set with an arrow (⇒) if they introduce important ideas taken up later in the text, or bracketed ( ) if they duplicate previous problems, and thus provide optional exercise only.

Our experience has been that this is about the right amount of material for a two-semester course; a single semester introduction is easily designed to include the first 7, 8, or 9 chapters. We have also found that majors in economics who may be pushed a bit harder can cover the first 10 chapters in one semester. This has allowed us in the second semester to use our forthcoming Econometrics text which provides more detailed coverage of the material in Chapters 11 to 15 of this book, plus additional material on serial correlation, identification, and other econometric problems.

So many have contributed to this book that it is impossible to thank them all individually. However, a special vote of thanks should go, without implication, to the following for their thoughtful reviews: Harvey J. Arnold, David A. Belsley, Ralph A. Bradley, Edward Greenberg, Leonard Kent, R. W. Pfouts, and especially Franklin M. Fisher. We are also indebted to our teaching assistants and the students in both mathematics and economics at the University of Western Ontario and Wesleyan (Connecticut) who suggested many improvements during a two-year classroom test.

London, Ontario, Canada
September, 1968

Thomas H. Wonnacott
Ronald J. Wonnacott

Contents

chapter I

Introduction

The word "statistics" originally meant the collection of population and economic information vital to the state. From that modest beginning, statistics has grown into a scientific method of analysis now applied to all the social and natural sciences, and one of the major branches of mathematics. The present aims and methods of statistics are best illustrated with a familiar example.

1-1 EXAMPLE

Before every presidential election, the pollsters try to pick the winner; specifically, they try to guess the proportion of the population that will vote for each candidate. Clearly, canvassing all voters would be a hopeless task. As the only alternative, they survey a sample of a few thousand in the hope that the sample proportion will be a good estimate of the total population proportion. This is a typical example of *statistical inference or statistical induction:* the (voting) characteristics of an unknown population are inferred from the (voting) characteristics of an observed sample.

As any pollster will admit, it is an uncertain business. To be *sure* of the population, one has to wait until election day when all votes are counted. Yet if the sampling is done fairly and adequately, we can have high hopes that the sample proportion will be close to the population proportion. This allows us to estimate the unknown population proportion π from the observed sample proportion (P), as follows:

$$\pi = P \pm \text{a small error} \tag{1-1}$$

with crucial questions being, "How small is this error?" and "How sure are we that we are right?"

Since this typifies the very core of the book, we state it more precisely in the language of Chapter 7 (where the reader will find the proof and a fuller understanding).

If the sampling is random and large enough, we can state with 95% confidence that

$$\pi = P \pm 1.96\sqrt{\frac{P(1-P)}{n}} \tag{1-2}$$

where π and P are the population and sample proportion, and n is the sample size.

As an illustration of how this formula works, suppose we have sampled 1,000 voters, with 600 choosing the Democratic candidate. With this sample proportion of .60, equation (1-2) becomes

$$\pi = .60 \pm 1.96\sqrt{\frac{.60(1-.60)}{1000}}$$

or approximately

$$\pi = .60 \pm .03 \tag{1-3}$$

Thus, with 95% confidence, we estimate the population proportion voting Democrat to be between .57 and .63.

This is referred to as a *confidence interval*, and making estimates of this kind will be one of our major objectives in this book. The other objective is to *test hypotheses*. For example, suppose we wish to test the hypothesis that the Republican candidate will win the election. On the basis of the information in equation (1-3) we would reject this claim; it is no surprise that a sample result that pointed to a Democratic majority of 57 to 63% of the vote will also allow us to reject the hypothesis of a Republican victory. In general, there is a very close association of this kind between confidence intervals and hypothesis tests; indeed, we will show that in many instances they are equivalent procedures.

We pause to make several other crucial observations about equation (1-3).

1. The estimate is *not* made *with certainty;* we are only 95% confident. We must concede the possibility that we are wrong—and wrong because we were unlucky enough to draw a misleading sample. Thus, even if less than half the population is in fact Democratic, it is still possible, although unlikely, for us to run into a string of Democrats in our sample. In such circumstances, our conclusion (1-3) would be dead wrong. Since this sort of bad luck is possible, but not likely, we can be 95% confident of our conclusion.

2. Luck becomes less of a factor as sample size increases; the more voters we canvass, the less likely we are to draw a predominantly Democratic

sample from a Republican population. Hence, the more precise our prediction. Formally, this is confirmed in equation (1-2); in this formula we note that the error term decreases with sample size. Thus, if we increased our sample to 10,000 voters, and continued to observe a Democratic proportion of .60, our 95% confidence interval would become the more precise:

$$.60 \pm .01 \tag{1-4}$$

3. Suppose our employer indicates that 95% confidence is not good enough. "Come back when you are 99% sure of your conclusion." We now have two options. One is to increase our sample size; as a result of this additional cost and effort we will be able to make an interval estimate with the precision of (1-4) but at a higher level of confidence. But if the additional resources for further sampling are not available, then we can increase our confidence only by making a less precise statement—i.e., that the proportion of Democrats is

$$.60 \pm .02$$

The less we commit ourselves to a precise prediction, the more confident we can be that we are right. In the limit, there are only two ways that we can be certain of avoiding an erroneous conclusion. One is to make a statement so imprecise that it cannot be contradicted.[1] The other is to sample the whole population[2]; but this is not statistics—it is just counting. Meaningful statistical conclusions must be prefaced by some degree of uncertainty.

1-2 INDUCTION AND DEDUCTION

Figure 1-1 illustrates the difference between inductive and deductive reasoning. Induction involves arguing from the specific to the general, or (in our case) from the sample to the population. Deduction is the reverse—arguing from the general to the specific, i.e., from the population to the sample.[3] Equation (1-1) represents inductive reasoning; we are arguing from a sample proportion to a population proportion. But this is only possible

[1] E.g., $\pi = .50 \pm .50$.

[2] Or, almost the whole population. Thus it would not be necessary to poll the whole population to determine the winner of an election; it would only be necessary to continue canvassing until one candidate comes up with a majority. (It is always possible, of course, that some people change their mind between the sample survey and their actual vote, but we don't deal with this issue here.)

[3] The student can easily keep these straight with the help of a little Latin, and recognition that the population is the point of reference. The prefix *in* means "into" or "towards." Thus *in*duction is arguing towards the population. The prefix *de* means "away from." Thus *de*duction is arguing away from the population. Finally, statistical *in*ference is based on *in*duction.

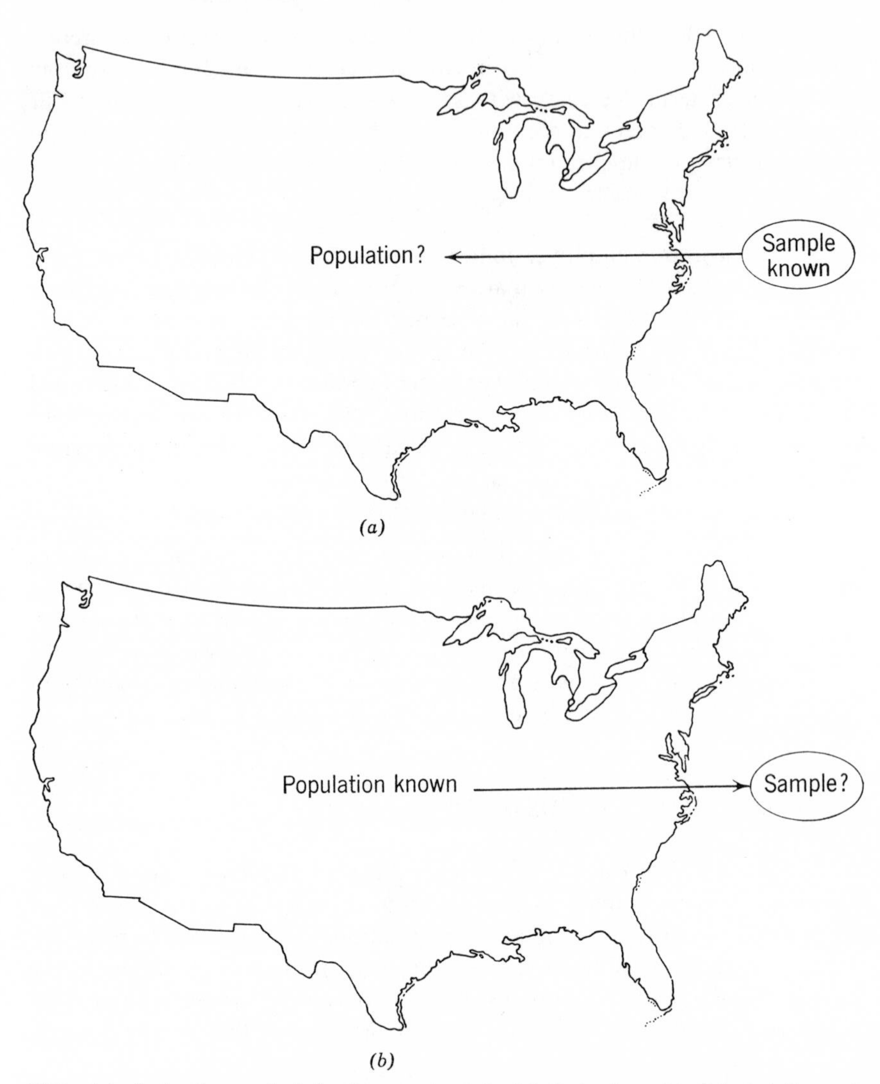

FIG. 1-1 Induction and deduction contrasted. (*a*) Induction (statistical inference). (*b*) Deduction (probability).

if we study the simpler problem of deduction first. Specifically, in equation (1-1), we note that the *inductive* statement (that the population proportion can be inferred from the sample proportion) is based on a prior *deduction* (that the sample proportion is likely to be close to the population proportion).

Chapters 2 through 5 are devoted to deduction. This involves, for example, the study of probability, which is useful for its own sake, (e.g., in

Game Theory); but it is even more useful as the basis for statistical induction dealt with in Chapters 7 through 10. In short, in the first 6 chapters we ask, "With a given population, how will a sample behave? Will the sample be 'on target'?" Only when this deductive issue is resolved can we move to questions of statistical inference. This involves, in the later chapters, turning the argument around and asking "How precisely can we make inferences about an unknown population from an observed sample?"

1-3 WHY SAMPLE?

We sample, rather than study the whole population, for any one of three reasons.

(1) Limited resources.
(2) Limited data available.
(3) Destructive testing.

1. Limited resources almost always play some part. In our example of preelection polls, funds were not available to observe the whole population; but this is not the only reason for sampling.

2. Sometimes there is only a small sample available, no matter what cost may be incurred. For example, an anthropologist may wish to test the theory that the two civilizations on islands A and B have developed independently, with their own distinctive characteristics of weight, height, etc. But there is no way in which he can compare the two civilizations *in toto*. Instead he must make an inference from the small sample of the 50 surviving inhabitants of island A and the 100 surviving inhabitants of island B. The sample size is fixed by nature, rather than by the researcher's budget.

There are many examples in business. An allegedly more efficient machine may be introduced for testing, with a view to the purchase of additional similar units. The manager of quality control simply cannot wait around to observe the entire population this machine will produce. Instead a sample run must be observed, with the decision on efficiency based on an inference from this sample.

3. Sampling may involve destructive testing. For example, suppose we have produced a thousand light bulbs and wish to know their average life. It would be folly to insist on observing the whole population of bulbs until they burn out.

1-4 HOW TO SAMPLE

In statistics, as in business or any other profession, it is essential to distinguish between bad luck and bad management. For example, suppose a man bets you $100 at even odds that you will get an ace (i.e., 1 dot) in rolling a die. You accept the challenge, roll an ace, and he wins. He's a bad manager and you're a good one; he has merely overcome his bad management with extremely good luck. Your only defense against this combination is to get him to keep playing the game—with your dice.

If we now return to our original example of preelection polls, we note that the sample proportion of Democrats may badly misrepresent the population proportion for either (or both) of these reasons. No matter how well managed and designed our sampling procedure may be, we may be unlucky enough to turn up a Democratic sample from a Republican population. Equation (1-2) relates to this case; it is assumed that the only complication is the luck of the draw, and not mismanagement. From that equation we confirm that the best defense against bad luck is to "keep playing"; by increasing our sample size, we improve the reliability of our estimate.

The other problem is that sampling can be badly mismanaged or biased. For example, in sampling a population of voters, it is a mistake to take their names from a phone book, since poor voters who often cannot afford telephones are badly underrepresented.

Other examples of biased samples are easy to find and often amusing. "Straw polls" of people on the street are often biased because the interviewer tends to select people that seem civil and well dressed; the surly worker or harassed mother is overlooked. A congressman can not rely on his mail as an unbiased sample of his constituency, for this is a sample of people with strong opinions, and includes an inordinate number of cranks and members of pressure groups.

The simplest way to ensure an unbiased sample is to give each member of the population an equal chance of being included in the sample. This, in fact, is our definition of a "random" sample.[4] For a sample to be random, it cannot be chosen in a sloppy or haphazard way; it must be carefully designed. A sample of the first thousand people encountered on a New York street corner will not be a random sample of the U.S. population. Instead, it is necessary to draw some of our sample from the West, some from the East, and so on. Only if our sample is randomized, will it be free of bias and, equally important, only then will it satisfy the assumptions of probability theory, and allow us to make scientific inferences of the form of (1-2).

In some circumstances, the only available sample will be a nonrandom one. While probability theory often cannot be strictly applied to such a sample, it still may provide the basis for a good educated guess—or what we might term the *art* of inference. Although this art is very important, it cannot be taught in an elementary text; we, therefore, consider only scientific

[4] Strictly speaking, this is called "simple random sampling," to distinguish it from more complex types of random sampling.

inference based on the assumption that samples are random. The techniques for ensuring this are discussed further in Chapter 6.

FURTHER READINGS

For readers who wish a more extensive introduction to Statistics, we highly recommend the following.

1. Huff, Darrell, "How to Lie with Statistics." New York: Norton, 1954.
2. Huff, Darrell, "How to Take a Chance." New York: Norton, 1957.
3. Wallis, W. A., and Roberts, H. V., "The Nature of Statistics." Free Press Paperback, 1956.
4. McDonald, J., and Osborn, R.,"Strategy in Poker, Business, and War." New York: Norton, 1950.
5. Slonim, M. J., "Sampling." Simon and Shuster Paperback, 1966.

chapter 2

Descriptive Statistics for Samples

2-1 INTRODUCTION

We have already discussed the primary purpose of statistics—to make an inference to the whole population from a sample. As a preliminary step, the sample must be simplified, and reduced to a few descriptive numbers; each is called a sample *statistic*.[1]

In the very simple example of Chapter 1, the pollster would record the answers of the 1000 people in his sample, obtaining a sequence such as D D R D R . . . . where D and R represent Democrat and Republican. The best way of describing this sample by a single number is the statistic P, the sample proportion of Democrats; this will be used to make an inference about π, the population proportion. Admittedly, this statistic is trivial to compute. In the sample of the previous chapter, computing the sample proportion (.60) required only a count of the number voting Democrat (600), followed by a division by sample size, ($n = 1{,}000$).

We now turn to the more substantial computations of statistics to describe two other samples.

(a) The results when a die is thrown 50 times.
(b) The average height of a sample of 200 American men.

2-2 FREQUENCY TABLES AND GRAPHS

(a) Discrete Example

Each time we toss the die, we record the number of dots X, which takes on the values $1, 2, \ldots, 6$. X is called a "discrete" random variable because it assumes only a finite (or countably infinite) number of values.

[1] Later, we shall have to define a statistic more rigorously; but for now, this will suffice.

TABLE 2-1 Results of Tossing a Die 50 Times

6, 2, 2, 3, 5, 1, 2, 6, ... 4, 2.

The 50 throws yield a string of 50 numbers such as given in Table 2-1.

To simplify, we keep a running tally of each of the six possible outcomes in Table 2-2. In column 3 we note that 9 is the frequency f (or total number of times) that we rolled a 1; i.e., we obtained this outcome on 9/50 of our tosses. Formally, this proportion (.18) is called relative frequency (f/n); it is computed in column 4.

TABLE 2-2 Calculation of the Frequency, and Relative Frequency of the Number of Dots in 50 Tosses of a Die

(1) Number of Dots	(2) Tally	(3) Frequency (f)	(4) Relative Frequency (f/n)
1	卌 ||||	9	.18
2	卌 卌 ||	12	.24
3	卌 |	6	.12
4	卌 |||	8	.16
5	卌 卌	10	.20
6	卌	5	.10
		$\sum f = 50 = n$	$\sum (f/n) = 1.00$

where $\sum f$ is "the sum of all f"

The information in column 3 is called a "frequency distribution," and is graphed in Figure 2-1. The "relative frequency distribution" in column 4 can be similarly graphed; the student who does so will note that the two graphs are identical except for the vertical scale. Hence, a simple change of vertical scale transforms Figure 2-1 into a relative frequency distribution. This now gives us an immediate picture of the sample result.

(b) Continuous Example

Suppose that a sample of 200 men is drawn from a certain population, with the height of each recorded in inches. The ultimate aim will be an inference about the average height of the whole population; but first we must efficiently summarize and describe our sample.

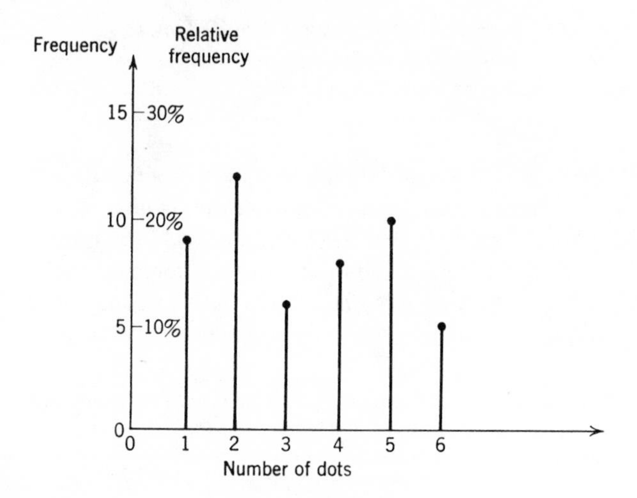

FIG. 2-1 Frequency and relative frequency distribution of the results of a sample of 50 tosses of a die.

In this example, height (in inches) is our random variable X. In this case, X is continuous; thus an individual's height might be any value, such as 64.328 inches.[2] It no longer makes sense to talk about the frequency of this specific value of X; chances are we'll never again observe anyone exactly 64.328 inches tall. Instead we can tally the frequency of heights within a

TABLE 2-3 Frequency, and Relative Frequency of the Heights of a Sample of 200 Men

Cell No.	(1) Cell Boundaries	(2) Cell Midpt	(3) Tally	(4) Frequency, f	(5) Relative Frequency f/n
1	55.5–58.5	57	||	2	.010
2	58.5–61.5	60	𝍸 ||	7	.035
3	61.5–64.5	63	𝍸 𝍸 𝍸 𝍸 ||	22	.110
4	.	66	.	13	.065
5	.	69	.	44	.220
6	.	72	.	36	.180
7	.	75	.	32	.160
8	.	78	.	13	.065
9	.	81	.	21	.105
10	82.5–85.5	84	.	10	.050
				$\sum f = 200 = n$	$\sum f/n = 1.00$

[2] We shall overlook the fact that although height is conceptually continuous, in practice the measured height is rounded to a few decimal places at most, and is therefore discrete.

class or cell, (e.g., 58.5″ to 61.5″) as in column 3 of Table 2-3. Then the frequency and relative frequency are tabulated as before.

The cells have been chosen somewhat arbitrarily, but with the following conveniences in mind.

1. The number of cells is a reasonable compromise between too much detail and too little.

2. Each cell midpoint, which hereafter will represent all sample values in the cell, is a convenient whole number.

The grouping of the 200 observations into cells is illustrated in Figure 2-2, where each observation is represented by a dot. For simplicity, we have assumed that the observations are recorded exactly, rather than being

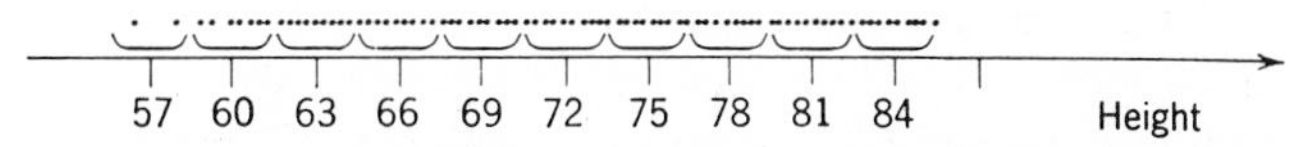

FIG. 2-2 The grouping of observations into cells, illustrating the first two columns of Table 2-3.

rounded off. (Rounding, to the nearest integer, for example, may in fact be regarded as a preliminary grouping into cells of width 1.)

The grouped data is then graphed in Figure 2-3. This frequency distribution, or so-called histogram, uses bars to represent frequencies as a reminder that the observations occurred throughout the cell, and not just at the midpoint.

We now turn to the question of how we may characterize a sample frequency distribution with a single descriptive measure, or sample statistic.

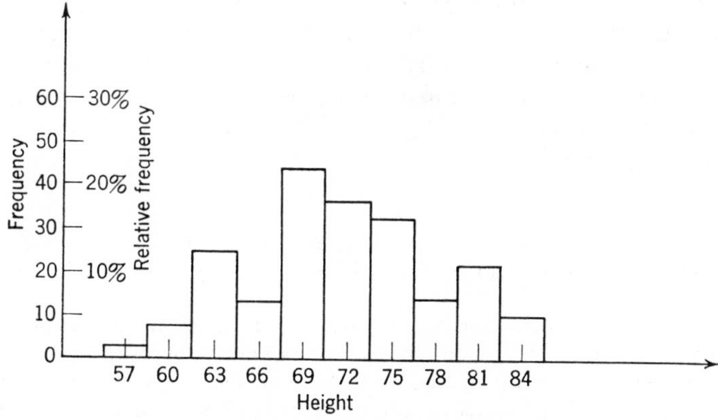

FIG. 2-3 The frequency and relative frequency distribution of a sample of 200 men.

In fact, there are two highly useful descriptions: the first is the central point of the distribution and the second is its spread.

2-3 CENTERS (MEASURES OF LOCATION)

There are several different concepts of the "center" of a frequency distribution. Three of these, the mode, the median, and the mean, are discussed below. We shall start with the simplest.

(a) The Mode[3]

This is defined as the most frequent value. In our example of heights, the mode is 69 inches, since this cell has the greatest frequency, or highest bar in Figure 2-3. Generally, the mode is *not* a good measure of central tendency, since it often depends on the arbitrary grouping of the data. (The student will note that, by redefining cell boundaries, the mode can be shifted up or down considerably.) It is also possible to draw a sample where the largest frequency (highest bar in the group) occurs at two (or even more) heights; this unfortunate ambiguity is left unresolved, and the distribution is "bimodal."

(b) The Median

This is the 50th percentile; i.e., the value below which half the values in the sample fall. Since it splits the observations into two halves, it is sometimes called the middle value. In the sample of 200 shown in Figure 2-2, the median (say, 71.46) is most easily derived by reading off the 100th value[4] from the left; but if the only information available is the frequency distribution in Figure 2-3, it must be calculated choosing an appropriate value within the median cell.[5]

---

[3] "Mode" means fashion, in French.

[4] Or 101st value. This ambiguity is best resolved by defining the median as the average of the 100th and 101st values. In a sample with an odd number of observations, this ambiguity does not arise.

[5] The median cell is clearly the 6th, since this leaves 44% (i.e., 88) of the sample values below and 38% (i.e., 76) above. The median value can be closely approximated by moving through this median cell from left to right to pick up another 6% of the observations. Since this cell includes 18% of the observations, we move 6/18 of the way through this cell. Thus our median approximation is $70.5 + (6/18 \times 3) = 71.5$.

(c) The Mean ($\bar{X}$)

This is sometimes called the arithmetic mean, or simply the average. This is the most common central measure. The original observations (X_1, $X_2, \ldots, X_n$) are simply summed, then divided by n. Thus

$$\bar{X} \triangleq \frac{1}{n}(X_1 + X_2 + \cdots + X_n)$$

Definition.

$$\boxed{\bar{X} \triangleq \frac{1}{n}\sum_{i=1}^{n} X_i} \tag{2-1a}$$

where X_i represents the ith value of X, and $\triangleq$ means "equals, by definition."

The average height of our sample could be computed by summing all 200 observations and dividing by 200. However, this tedious calculation can be greatly simplified by using the grouped data in Table 2-3. Let f_1 represent the number of observations in cell 1, where each observation may be approximated[6] by the cell midpoint, x_1. Similar approximations hold for all the other cells too, so that

$$\bar{X} \simeq \frac{1}{n}\Big\{\underbrace{(x_1 + x_1 + \cdots + x_1)}_{f_1 \text{ times}} + \underbrace{(x_2 + x_2 + \cdots x_2)}_{f_2 \text{ times}} + \cdots + \underbrace{(x_{10} + \cdots x_{10})}_{f_{10} \text{ times}}\Big\}$$

where $\simeq$ represents approximate equality; it follows that

$$\begin{aligned}\bar{X} &\simeq \frac{1}{n}\{f_1 x_1 + f_2 x_2 + \cdots f_{10} x_{10}\} \\ &\simeq \frac{f_1}{n} x_1 + \frac{f_2}{n} x_2 + \cdots \frac{f_{10}}{n} x_{10} \\ &\simeq \sum_{i=1}^{10} \left(\frac{f_i}{n}\right) x_i\end{aligned}$$

In general

$$\bar{X} \simeq \sum_{i=1}^{m} x_i \left(\frac{f_i}{n}\right) \tag{2-1b}$$

[6] In approximating each observed value by the midpoint of its cell, we sometimes err positively, sometimes negatively; but unless we are very unlucky, these errors will tend to cancel. Even in the unluckiest case, however, the error must be smaller than half the cell width. Note that cell midpoints are designated by the small x_i, to distinguish them from the observed values X_i.

TABLE 2-4 Calculation of Mean and Mean Squared Deviation of a Sample of 200 Men's Heights[a]

Given		Calculation of $\overline{X}$ using (2-1b)	Calculation of MSD using 2-5(b), and s^2			Easier Calculation of s^2 using (2-7b)	
(1) x_i	(2) f_i/n	(3) $x_i(f_i/n)$	(4) $(x_i - \overline{X})$	(5) $(x_i - \overline{X})^2$	(6) $(x_i - \overline{X})^2(f_i/n)$	(7) x_i^2	(8) $x_i^2 f_i$
57	.010	.570	−14.80	218.	2.18	3,249	6,498
60	.035	2.100	−11.80	139.	4.88	3,600	25,200
63	.110	6.930	−8.80	77.	8.47	3,969	873,180
.	.	.	.	.			
.	.	.	.	.			
.	.	.	.	.			
84	.050	4.200	+12.20	149.	7.45	7,056	70,560

$$\overline{X} \simeq \sum x_i\left(\frac{f_i}{n}\right) = \boxed{71.80}$$

$$\sum (x_i - \overline{X})\left(\frac{f_i}{n}\right) = 0$$

$$MSD \simeq \sum (x_i - \overline{X})^2\left(\frac{f_i}{n}\right) = \boxed{40.0}$$

Comparing (2-5a) and (2-6):

$$s^2 = MSD\left(\frac{n}{n-1}\right) = 40\left(\frac{200}{199}\right) = \boxed{40.2}$$

$$s = \sqrt{40.2} = \boxed{6.35}$$

$$\sum x_i^2 f_i = 1,039,000$$

$$n\overline{X}^2 = 1,031,000$$

$$s^2 = \frac{1}{199}\{8,000\} = \boxed{40.2}$$

[a] *Warning*. This computation is very tedious. Just verify a few calculations rather than all the detail, since an easier method is shown later in Table 2-5.

where (f_i/n) = relative frequency in the ith cell, and m = number of cells. We number this equation (2-1b) to emphasize that it is the equivalent formulation of (2-1a), appropriate for grouped data. In our example, the calculation of (2-1b) is based on the data in Table 2-3, and is shown in column 3 of Table 2-4. We can think of this as a "weighted" average, with each x value weighted appropriately by its relative frequency.

(d) Comparison of Mean, Median and Mode

These three measures of center are compared in Figure 2-4. In part *a* we show a distribution which has a single peak and is symmetric (i.e., one half is the mirror image of the other); in this case all three central measures coincide. But when the distribution is skewed to the right as in *b*, the median

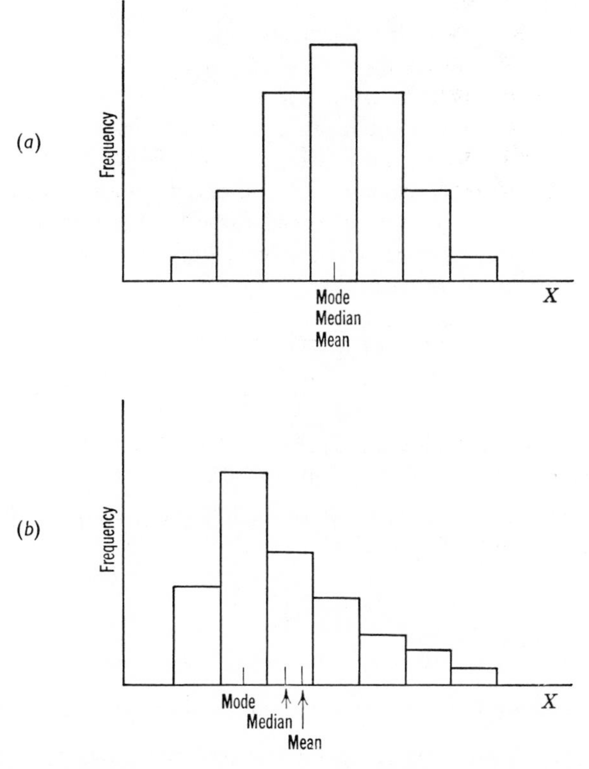

FIG. 2-4(*a*) A symmetric distribution with a single peak. The mode, median, and mean coincide at the point of symmetry. (*b*) A right-skewed distribution, showing mode < median < mean.

falls to the right of the mode; with the long scatter of observed values strung out in the right hand tail, it is generally necessary to move from the mode to the right to pick up half the observations. Moreover, the mean will generally lie even further to the right, as explained in the next section.

Interpreting the Mean by an Analogy from Physics. The 200 observations in the sample of heights appear in Figure 2-2 as points along the X-axis. If we think of these observations as masses (each observation a one pound mass, for example), and the X-axis as a weightless supporting rod, we might ask where this rod balances. Our intuition suggests "the center."

The precise balancing point, also called the center of gravity, is given by the formula

$$\frac{1}{n}\sum X_i$$

which is exactly the formula for the mean. Thus we are quite justified in thinking of the sample mean as the "balancing point" of the data, and representing it in graphs as a fulcrum ▲.

It can easily be seen why the mean lies to the right of the median in a right-skewed distribution, as shown in Figure 2-4*b*. Experiment by trying to balance at the median. Fifty percent of the observed values now lie on either side, but the observations to the right tend to be further distant, tilting the distribution to the right. Balance can be achieved only by placing the fulcrum (mean) to the right of the median.

PROBLEMS

2-1 Show the mean, mode, and median in our example in Figure 2-3. Is the mode a good central measure in this case?

2-2 Find the mean, median, and mode of the following sample of litter sizes. Graph the frequency distribution.

7	4	10	9	15	12	7
8	11	4	14	10	5	14
1	10	8	12	6	5	

2-3 Sort the following data into 8 cells, whose midpoints are 55, 60 . . . 90.

55.31	81.47	64.90	70.88	86.02	77.25	76.73	84.21	56.02
84.92	90.23	78.01	88.05	73.37	87.09	57.41	85.43	
74.76	86.51	86.37	76.15	88.64	84.71	66.05	83.91	

Approximately what are the mean, median, and mode? Graph the frequency distribution.

2-4 Sort the data of Problem 2-3 into 4 cells, whose midpoints are 60, 70, 80, 90. Then answer the same questions as in Problem 2-3.

2-5 Summarize the answers to the previous two problems in the following table.

	Mean	Median	Mode
Original data (exact values)	77.78	81.47	Not defined
—Fine grouping (Problem 2-3)			
—Coarse grouping (Problem 2-4)			

(a) Do you see why the mode is not a good measure?

(b) Will coarse grouping *always* give worse approximations (for the mean and median) than fine grouping, or will it do so usually?

2-4 DEVIATIONS (MEASURES OF SPREAD)

Although average height may be the most important characteristic (statistic) of the sample, it is also important to know how spread out or varied are the sample observations.

As with measures of center, we find that there are several measures of spread; we start with the simplest.

(a) *The range* is simply the distance between the largest and smallest value.

$$\text{Range} \triangleq \text{largest-smallest observation}$$

For men's heights, the range is 30. It may be fairly criticized on the grounds that it tells us nothing about the distribution except where it ends. And these two extreme values may be very unreliable. We therefore turn to measures of spread which take account of all observations.

The average deviation, as its name implies, is found by calculating the deviation of each observed value (X_i) from the mean ($\bar{X}$); these deviations ($X_i - \bar{X}$) are then averaged by summing and dividing by n. Although this sounds like a promising measure, in fact it is worthless; positive deviations always cancel negative deviations, leaving an average of zero.[7] This sign problem can be avoided by ignoring all negative signs and taking the average of the *absolute* values of the deviations, as follows.

$$\text{(b) The Mean Absolute Deviation} \triangleq \frac{1}{n}\sum_{i=1}^{n} |X_i - \bar{X}| \qquad (2\text{-}4)$$

Intuitively, this is a good measure of spread; the problem is that it is mathematically intractable.[8] We therefore turn to an alternative means of avoiding the sign problem—namely, squaring each deviation.

$$\text{(c)} \quad \textit{Mean Squared Deviation}\ \text{(MSD)} \triangleq \frac{1}{n}\sum_{i=1}^{n} (X_i - \bar{X})^2 \qquad (2\text{-}5a)$$

If we use grouped data as in Table 2-4, this formula becomes

$$\text{Mean Squared Deviation} \simeq \sum_{i=1}^{m} (x_i - \bar{X})^2 \left(\frac{f_i}{n}\right) \qquad (2\text{-}5b)$$

This is a good measure, provided we wish only to describe the sample. But typically we shall want to go one step further, and use this to make a statistical inference about the population. For this purpose it is better to use the divisor $n - 1$ rather than[9] n. The resulting sample statistic is referred to as the variance.

$$\text{(d)} \quad \textit{Variance},\ s^2 \triangleq \frac{1}{n-1}\sum_{i=1}^{n} (X_i - \bar{X})^2 \qquad (2\text{-}6)$$

---

[7] This may be proved as follows:

$$\text{Average deviation} \triangleq \frac{1}{n}\sum_{i=1}^{n}(X_i - \bar{X}) \qquad (2\text{–}2)$$

$$= \frac{1}{n}\left(\sum X_i - \sum \bar{X}\right)$$

$$= \frac{1}{n}\sum X_i - \frac{1}{n}(n\bar{X})$$

$$= \bar{X} - \bar{X} = 0$$

$$\text{Average deviation} = 0 \qquad (2\text{-}3)$$

[8] One difficulty is the problem of differentiating the absolute value function.

[9] Technically, this makes the sample variance an unbiased estimator of the population variance. See Chapter 7.

TABLE 2-5 Coded Computation of Mean and Standard Deviation of a Sample of 200 Men's Heights (Compare with Table 2-4)

	Coding		For $\overline{Y}$	For s_Y^2, using easy calculation (2-7)
(1)	(2)	(3)	(4)	(5)
x_i	$y_i = \dfrac{x_i - 69}{3}$	f_i	$f_i y_i$	$f_i y_i^2$
57	−4	2	−8	+32
60	−3	7	−21	+63
63	−2	22	−44	+88
66	−1	13	−13	+13
69	0	44	0	0
72	1	36	36	36
75	2	32	64	128
78	3	13	39	117
81	4	21	84	336
84	5	10	50	250
			$\sum f_i y_i = 187$	$\sum f_i y_i^2 = 1063$
			$\overline{Y} = \dfrac{\sum f_i y_i}{n} = \dfrac{187}{200}$	$n\overline{Y}^2 = 175$
			$= .935$	$s_Y^2 = \dfrac{1}{199}(888)$
			$\overline{X} = 3\overline{Y} + 69$	$= 4.46$
			$\overline{X} = \boxed{71.80}$	$s_X = 3s_Y$
				$= 3\sqrt{4.46}$
				$s_X = \boxed{6.35}$

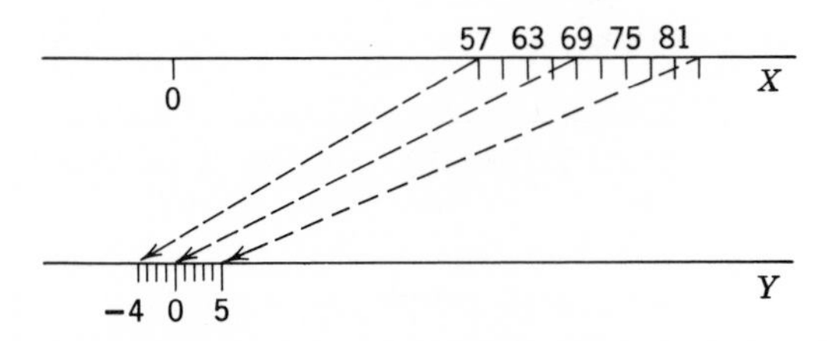

FIG. 2-7 Coding from inches (X) into trintals (Y), involving both a change of origin and of scale.

Moreover, it is evident that when $X_i = 69$, $Y_i = 0$. Furthermore, as X_i progresses by steps of 3, Y_i progresses in unit steps. With these guidelines we can fill in the appropriate Y values in column 2 of Table 2-5; diagrammatically, this coding is illustrated in Figure 2-7.

2. Compute the mean and standard deviation of the Y values. We note in the successive columns of Table 2-5 how easily this is now done. With $\overline{Y}$ and s_Y now in hand, we are in a position to:

3. Translate this mean and standard deviation back into X values. This involves applying the theory of linear transformations (Theorem III) to (2-27).

$$\overline{Y} = \left(\frac{-69}{3}\right) + \tfrac{1}{3}\overline{X} = \frac{\overline{X} - 69}{3} \tag{2-28}$$

and

$$s_Y = \tfrac{1}{3}s_X \tag{2-29}$$

From (2-28)

$$\begin{aligned} \overline{X} &= 3\overline{Y} + 69 \\ &= 71.80 \end{aligned} \tag{2-30}$$

From (2-29)

$$\begin{aligned} s_X &= 3s_Y \\ &= 6.35 \end{aligned} \tag{2-31}$$

Thus the simple coded computation of $\overline{X}$ and s_X is complete.

PROBLEMS

(2-9) By coding the heights shown in Table 2-5 from inches (X) into feet (Y), compute $\overline{X}$ and s_X. Show your linear transformation with a diagram similar to Figure 2-7. Why is the coding used in the text preferred?

2-10 Use coding to find the mean and standard deviation of the data in Problem 2-4.

2-11 Find the mean of the following:

239510	239250	239860	239360
239480	239430	239230	239680
239370	239290	239850	

(*Hint.* It is natural to simply drop the first 3 digits of every number, and just work with the numbers 510, 250, ... This is mathematically justified—it is just the linear transformation $Y = X - 239{,}000$)

2-12 To show that nonlinear transformations are trickier, see if this is true. If $Y = X^2$, then $\bar{Y} = \bar{X}^2$, when there are three values of X_i—1, 3, 5.

(2-13) Using coding, find the mean and standard deviation of
(a) The data of Problem 2-3.
(b) The data of Problem 2-2.

(2-14) Find the mean and standard deviation of the following sample of 50 executive ages. Graph the relative frequency distribution.

35	46	63	69	54	50	62	68	38	40
55	43	42	59	45	44	57	47	48	46
43	64	49	36	59	60	42	60	42	38
51	50	66	63	57	56	51	38	61	54
50	44	48	69	64	37	56	53	62	52

Review Problems

2-15 The weekly wage rates for 5 major industrial groupings are listed below. Find the average weekly wage.

Industry	A	B	C	D	E
% of employment	30%	25	20	20	5
Weekly wage	$120	150	120	100	80

2-16 Suppose the number of children was recorded for each of 25 families, obtaining the following data:

2, 4, 1, 0, 1, 3, 0, 4, 2, 6, 0, 0, 2, 3, 1, 5, 4,
0, 3, 1, 2, 5, 3, 4, 1.

(a) Construct a frequency table and graph.
(b) Find the sample mean and standard deviation.

(2-17) The following table* gives the actual percent of farmland that was harvested (as opposed to pasture, woodlot, etc.) in the U.S.A. in 1959, according to region. Compute the percentage harvested in the U.S.A. as a whole.

* *Source.* Statistical Abstract of the United States, 1963, pp. 625, 614.

Region	Amount of Farmland (millions of acres)	Percent Harvested
North	421	46.7
South	357	21.0
Mountain	264	8.7
Pacific	80	18.8
U.S.A.	1,122	?

2-18 A certain species of beetle was sampled, yielding the following 10 lengths, in centimeters: 1.5, 1.0, 1.2, 1.0, 1.1, 1.0, 1.6, 1.2, 1.4, 2.0. Find the median, mean, range, variance, and standard deviation

(a) For the original lengths.

(b) If the lengths are expressed in mm (1 cm = 10 mm).

(c) If the lengths are expressed as "centimeters above a standard beetle height of 1.1," (i.e., the sample values become +.4, −.1, .1, −.1, . . .).

2-19 Throw a die 100 times (or else simulate this by consulting the random numbers in Appendix Table IIa). Graph the relative frequency distribution, and calculate the sample mean

(a) After 10 throws;

(b) After 25 throws;

(c) After 100 throws;

(d) After millions of throws (guess).

chapter 3

Probability

3-1 INTRODUCTION

In the next four chapters we make deductions about a sample from a known population; this is a necessary prelude to the induction involved in Chapters 7 to 10, where we shall make inferences about an unknown population from an observed sample.

If the population of American voters is 55% Democrat, we cannot be certain that exactly the same percentage of Democrats will occur in a random sample. Nevertheless, it is "likely" that "close to" this percentage will turn up in our sample. Our objective is to define "likely" and "close to" more precisely; in this way we shall be able to make useful predictions. First, however, we must lay a good deal of ground work. Predictions in the face of uncertainty or chance require a knowledge of the laws of *probability*, and this chapter is devoted exclusively to their development. We start with the simplest examples—tossing coins and rolling dice.

Consider again our example in Chapter 1, in which the reader gambled against rolling an ace on a die. This gamble was based on the judgement that this outcome was unlikely. Now let's be more specific, and try to define its probability precisely. Intuitively, since this is but one of six equally probable outcomes, we might (correctly) guess its probability to be one in six, or one-sixth—provided it is an honest die. Alternatively we might say that if the die were thrown a large number of times, the relative frequency (of rolling an ace) would approach one-sixth (as in Problem 2-19). This is a useful operational approach; thus, if we suspect that this die is not, in fact, a fair one, we could test by tossing it many times, and observing whether or not the relative frequency of this outcome approached one sixth.

This definition of probability as "the limit of relative frequency," is formally stated as:

Definition

$$\boxed{\Pr(e_1) \triangleq \lim_{n\to\infty} \frac{n_1}{n}} \tag{3-1}$$

where e_1 is the outcome ("ace")
n is the total number of times that the trial is repeated (die is thrown)
n_1 is the number of times that the outcome e_1 occurs, [also called $n(e_1)$ or the frequency f of e_1]
$\frac{n_1}{n}$ is therefore the relative frequency of e_1

We shall use this definition of probability because it provides the clearest intuitive idea. However, you will find in Section 3-6 that it involves conceptual difficulties; thus, if you choose to study probability further you will soon be forced to turn to the axiomatic approach.

PROBLEMS

3-1 (a) Throw a thumbtack 50 times. Define tossing "the point up" as e_1. Record your results as in the following table, and keep a permanent record for future reference.

Trial Number (n)	Point Up?	Frequency of "Ups" (n_1) Accumulated	Relative Frequency (n_1/n)
1	No	0	.00
2	Yes	1	.50
3	Yes	2	.67
4	No	2	.50
5	Yes	3	.60
.	.	.	.
.	.	.	.
.	.	.	.
10			
20			
30			
40			
50			

(b) Show your results on the following graph:

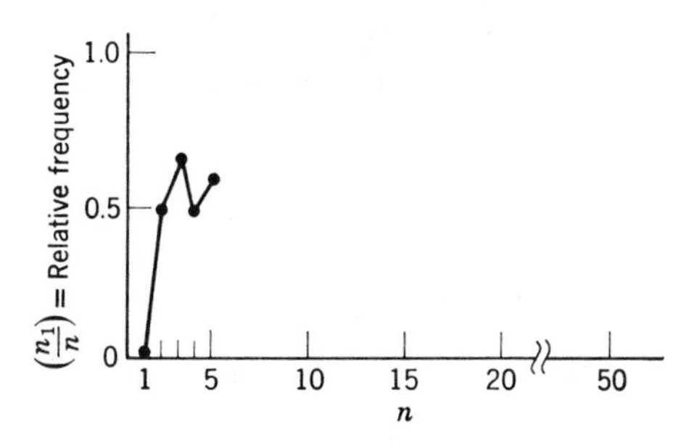

(c) What is your best guess of the probability of tossing the point up? Of tossing the point down?

3-2 In tossing a coin, define a "head" as e_1; and proceed as in 3-1(a) and (b), tossing it 100 times. (Record your results for use in Chapter 9.)

3-3 Roll a die 100 times. Define rolling a four as e_4, and proceed as in 3-1(a) and (b). (Record your results for future use. You may use the same data as in Problem 2-19.)

3-4 Roll a pair of dice, and define the event E to occur if you get a total of 7 or 11. Repeat 50 times, as in 3-1(a) and (b). What is your estimate of Pr (E)? Can you derive Pr (E) theoretically, in order to be exact, and also save the empirical work?

3-2 ELEMENTARY PROPERTIES OF PROBABILITY

We generalize by considering an experiment with N elementary outcomes $(e_1, e_2, \ldots, e_i, \ldots, e_N)$. The relative frequency n_i/n of any outcome e_i must be positive, since both the numerator and denominator are positive; moreover, since the numerator cannot exceed the denominator, relative frequency cannot exceed 1. Thus

$$0 \le \frac{n_i}{n} \le 1$$

The same relations are true in the limit, so that from (3-1)

$$\boxed{0 \le \Pr(e_i)} \tag{3-2}$$

and

$$\Pr(e_i) \le 1 \tag{3-3}$$

Next we note that the frequencies of all possible outcomes sum to n.

$$n_1 + n_2 + \cdots n_N = n$$

Dividing this equation by n, we find that all the *relative* frequencies sum to 1.

$$\frac{n_1}{n} + \frac{n_2}{n} + \cdots + \frac{n_N}{n} = 1$$

This same relation is true in the limit, so that

$$\boxed{\Pr(e_1) + \Pr(e_2) + \cdots \Pr(e_N) = 1} \tag{3-4}$$

3-3 EVENTS AND THEIR PROBABILITIES

(a) The Outcome Set—An Example

In the previous section, the die example was an experiment where the outcomes $e_1, e_2, \ldots, e_6$ were numerical, and involved no complications. Usually, an experiment will have a more complex set of outcomes.

For example, suppose the experiment consists of flipping a coin three times (or, equivalently, flipping three coins at once). A typical outcome (designated as e_4) is the sequence H, T, T. The list of all possible outcomes, or outcome set, is shown in Figure 3-1. Since most experiments of interest to the practical statistician are sampling experiments, the outcome set is also often known as the *sample space S*.

- (H, H, H) = e_1
- (H, H, T) = e_2
- (H, T, H) = e_3
- (H, T, T) = e_4
- (T, H, H) = e_5
- (T, H, T) = e_6
- (T, T, H) = e_7
- (T, T, T) = e_8

FIG. 3-1 Outcome set in flipping a coin three times.

We note several features. The order in which the set of eight outcomes $\{e_1, e_2, \ldots, e_8\}$ is listed doesn't matter. Whenever this is the case, it is a mathematical convention to use curly brackets. Thus the two outcome sets $\{e_1, e_2, \ldots, e_8\}$ and $\{e_2, e_1, e_8, \ldots, e_5\}$ are the *same* set.

However, since (H, H, T) and (H, T, H) are separate and distinct outcomes, the order in which H and T appear is an essential feature; in this case we use *round* brackets and call the result an *ordered* triple.

Finally, we note that an experimental outcome involves an entire ordered triple. It is tempting to try to tear each triple into three parts, and think of 24 outcomes. This mistake is avoided by writing down a dot for each of the 8 elementary outcomes. (Hereafter, we shall often refer to outcomes as "points" for short.)

To simplify calculations without restricting our concepts in any way, let us suppose that the coin is fairly tossed, so that all 8 outcomes are equally probable. Since all 8 probabilities must sum to 1 according to (3-4) we have

$$P(e_1) = P(e_2) = \cdots = P(e_8) = \tfrac{1}{8} \tag{3-5}$$

(b) Events

Continuing the example of 3 coins, consider the event

E: at least 2 heads

This event includes outcomes e_1, e_2, e_3, and e_5 in Figure 3-1. We might say

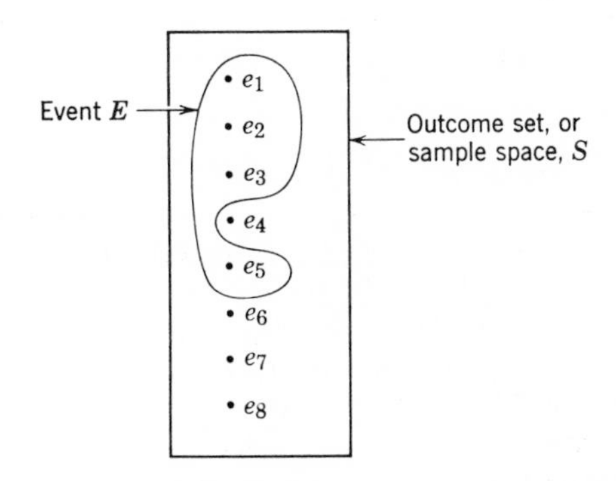

FIG. 3-2 An event as a subset of points within an outcome set.

the event E is the collection of points $\{e_1, e_2, e_3, e_5\}$ as in Figure 3-2. In fact, this is a convenient way to generally define an event:

Definition

An event E is a subset of the outcome set S	(3-6)

We now ask "What is the probability of E?" Using the definition of limiting relative frequency, we may write

$$\Pr(E) = \lim_{n \to \infty} \frac{n_E}{n} \tag{3-7}$$

where n_E = frequency of E. But of course E occurs whenever the outcomes e_1, e_2, e_3, or e_5 occur. Thus

$$n_E = n_1 + n_2 + n_3 + n_5$$

and from (3-7)

$$\Pr(E) = \lim \frac{n_1 + n_2 + n_3 + n_5}{n}$$

$$= \lim \left(\frac{n_1}{n} + \frac{n_2}{n} + \frac{n_3}{n} + \frac{n_5}{n}\right)$$

$$= \Pr(e_1) + \Pr(e_2) + \Pr(e_3) + \Pr(e_5) \tag{3-8}$$

$$= \tfrac{1}{8} + \tfrac{1}{8} + \tfrac{1}{8} + \tfrac{1}{8} = \tfrac{1}{2}$$

TABLE 3-1 Several Events in the Experiment of Figure 3-1 (Tossing 3 Coins)

Three alternative ways of naming an event			
(1) Arbitrary Symbol for Event	(2) Description	(3) Outcome List	(4) Probability
E	At least 2 heads	$\{e_1, e_2, e_3, e_5\}$	1/8 + 1/8 + 1/8 + 1/8 = 1/2
F	Second coin head, followed by tail	$\{e_2, e_6\}$	1/4
G	Fewer than 2 heads	$\{e_4, e_6, e_7, e_8\}$	1/2
H	All coins the same	$\{e_1, e_8\}$	1/4
I	No heads	$\{e_8\}$	1/8
I_1	Exactly 1 head	$\{e_4, e_6, e_7\}$	3/8
I_2	Exactly 2 heads	$\{e_2, e_3, e_5\}$	3/8
I_3	Exactly 3 heads	$\{e_1\}$	1/8
J	Less than 2 tails	$\{e_1, e_2, e_3, e_5\}$	4/8

The obvious generalization of (3-8) is that the probability of an event is the sum of the probabilities of all the points (or outcomes) included in that event, that is

$$\boxed{\Pr(E) = \sum \Pr(e_i)} \tag{3-9}$$

summing over just those outcomes e_i which are in E. We note an analogy between mass (in physics) and probability: the mass of an object is the sum of the masses of all the atoms in that object; the probability of an event is the sum of the probabilities of all the outcomes included in that event.

Various events are considered in Table 3-1; all the outcomes included in each event are listed in column 3. Since the probability of each outcome is 1/8, the calculation of the probability of each event in column 4 is very simple. The value of the list is evident when we consider the first and last events in this table. In fact, they are the same event; although this may not have been clear immediately from the description, the list makes it obvious.

(c) Combining Events

As an example, we might ask for the probability of "G or H," that is, that there will be less than 2 heads *or* all coins the same (or both). This

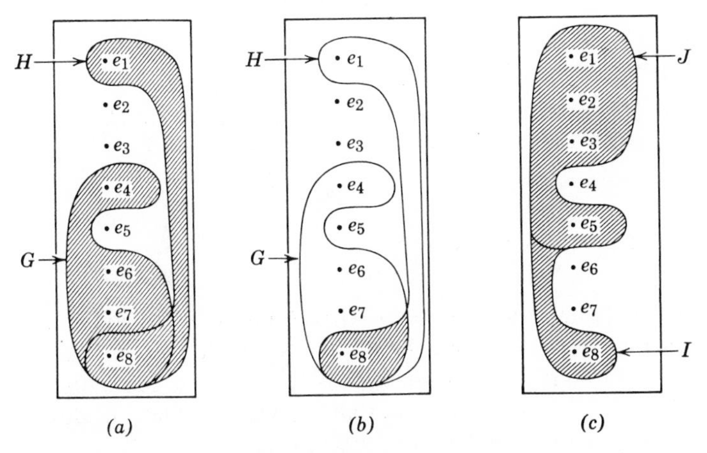

FIG. 3-3 Venn diagrams, illustrating probability of combined events. (The rectangle in each case represents the whole sample space; hence the probability of all points (or outcomes) within a rectangle sum to 1.) (*a*) $G \cup H$ shaded, "G or H"; (*b*) $G \cap H$ shaded, "G and H"; (*c*) $I \cup J$ shaded.

combined event is denoted by "$G \cup H$," and may be read "G union H" as well as "G or H." From the lists of Table 3-1 it can be seen that

$$G \cup H = \{e_4, e_6, e_7, e_8, e_1\}.$$

In general, for any two events G, H:

Definition.

$$\boxed{G \cup H \triangleq \text{set of points which are in } G\text{, or in } H\text{, or in both.}} \qquad (3\text{-}10)$$

A little abstract art in Figure 3-3*a*, called a Venn diagram, illustrates this definition. Since five of the eight equiprobable outcomes are included in $G \cup H$, its probability is 5/8.

Similarly, we might be interested in the event "G and H," that is, that there will be fewer than 2 heads, *and* all coins the same. This is clearly a much more restricted combined event; any outcome must satisfy *both* G and H, rather than either G or H. Again, we can use a Venn diagram as in Figure 3-3*b*; this shows clearly that there is only one outcome (3 tails) that qualifies. This combined event is denoted by $G \cap H$, and may be read[1] "G intersect H" as well as "G and H." The lists of G and H in Table 3-1 confirm that

$$G \cap H = \{e_8\}$$

since the only outcome appearing in both lists is e_8. Hence the probability of $G \cap H$ is 1/8. In general, for any 2 events G, H

Definition.

$$\boxed{G \cap H \triangleq \text{set of points which are in both } G \text{ and } H.} \tag{3-11}$$

(d) Probabilities of Combined Events

We have already shown how $\Pr(G \cup H)$ may be found from the Venn diagram in Figure 3-3. Now we should like to develop a formula. First consider a pair of events that do not have any points in common, such as I and J from Table 3-1. (We also say that they are mutually exclusive, or do not overlap). From Figure 3-3*c* it is obvious that

$$\Pr(I \cup J) = \Pr(I) + \Pr(J) \tag{3-12}$$

$$\tfrac{5}{8} = \tfrac{1}{8} + \tfrac{4}{8}$$

But this simple addition does not always work. For example

$$\Pr(G \cup H) \neq \Pr(G) + \Pr(H) \tag{3-13}$$

$$\tfrac{5}{8} \neq \tfrac{2}{8} + \tfrac{4}{8}$$

What has gone wrong in this case? Since G and H overlap, in summing $\Pr(G)$ and $\Pr(H)$ we count the intersection $G \cap H$ *twice*; this is why (3-13) overestimates. This is easily corrected; subtracting $\Pr(G \cap H)$ eliminates

[1] To remember when $\cup$ or $\cap$ is used, it may help to recall that $\cup$ stands for "*u*nion," and that $\cap$ resembles the letter "A" in the word "*a*nd." These technical symbols are used to avoid the ambiguity that might occur if we used ordinary English. For example, the sentence "$E \cup F$ has 5 points" has a precise meaning, but the informal "E or F has 5 points" is ambiguous.

this double counting. Thus, we have shown

Theorem.

$$\Pr(G \cup H) = \Pr(G) + \Pr(H) - \Pr(G \cap H) \tag{3-14}$$

In our example

$$\tfrac{5}{8} = \tfrac{4}{8} + \tfrac{2}{8} - \tfrac{1}{8}$$

Formula (3-14) is in fact quite general, and applies to any two events. It works in this example where G and H overlap. It also applies in cases like (3-12) where I and J do not overlap; hence $\Pr(I \cap J) = 0$, and this last term disappears when (3-14) is applied. For emphasis we may write in general,

Theorem.

$$\Pr(I \cup J) = \Pr(I) + \Pr(J) \quad \text{if } I \text{ and } J \text{ are mutually exclusive.} \tag{3-15}$$

But it must be recognized that this is just a special case of (3-14).

A collection of several events is defined as mutually exclusive if there is no overlap, i.e., if no outcome e_i belongs to more than one event. For example, in Table 3-1, events I, I_1, and I_2 are mutually exclusive; but E, F, and I are not, because E and F overlap at e_2.

The collection of events $\{I, I_1, I_2, I_3\}$ is mutually exclusive, and also "covers" the whole sample space S. We therefore call it a partition of S. In general,

Definition.

A partition of a sample space S is a collection of mutually exclusive events $\{I, \ldots I_n\}$ whose union is the whole sample space S. (3-16)

$$I \cup I_1 \cup I_2 \cdots \cup I_n = S$$

Thus a partition completely divides the sample space into nonoverlapping events, as illustrated in Figure 3-4*b*.

In Table 3-1 note that G consists of exactly those points which are not in E. We therefore could call G the "*complement* of E," or "not E," and denote it by $\bar{E}$. And in general, for any event E

Definition.

$$\bar{E} \triangleq \text{points in sample space not in } E. \tag{3-17}$$

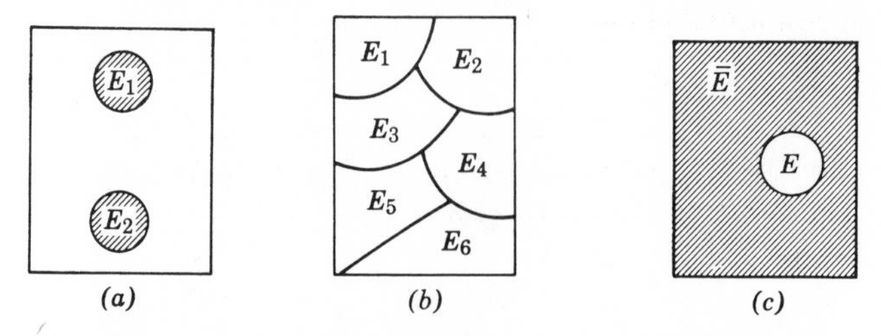

FIG. 3-4 Venn diagrams to illustrate definitions. (*a*) E_1 and E_2 are mutually exclusive; (*b*) $E_1, E_2, \ldots, E_6$ form a partition; (*c*) $\bar{E}$ shaded, (*Note.* $\{E, \bar{E}\}$ form a partition). Sample space S in each case is represented by rectangle.

An event and its complement $\{E, \bar{E}\}$ form a very simple partition. Because these events are mutually exclusive, by (3-15)

$$\Pr(E \cup \bar{E}) = \Pr(E) + \Pr(\bar{E}) \tag{3-18}$$

and since $\{E, \bar{E}\}$ form a partition

$$\Pr(E \cup \bar{E}) = 1 \tag{3-19}$$

Substituting (3-19) into (3-18)

$$1 = \Pr(E) + \Pr(\bar{E})$$

This yields a solution for $\Pr(\bar{E})$ in terms of $\Pr(E)$:

Theorem.

$$\boxed{\Pr(\bar{E}) = 1 - \Pr(E)} \tag{3-20}$$

As an example, consider the probability of getting at least one head. The complement is "no heads," and is very simple to calculate. Thus

$$\begin{aligned} \Pr(\text{at least one head}) &= 1 - \Pr(\text{no heads}) \\ &= 1 - \tfrac{1}{8} \\ &= \tfrac{7}{8} \end{aligned}$$

This is not the only way to answer this question, but it is by far the simplest, since Pr (no heads) is so easy to evaluate. The student should be on the alert for similar problems: the key words to watch for are "at least," "more than," "less than," "no more than," etc.

PROBLEMS

3-5 Suppose a penny and nickel are thrown on the table.

(a) The outcome set may be listed conveniently:

(penny, nickel)

· (H, H) $= e_1$
· (H, T) $= e_2$
· (T, H) $= e_3$
· (T, T) $= e_4$

Satisfy yourself that all 4 outcomes are equally likely, in 2 ways:

1. *Philosophical argument.* Obviously e_1 and e_2 are equally likely, because they differ only in what happens to the symmetric nickel. Similarly e_3 and e_4 are equally likely. Finally, e_1 and e_3 are equally likely, because they differ only in what happens to the symmetric penny. Thus all 4 outcomes are equally likely.

2. *Empirical argument.* Have everyone in the class repeat the experiment 10 times, so that a large amount of data can be pooled. Is the relative frequency of each outcome about 1/4?

(b) Consider the following alternate outcome set [which is recognized as just a reduction of the outcome set in (a)].

· Both heads
· Both tails
· One of each

Are these three outcomes equally likely? What are their probabilities?

(c) What is the probability of at least one head? Answer using the two alternate outcome sets, and verify that you get the same answer.

3-6 The outcome set of Problem 3-5(a) could alternatively be written as

Penny \ Nickel	H	T
H	· (H, H)	· (H, T)
T	· (T, H)	· (T, T)

In the same way, list the outcome set when a pair of dice are thrown—one red, one white. Then calculate the probability of
(1) A total of 4 dots.
(2) A total of 7 dots.
(3) A total of 7 or 11 dots (as in Problem 3-4).
(4) A double.
(5) A total of at least 8 dots.
(6) A double 3.
(7) A 1 on one die, 5 on the other.
(8) Would you get the same answers to (1)–(7) if the dice were both painted white? In particular, compare the chance of a $\{3, 3\}$ combination to the chance of a $\{1, 5\}$ combination.

3-7 Suppose the coin of Figure 3-1 were not fairly thrown, and that over the long run, the following relative frequencies were observed

e	Pr (e)
· (H H H)	.15
· (H H T)	.10
· (H T H)	.10
· (H T T)	.15
· (T H H)	.15
· (T H T)	.10
· (T T H)	.10
· (T T T)	.15

Recalling the definitions of Table 3-1,

G: fewer than 2 heads
H: all coins the same

find the following probabilities. (*Hint*. Use (3-9) and a Venn diagram.)
(a) Pr (G); Pr (H); Pr $(G \cup H)$; Pr $(G \cap H)$
(b) Verify that (3-14) holds true.
Let us further define

K: fewer than 2 tails
L: some coins different

Then find
(c) Pr (K) ; Pr (L); Pr $(K \cup L)$; Pr $(K \cap L)$
(d) Verify that (3-14) holds true.

3-8 (a) List the sample space of 4 coins tossed simultaneously.
(b) Define events A: all coins the same
B: precisely 1 head
C: at least 2 heads
Evaluate $\Pr(A) + \Pr(B) + \Pr(C)$. Do these events form a partition?
(c) Redefine A as "all tails." Do A, B, C now form a partition? What is $\Pr(A) + \Pr(B) + \Pr(C)$?

3-9 When a coin is fairly tossed 4 times, let Y denote the number of changes in sequence. For example, the outcome H T H H may be written H/T/HH, where the two changes in sequence are indicated by slashes; similarly, the outcome H/TTT has only 1 change. What is
(a) $\Pr(Y = 1)$
(b) $\Pr(Y = 2)$
(c) Do the events of (a) and (b) form a partition?

3-10 (a) What is the probability of at least one head when 4 coins are tossed?
(b) What is the probability of at least one head when 10 coins are tossed?

⇒ 3-11[1] Suppose a class of 100 students consists of several groups, in the following proportions:

	Men	Women
Taking math	$\frac{17}{100}$	$\frac{38}{100}$
Not taking math	$\frac{23}{100}$	$\frac{22}{100}$

If a student is chosen by lot to be class president, what is the chance the student will be:
(a) A man?
(b) A woman?
(c) Taking math?
(d) A man, or taking math?
(e) A man, and taking math?
(f) If the class president in fact turned out to be a man, what is the chance that he is taking math? Not taking math?

---

[1] Problems preceded by arrows are important, because they introduce a later section in the text.

⇒ 3-12 The students of a certain school engage in various sports in the following proportions:

Football, 30% of all students.
Basketball, 20%.
Baseball, 20%.
Both football and basketball, 5%.
Both football and baseball, 10%.
Both basketball and baseball, 5%.
All three sports, 2%.

If a student is chosen by lot for an interview, what is the chance that he will be:

(a) An athlete (playing at least one sport)?
(b) A football player only?
(c) A football player or a baseball player?

If an *athlete* is chosen by lot, what is the chance that he will be:

(d) A football player only?
(e) A football player or a baseball player?
Hint. Use a Venn diagram.
(f) Use your result in (a) to generalize (3-14).

3-4 CONDITIONAL PROBABILITY

Continuing with the experiment of fairly tossing 3 coins, suppose that the tossing is completed, and we are informed that there were fewer than 2 heads, i.e., that event G had occurred. Given this condition, what is the probability that event I (no heads) occurred? This is an example of "conditional probability," and is denoted as $\Pr(I/G)$, or "the probability of I, given G."

The problem may be solved by keeping in mind that our relevant outcome set is reduced to G. From Figure 3-5 it is evident that $\Pr(I/G) = 1/4$.

The second illustration in this figure shows the conditional probability of H (all coins the same), given G (less than 2 heads). Our knowledge of G means that the only relevant part of H is $H \cap G$ ("no heads" $= I$) and thus $\Pr(H/G) = 1/4$. This example is immediately recognized as equivalent to the preceding one; we are just asking the same question in two different ways.

Suppose $\Pr(G)$, $\Pr(H)$, and $\Pr(G \cap H)$ have already been computed for the original sample space S. It may be convenient to have a formula for $\Pr(H/G)$ in terms of them. We therefore turn to the definition (3-1) of probability as relative frequency. We imagine repeating the experiment n times, with G occurring $n(G)$ times, of which H also occurs $n(H \cap G)$ times.

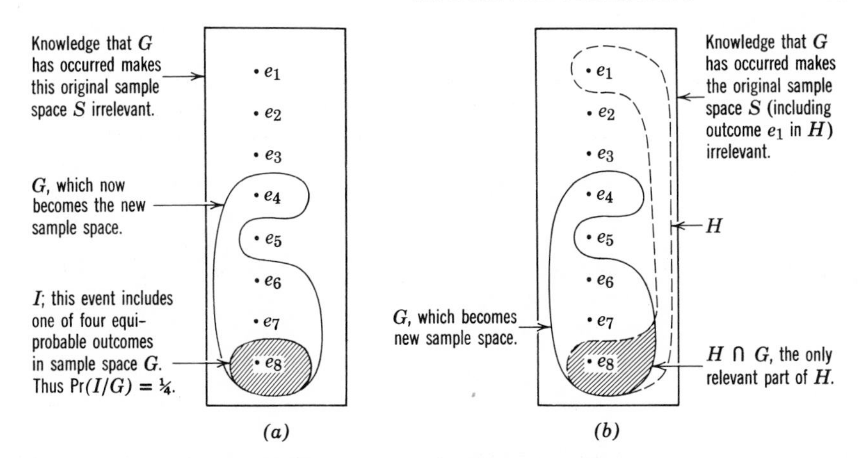

FIG. 3-5 Venn diagrams to illustrate conditional probability. (*a*) Pr(I/G). (*b*) Pr(H/G). Note Pr(H/G) is identical to Pr(I/G).

The ratio is the conditional relative frequency, and in the limit

$$\Pr(H/G) \triangleq \lim_{n\to\infty} \frac{n(H \cap G)}{n(G)} \tag{3-21}$$

On dividing numerator and denominator by n, we obtain

$$\Pr(H/G) = \lim_{n\to\infty} \frac{n(H \cap G)/n}{n(G)/n}$$

$$\boxed{\Pr(H/G) = \frac{\Pr(H \cap G)}{\Pr(G)}} \tag{3-22}$$

This formula[2] is often used in a slightly different form, obtained by cross multiplying by Pr (G)

$$\boxed{\Pr(H \cap G) = \Pr(G)\Pr(H/G)} \tag{3-23}$$

PROBLEMS

(3-13) Flip 3 coins over and over again, recording your results as in the following table.

[2] In this section and the next, we shall assume all events under consideration have nonzero probabilities. This permits us to divide legitimately by various probabilities at will.

Trial Number n	G Occurs?	Accumulated Frequency $n(G)$	If G Occurs, Then H Also Occurs?	Accumulated Frequency $n(H \cap G)$	Conditional Relative Frequency $n(H \cap G)/n(G)$
1	No	0			
2	Yes	1	Yes	1	1.00
3	No	1			
4	Yes	2	No	1	.50
5	Yes	3	Yes	2	.67
.					
.					
.					

After 50 trials, is the relative frequency $n(H \cap G)/n(G)$ close to the probability calculated theoretically in the previous section? (If not, it is because of insufficient trials, so pool the data from the whole class.)

3-14 Using the unfair coins and definitions of Problem 3-7, calculate
(a) $\Pr(G/H)$
(b) $\Pr(H/G)$
(c) $\Pr(K/L)$
(d) $\Pr(\bar{K}/L)$

3-15 (a) A consumer may buy brand X or brand Y but not both. The probability of buying brand X is .06, and brand Y is .15. Given that the consumer bought either X or Y, what is the probability that he bought brand X?
(b) If events A and B are mutually exclusive (and of course non-empty, i.e., include at least one possible outcome), is it always true that

$$\Pr(A/A \cup B) = [\Pr(A)]/[\Pr(A) + \Pr(B)]?$$

3-16 A bowl contains 3 red chips (numbered R_1, R_2, R_3) and 2 white chips (numbered W_1, W_2). A sample of 2 chips is drawn, one after the other. List the sample space. For each of the following events, diagram the subset of outcomes included and find its probability.
(a) Second chip is red.
(b) First chip is red.
(c) Second chip is red, given the first chip is red.
(d) First chip is red, given the second chip is red.
(e) Both chips are red.

Then note the following features, which are perhaps intuitively obvious also:

(1) The answers to (a) and (b) agree, as do the answers to (c) and (d).
(2) Show that the answer to (e) can be found alternatively by applying (3-23) to parts (b) and (c).
(3) Extension of part (2): if 3 chips are drawn what is the probability that all 3 are red? Can you now generalize Theorem (3-23)?

(3-17) Two cards are drawn from an ordinary deck. What is the probability that:
(a) They are both aces?
(b) They are the two black aces?
(c) They are both honor cards (ace, king, queen, jack or ten)?

3-18 A poker hand (5 cards) is drawn from an ordinary deck of cards. What is the chance of drawing, in order,
(a) 2 aces, then 3 kings?
(b) 2 aces, then 2 kings, finally a queen?
(c) 4 aces, then a king?
What is the chance of drawing, in any order whatsoever,
(d) 4 aces and a king?
(e) 4 aces?
(f) "Four of a kind" (i.e., 4 aces, or 4 kings, or 4 jacks, etc.)?
If the 5 cards are drawn with replacement (i.e., each card is replaced in the deck before drawing the next card, so that it is no longer a real poker deal), what is the probability of drawing, in any order,
(g) Exactly 4 aces?

3-19 A supply of 10 light bulbs contains 2 defective bulbs. If the bulbs are picked up in random order, what is the chance that
(a) The first two bulbs are good?
(b) The first defective bulb was picked 6th?
(c) The first defective bulb was not picked until the 9th?

⇒ 3-20 Two dice are thrown. Let
E: first die is 5
F: total is 7
G: total is 10
Compute the relevant probabilities using Venn diagrams. Show that:
(a) $\Pr(F/E) = \Pr(F)$.
(b) $\Pr(G/E) \neq \Pr(G)$.
(c) Is it true that $\Pr(E/F) = \Pr(E)$? Do you think this is closely related to (a), or just an accident?

3-21 If E and F are any 2 mutually exclusive events (and both are non-empty, of course), what can be said about $\Pr(E/F)$?

3-22 A company employs 100 persons—75 men and 25 women. The accounting department provides jobs for 12% of the men and 20%

of the women. If a name is chosen at random from the accounting department, what is the probability that it is a man? That it is a woman?

⇒ 3-23 (Bayes' Theorem). In a population of workers, suppose 40% are grade school graduates, 50% are high school graduates, and 10% are college graduates. Among the grade school graduates, 10% are unemployed, among the high school graduates, 5% are unemployed, and among the college graduates 2% are unemployed.

If a worker is chosen at random and found to be unemployed, what is the probability that he is

(a) A grade school graduate?
(b) A high school graduate?
(c) A college graduate?

(This problem is important as an introduction to Chapter 15; therefore its answer is given in full.)

Answer. Think of probability as proportion of the population, if you like.

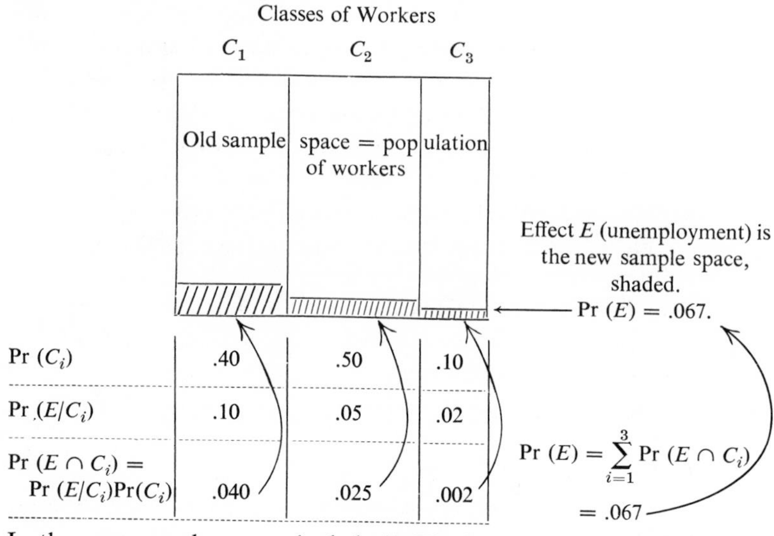

In the new sample space shaded, (3-22) gives

$$\text{(a)} \quad \Pr(C_1/E) = \frac{.040}{.067} = .597$$

$$\text{(b)} \quad \Pr(C_2/E) = \frac{.025}{.067} = .373$$

$$\text{(c)} \quad \Pr(C_3/E) = \frac{.002}{.067} = .030$$

$$\text{check, sum} = 1 \qquad 1.000$$

Notes on Bayes' Theorem. Problem 3-23 is an example of Bayes' Theorem, which may be stated as follows:

Certain "*causes*" (education levels) $C_1, C_2, \ldots C_n$, have *prior probabilities* $\Pr(C_i)$. In a sense the causes produce an "*effect*" E (unemployment) not with certainty, but with conditional probabilities $\Pr(E/C_i)$. Using conditional probability manipulations, one calculates eventually the probability of a cause given the effect, $\Pr(C_i/E)$:

$$\begin{array}{cc} \textit{Given} & \textit{Deduced} \\ \left.\begin{array}{r} \Pr(C_i) \\ \Pr(E/C_i) \end{array}\right\} & \rightarrow \Pr(C_i/E) \end{array}$$

⇒ 3-24 In a certain country, it rains 40% of the days and shines 60% of the days. A barometer manufacturer, in testing his instrument in the lab, has found that it sometimes errs: on rainy days it erroneously predicts "shine" 10% of the time, and on shiny days it erroneously predicts "rain" 30% of the time.

(a) In predicting tomorrow's weather before looking at the barometer, the (prior) chance of rain is 40%. *After* looking at the barometer and seeing it predict "rain," what is the (posterior) chance of rain?

(b) What is the posterior chance of rain if an improved barometer (error rates of 10 and 20% respectively) predicts "rain"?

(c) What is the posterior chance of shine if the improved barometer predicts "rain"?

3-5 INDEPENDENCE

In Problem 3-20 we noticed that $\Pr(F/E) = \Pr(F)$. This means that the chance of F, knowing E, is exactly the same as the chance of F, without knowing E; or, knowledge of E does not change the probability of F at all. It seems reasonable, therefore, to call F *statistically independent* of E. In fact, this is the basis for the general definition:

Definition.

An event F is called *statistically independent* of an event E if $\boxed{\Pr(F/E) = \Pr(F)}$ (3-24)

Of course, in the case of events G and E, where $P(G/E) \neq P(G)$, we would say that G was statistically *dependent* on E. In this case, knowledge of E changes the probability of G.

We now develop the consequences of F being independent of E. Substituting (3-22) in (3-24), we obtain

$$\frac{\Pr(F \cap E)}{\Pr(E)} = \Pr(F)$$

hence

$$\boxed{\Pr(F \cap E) = \Pr(F)\Pr(E)} \tag{3-25}$$

We can reverse this argument, and work backwards from (3-25) as follows:

$$\frac{\Pr(F \cap E)}{\Pr(F)} = \Pr(E)$$

$$\boxed{\Pr(E/F) = \Pr(E)} \tag{3-26}$$

That is, E is independent of F whenever F is independent of E. In other words, the result in Problem 3-20(c) above was no accident. In view of this symmetry, we may henceforth simply state that E and F are statistically independent of each other, whenever any of the three logically equivalent statements (3-24), (3-25), or (3-26) is true. Usually, statement (3-25) is the preferred form, in view of its symmetry. Sometimes, in fact, this "multiplication formula" is taken as the definition of statistical independence. But this is just a matter of taste.

Notice that so far we have insisted on the phrase "*statistical* independence," in order to distinguish it from other forms of independence—philosophical, logical, or whatever. For example, we might be tempted to say that in our dice problem, F was "somehow" dependent on E because the total of the two tosses depends on the first die. This vague notion of dependence is of no use to the statistician, and will be considered no further. But let it serve as a warning that *statistical* independence is a very precise concept, defined by (3-24), (3-25), or (3-26) above.

Now that we clearly understand statistical independence, and agree that this is the only kind of independence we shall consider, we shall run no risk of confusion if we are lazy and drop the word "statistical."

Our results so far are summarized as follows:

	$\Pr(E \cup F)$	$\Pr(E \cap F)$
General Theorem	$= \Pr(E) + \Pr(F) - \Pr(E \cap F)$	$= \Pr(F) \cdot \Pr(E/F)$
Special Case	$= \Pr(E) + \Pr(F)$ if E and F mutually exclusive; i.e., if $\Pr(E \cap F) = 0$	$= \Pr(F) \cdot \Pr(E)$ if E and F are independent; i.e., if $\Pr(E/F) = \Pr(E)$

PROBLEMS

3-25 Three coins are fairly tossed.

E_1: first two coins are heads;
E_2: last coin is a head;
E_3: all three coins are heads.

Try to answer the following questions intuitively (does knowledge of the condition affect your betting odds?). Then verify by drawing the sample space and calculating the relevant probabilities for (3-24).
(a) Are E_1 and E_2 independent?
(b) Are E_1 and E_3 independent?

3-26 Repeat Problem 3-25 using the three unfair coins whose sample space is as follows (compare Problem 3-7).

e	Pr (e)
· (H H H)	.15
· (H H T)	.10
· (H T H)	.10
· (H T T)	.15
· (T H H)	.15
· (T H T)	.10
· (T T H)	.10
· (T T T)	.15

3-27 A certain electronic mechanism has 2 bulbs which have been observed on or off with the following long-run relative frequencies:

Bulb 1 \ Bulb 2	On	Off
On	.15	.45
Off	.10	.30

This table means, for example, that both bulbs were simultaneously off 30 percent of the time.
(a) Is "bulb 1 on" independent of "bulb 2 on"?
(b) Is "bulb 1 off" independent of "bulb 2 on"?

3-28 A single card is drawn from a deck of cards, and let

E: it is an ace
F: it is a heart.

Are E and F independent, when we use
(a) An ordinary 52-card deck.
(b) An ordinary deck, with all the spades deleted.
(c) An ordinary deck, with all the spades from 2 to 9 deleted.

3-6 OTHER VIEWS OF PROBABILITY

In Section 3-1 we defined probability as the limit of relative frequency. There are several other approaches, including *symmetric* probability, *axiometric* probability, and *subjective* probability.

(a) Symmetric Probability

The physical symmetry of a fair die assures us that all six of its outcomes are equally probable. Thus

$$\Pr(e_1) = \Pr(e_2) = \cdots = \Pr(e_6)$$

In order that these six probabilities sum to one, each must be 1/6, (compare to (3-5)).

In general, for an experiment having N equally likely outcomes or points, for each point e_j

$$\Pr(e_j) = \frac{1}{N}$$

Then, for an event E consisting of N_E points, the probability is given by (3-9) as

$$\Pr(E) = \sum \Pr(e_j) = N_E \cdot \left(\frac{1}{N}\right)$$

where the summation $\sum$ extends only over points e_j in E (N_E in number). Thus, for equally probable outcomes

$$\boxed{\Pr(E) = \frac{N_E}{N}} \tag{3-27}$$

For example, in rolling a fair die consider the event

E: number of dots is an even number.

E consists of three of the six equiprobable elementary outcomes (2, 4, or 6 dots); thus its probability is 3/6.

Symmetric probability theory begins with (3-27) as the definition of probability, and gives a simpler development than our earlier relative frequency approach. However, our earlier analysis was more general; although the examples we cited often involved equiprobable outcomes, the theory we developed was in no way limited to such cases. In reviewing it, you should confirm that it may be applied whether or not outcomes are equiprobable; special attention should be given to those cases (e.g., Problem 3-26) where outcomes were not equiprobable.

Not only is symmetric probability limited because it lacks generality; it also has a major philosophical weakness. Note how the definition of probability in (3-27) involves the phrase "equally probable"; we are guilty of circular reasoning.

Our own relative frequency approach to probability suffers from the same philosophical weakness. We might ask what sort of limit is meant in equation (3-1)? It is logically possible that the relative frequency n_1/n behaves badly, even in the limit; for example, no matter how often we toss a die, it is just conceivable that the ace will keep turning up every time, making $\lim n_1/n = 1$. Therefore, we should qualify equation (3-1) by stating that the limit occurs with high *probability*, not logical certainty. In using the concept of probability in the definition of probability, we are again guilty of circular reasoning.

(b) Axiomatic Objective Probability

The only philosophically sound approach, in fact, is an abstract axiomatic approach. In a simplified version, the following properties are taken as axioms:

Axioms.

$$\Pr(e_i) \geq 0 \qquad \text{(3-2) repeated}$$

$$\Pr(e_1) + \Pr(e_2) \cdots + \Pr(e_N) = 1 \qquad \text{(3-4) repeated}$$

$$\Pr(E) = \sum \Pr(e_i) \qquad \text{(3-9) repeated}$$

Then the other properties, such as (3-1), (3-3), and (3-20) are theorems derived from these axioms—with axioms and theorems together comprising a system of analysis that appropriately describes probability situations such as die tossing, etc.

Equation (3-1) is particularly important, and is known as the *law of large numbers*. Equations (3-3) and (3-20) may be proved very easily, so easily in fact that we shall give the proof to illustrate how nicely this axiomatic theory can be developed. We can prove even stronger results: for any event E,

Theorems.

$$0 \leq \Pr(E) \quad \text{(3-28), like (3-2)}$$

$$\Pr(E) \leq 1 \quad \text{(3-29), like (3-3)}$$

$$\Pr(\bar{E}) = 1 - \Pr(E) \quad \text{(3-30), repeating (3-20)}$$

Proof. According to axioms (3-9) and (3-2), Pr (E) is the sum of positive terms, and is therefore positive; thus (3-28) is proved.

To prove (3-30), we write out axiom (3-4):

$$\underbrace{\Pr(e_1) + \Pr(e_2) + \cdots}_{\text{Terms for } E} \underbrace{+ \Pr(e_N)}_{\text{Terms for } \bar{E}} = 1$$

According to (3-9), this is just

$$\Pr(E) + \Pr(\bar{E}) = 1 \tag{3-31}$$

from which (3-30) follows.

In (3-28) we proved that every probability is positive or zero. In particular Pr $(\bar{E})$ is positive or zero; substituting this into (3-31) ensures that:

$$\Pr(E) \leq 1 \quad \text{(3-29) proved.}$$

Thus our above theorems are established; other theorems may similarly be derived.

(c) Subjective Probability

Sometimes called personal probability, this is an attempt to deal with events that cannot be repeated, even conceptually, and hence cannot be given any frequency interpretation. For example, consider events such as an increase in the stock market average tomorrow, or the overthrow of a certain government within the next month. These events are described by the layman as "likely" or "unlikely," even though there is no hope of estimating this by observing their relative frequency. Nevertheless, their likelihood vitally influences policy decisions, and as a consequence must be estimated in some rough-and-ready way. It is only then that decisions can be made on what risks are worth taking.

To answer this practical need, an axiomatic theory of personal probability has been developed. Roughly speaking, personal probability is defined by the odds one would give in betting on an event; we shall find this a useful concept later in decision theory (Chapter 15).

Review Problems

3-29 A tetrahedral (four-sided) die has been loaded. Find Pr (e_4) if possible, given the following conditions. (If the problem is impossible, state so.)
(a) Pr $(e_1) = .2$; Pr $(e_2) = .4$; Pr $(e_3) = .1$
(b) Pr $(e_1) = .4$; Pr $(e_2) = .4$; Pr $(e_3) = .3$
(c) Pr $(e_1) = .6$; Pr $(e_3) = .2$
(d) Pr $(e_1) = .7$; Pr $(e_2) = .5$

3-30 In a family of 3 children, what is the chance of
(a) At least one boy?
(b) At least 2 boys?
(c) At least 2 boys, given at least one boy?
(d) At least 2 boys, given that the eldest is a boy?

3-31 Suppose that the last 3 customers out of a restaurant all lose their hat-checks, so that the girl has to hand back their 3 hats in random order. What is the probability
(a) That no man will get the right hat?
(b) That exactly 1 man will?
(c) That exactly 2 men will?
(d) That all 3 men will?

3-32 What is the probability that
(a) 3 people picked at random have different birthdays?
(b) A roomful of 30 people all have different birthdays?
(c) In a roomful of 30 people there is at least one pair with the same birthday?

3-33 A bag contains a thousand coins, one of which has heads on both sides. A coin is drawn at random. What is the probability that it is the loaded coin, if it is flipped and turns up heads without fail
(a) 3 times in a row
(b) 10 times in a row
(c) 20 times in a row.

3-34 Repeat Problem 3-33 when the loaded coin in the bag has both H and T faces, but is biased so that the probability of H is 3/4.

chapter 4

Random Variables and Their Distributions

4-1 DISCRETE RANDOM VARIABLES

Again consider the experiment of fairly tossing 3 coins. Suppose that our only interest is the total number of heads. This is an example of a *random variable* or *variate* and is customarily denoted by a capital letter thus:

$$X = \text{the total number of heads} \tag{4-1}$$

The possible values of X are 0, 1, 2, 3; however, they are not equally likely. To find what the probabilities are, it is necessary to examine the original sample space in Figure 4-1. Thus, for example, the event "two heads" ($X = 2$) consists of 3 of the 8 equiprobable outcomes; hence its probability is 3/8. Similarly, the probability of each of the other events is computed. Thus in Figure 4-1 we obtain the probability function of X.

The mathematical definition of a random variable is "a numerical-valued function defined over a sample space." But for our purposes we can be less abstract; it is sufficient to observe that:

A discrete random variable takes on various values with probabilities specified in its probability function.	(4-2)

In our specific example, the random variable X (number of heads) takes on the values 0, 1, 2, 3, with probabilities specified by the probability function in Figure 4-1*b*.[1]

[1] Although the intuitive definition (4-2) will serve our purposes well enough, it is not always as satisfactory as the more rigorous mathematical definition which stresses the random variable's relation to the original sample space. Thus, for example, in tossing 3 coins, the random variable Y = total number of tails, is seen to be a different random variable from X = total number of heads. Yet X and Y have the *same probability distribution*, (*cont'd*)

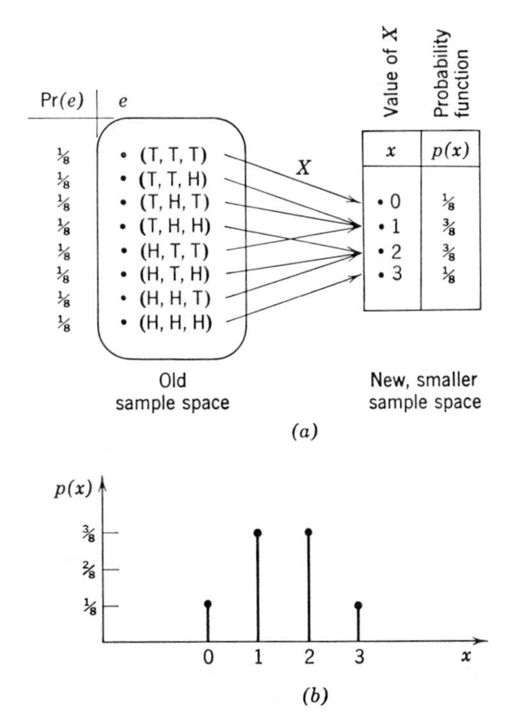

FIG. 4-1 (*a*) X, the random variable "number of heads in three tosses." (*b*) Graph of the probability function.

In the general case of defining a probability function, as in Figure 4-2, we begin by considering in the original sample space events such as $(X = 0)$, $(X = 1), \ldots$, in general $(X = x)$; (note that capital X represents the random variable, and small x a specific value it may take). For these events we calculate the probabilities and denote[2] them $p(0)$, $p(1), \ldots p(x) \ldots$. This probability function $p(x)$ may be presented equally well in any of 3 ways:

1. Table form, as in Figure 4-1*a*.
2. Graph form, as in Figure 4-1*b*.
3. By formula, as in Equation (4-7) given later on.

The *purpose* of a random variable is clear from Figures 4-1 and 4-2:

---

and anyone who used the loose definition (4-2) might be deceived into thinking that they were the *same random variable*. In conclusion, there is more to a random variable than its probability function.

[2] This notation, like any other, may be regarded simply as an abbreviation for convenience. Thus, for example, $p(3)$ is short for Pr $(X = 3)$, which in turn is short for "the probability that the number of heads is three." Note that when $X = 3$ is abbreviated to 3, Pr is correspondingly abbreviated to p.

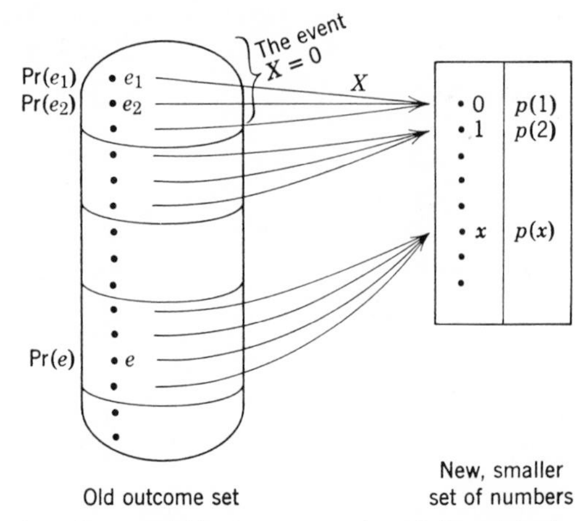

FIG. 4-2 A general random variable X as a mapping of the original outcome set onto a condensed set of numbers. (The set of numbers illustrated is 0, 1, 2, . . . , the set of positive integers. We really ought to be more general, however, allowing both negative values and fractional (or even irrational) values as well. Thus our notation, strictly speaking, should be $x_1, x_2, \ldots, x_i, \ldots$ rather than $0, 1, 2, \ldots, x, \ldots$.)

a complicated sample space (outcome set) is reduced to a much smaller, numerical sample space. The original sample space is introduced to enable us to calculate the probability function $p(x)$ for the new space; having served its purpose, the old unwieldy space is then forgotten. The interesting questions can be answered very easily in the new space. For example, referring to Figure 4-3, what is the probability of 1 head or fewer? We simply add up the relevant probabilities in the new sample space

$$\Pr\ (X \leq 1) = p(0) + p(1) = \tfrac{1}{8} + \tfrac{3}{8} = \tfrac{1}{2} \tag{4-2}$$

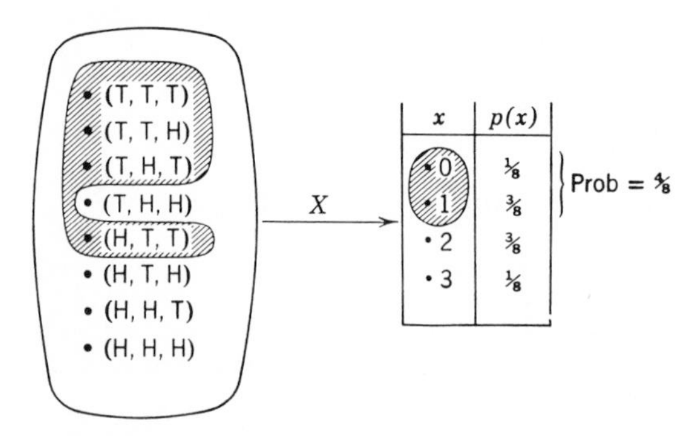

FIG. 4-3 The event $X \leq 1$ in both sample spaces, illustrating the easier calculation in the new sample space.

The answer could have been found, but with more trouble, in the original sample space.

EXAMPLE

In the same experiment of 3 fair tosses of a coin, let $Y =$ number of changes in the sequence. For example, for the sequence HTT, the value of Y is 1, because there is one changeover from H to T. In Figure 4-4 we use the

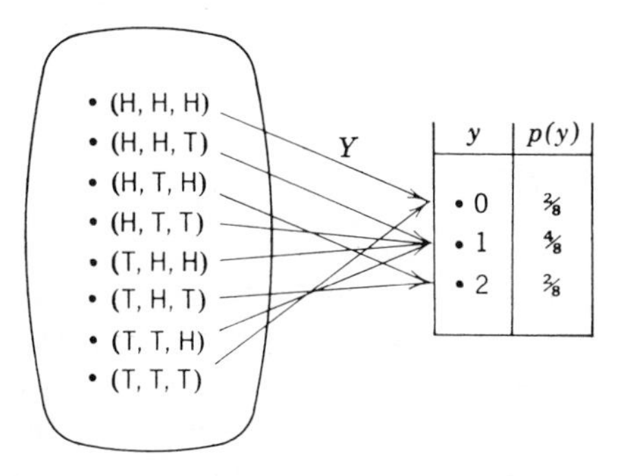

FIG. 4-4 The random variable Y ("number of changes in sequence of 3 tosses of a coin") and its probability distribution.

technique developed above to define this random variable and its probability function $p(y)$.

PROBLEMS

In each case, tabulate the probability function of the random variable, by first constructing a sample space of the experimental outcomes.

4-1 In 4 fair tosses of a coin, let
   (a) $X =$ number of heads.
   (b) $Y =$ number of changes in sequence.

4-2 Let X be the total number of dots showing when two fair dice are tossed.

4-3 Two boxes each contain 6 slips of paper numbered 1 to 6. Two slips of paper are drawn, one from each of the boxes. Let X be the difference between the numbers drawn (absolute value).

⇒ 4-4 To review Chapter 2, consider the experiment of tossing 3 coins; the number of heads X may be 0, 1, 2, or 3. Repeat this experiment 50 times[3] to obtain 50 values of X, so that you can

(a) Construct a relative frequency table of X.
(b) Graph this relative frequency table.
(c) Calculate the sample $\bar{X}$ from (2-1b).
(d) Calculate the mean squared deviation from (2-5b).
(e) If the experiment were repeated millions of times, to what value would

1. The relative frequencies tend?
2. $\bar{X}$ tend?
3. MSD (mean squared deviation) tend?
4. s^2 tend?

4-2 MEAN AND VARIANCE

In Chapter 3 we defined probability as limiting relative frequency. Now we notice the close relation between the relative frequency table observed in Problem 4-4 and the probability table calculated in Figure 4-1, for tossing 3 coins. If the sample size were increased without limit, (i.e., if we continued to toss ad infinitum), the relative frequency table would settle down to the probability table.

From the relative frequency table (Problem 4-4), we calculated the mean $\bar{X}$ and variance s^2 of our *sample*[4]. It is natural to calculate analogous population values from the probability table, and call them the mean μ and variance σ^2 of the probability distribution $p(x)$, or of the random variable X itself. Thus

Population mean, $\mu \triangleq \sum_x x\, p(x)$	(4-3) cf. (2-1b)
Population variance, $\sigma^2 \triangleq \sum_x (x - \mu)^2 p(x)$	(4-4) cf. (2-5b)

[3] Or simulate this by consulting the random numbers in Appendix Table II*a*, (with an even number representing a head, and an odd number a tail); or else use the authors' results, as follows:

03220 11232 11221 22213 13332
12212 12121 11233 21112 11213

[4] Strictly speaking, we calculated the mean squared deviation, rather than s^2. However, as $n \to \infty$, they become practically indistinguishable.

For our example of tossing three coins, we compute μ and σ^2 in Table 4-1.[5] Note the analogy here to our calculations in Table 2-4.

We call μ the "population mean," since it is based on the population of all possible tosses of the coins. On the other hand, the mean $\bar{X}$ is called a "sample mean," since it is based on a mere sample of tosses drawn from the parent population of all possible tosses. Similarly σ^2 and s^2 represent population and sample variance, respectively. A clear distinction between population and sample values is crucial; we return to this point in Chapters 6 and 7.

TABLE 4-1 Calculation of the Mean and Variance of a Random Variable

Given Probability Function		Calculation of μ from (4-3)	Calculation of σ^2 from (4-4)			Easier Calculation of σ^2, Using (4-5)
x	$p(x)$	$x\,p(x)$	$(x-\mu)$	$(x-\mu)^2$	$(x-\mu)^2\,p(x)$	$x^2\,p(x)$
0	1/8	0	−3/2	9/4	9/32	0
1	3/8	3/8	−1/2	1/4	3/32	3/8
2	3/8	6/8	+1/2	1/4	3/32	12/8
3	1/8	3/8	+3/2	9/4	9/32	9/8
		$\mu = 12/8$ $= 1.50$			$\sigma^2 = 24/32$ $= .75$	$\sum x^2\,p(x) = 24/8$ $\mu^2 = 18/8$ $\sigma^2 = 6/8$

Since the definitions of μ and σ^2 parallel those of $\bar{X}$ and s^2, we find parallel interpretations. We continue to think of the mean μ as a weighted average, using probability weights rather than relative frequency weights. The mean is also a fulcrum and center. The standard deviation is a measure of spread.

[5] The computation of σ^2 is often simplified by using:

$$\sigma^2 = \sum x^2\,p(x) - \mu^2 \tag{4-5}$$

This formula, with its proof, is analogous to (2-7). The computation is illustrated in the last column of Table 4-1.

Proof that (4-5) *is equivalent to* (4-4). Reexpress (4-4) as:

$$\sigma^2 = \sum (x^2 - 2\mu x + \mu^2)\,p(x)$$

and noting that μ is a constant:

$$= \sum x^2\,p(x) - 2\mu \sum x\,p(x) + \mu^2 \sum p(x)$$

Since $\sum x\,p(x) = \mu$ and $\sum p(x) = 1$, we have

$$\sigma^2 = \sum x^2\,p(x) - 2\mu(\mu) + \mu^2(1)$$

$$= \sum x^2\,p(x) - \mu^2 \qquad \text{(4-5) proved}$$

When a random variable is linearly transformed, the new mean and variance behave in exactly the same way as when sample observations were transformed in Section 2-5 (the proof is quite analogous and is left as an exercise). For future reference, we state these results in Table 4-2.

We could write out verbally all the information in this table, working across the rows, as follows:

TABLE 4-2 Linear Transformation (Y) of a Random Variable (X)

Random Variable	Mean	Variance	Standard Deviation
X	μ_X	σ_X^2	σ_X
$Y \triangleq a + bX$	$\mu_Y = a + b\mu_X$	$\sigma_Y^2 = b^2\sigma_X^2$	$\sigma_Y = \lvert b \rvert\, \sigma_X$

"Consider the random variable X, with mean μ_X and variance σ_X^2. If we define a new random variable Y as a linear function of X (specifically $Y = a + bX$), then the mean of Y will be $a + b\mu_X$, and its variance will be $b^2\sigma_X^2$."

PROBLEMS

4-5 Compute μ and σ^2 for the probability distributions in Problem 4-1. As a check, compute σ^2 in 2 ways—from the definition (4-4), and from the easy formula (4-5).

(4-6) Compute μ and σ^2 for the random variables of
(a) Problem 4-2.
(b) Problem 4-3.

4-7 Letting $X =$ the number of dots rolled on a fair die, find μ_X and σ_X. If $Y = 2X + 4$, calculate μ_Y and σ_Y in 2 ways:
(a) By tabulating the probability function of Y, then using (4-3) and (4-5).
(b) By Table 4-2.

(4-8) A bowl contains tags numbered from 1 to 10. There are ten 10's, nine 9's, etc. Let X denote the number on a tag drawn at random.
(a) Make a table of its probability function.
(b) Find μ_X and σ_X.

4-9 A student is given 4 questions, each with a choice of 3 answers. Let X be the number of correct answers when the student has to guess each answer. Compute the probability function and the mean and variance of X.

⇒ 4-10 Let X be a random variable with mean μ and standard deviation σ. What are the mean and standard deviation of Z, where $Z = \frac{X - \mu}{\sigma}$? (This introduces section 4-5.)

4-11 Suppose that the whole population of American families yields the following table for family size. (For simplicity, the data is slightly altered by truncating at 6.)

No. Children	0	1	2	3	4	5	6
Proportion of families	.43	.18	.17	.11	.06	.03	.02

Source. Statistical Abstract of U.S., 1963, p. 41.

(a) Let X be the number of children in a family selected at random. (This selection may be done by lots: imagine each family being represented on a slip of paper, the slips well mixed, and then one slip drawn.) The probability function of X is given in the table, of course. Find μ_X and σ_X.
(b) Now let a child be selected at random (rather than a family), and let Y be the number of children in his family. (This selection may be done by a teacher, for example, who picks a child by lot from the register of children.) What are the possible values of Y? Complete the probability table, and compute μ_Y and σ_Y.
(c) Is μ_X or μ_Y more properly called the "average family size"?

4-3 BINOMIAL DISTRIBUTION

There are many types of discrete random variables. We shall study one—the binomial—as an example of how a general formula can be developed for the probability function $p(x)$.

The classical example of a binomial variable is

$$X = \text{number of heads in } n \text{ tosses of a coin}$$

In order to generalize, we shall speak of n independent "trials," each resulting in either "success" or "failure," with respective probabilities π and $(1 - \pi)$. Then the total number of successes X is defined as a binomial random variable.

There are many practical random variables of this type, some of which are listed in Table 4-3. We shall now derive a simple formula for the probability

TABLE 4-3 Examples of Binomial Variables

"Trial"	"Success"	"Failure"	π	n	X
Tossing a fair coin	Head	Tail	1/2	n tosses	Number of heads
Birth of a child	Boy	Girl	Practically 1/2	Family size	Number of boys in family
Throwing 2 dice	7 dots	Anything else	6/36	n throws	Number of sevens
Drawing a voter in a poll	Democrat	Republican	Proportion of Democrats in the population	Sample size	Number of Democrats in the sample
The history of one atom which may radioactively decay during a certain time period	Decay	No change	Very small	Very large, the number of atoms in the sample	Number of radioactive decays

function $p(x)$. First, consider the special case in which we compute the probability of getting 3 heads in tossing 5 coins (Figure 4-5a). Each point in our outcome set is represented as a sequence of five of the letters S (success) and F (failure). We concentrate on the event three heads ($X = 3$), and show all outcomes that comprise this event. In each of these outcomes S appears

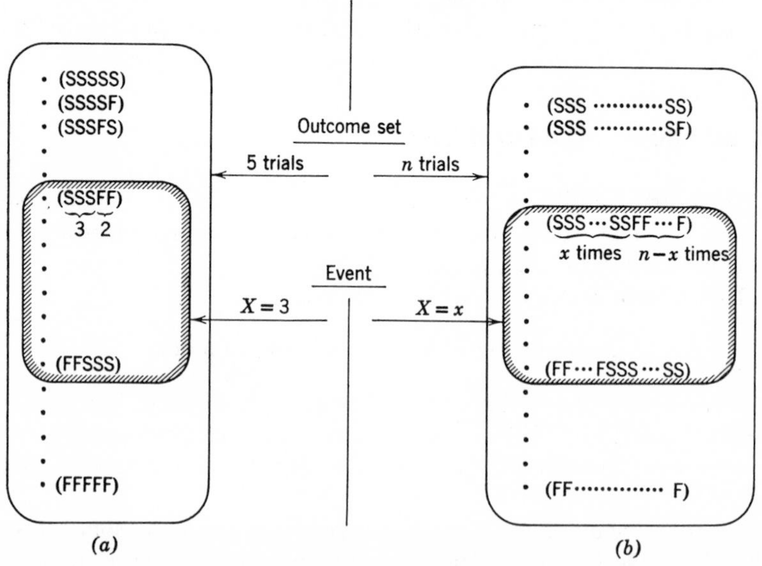

FIG. 4-5 Computing binomial probability. (a) Special case: 3 heads in 5 tosses of a coin. (b) General case: x successes in n trials.

three times, and F twice. Since the probability of S is π and F is $(1 - \pi)$,

The probability of the sequence SSSFF is

$$\pi \cdot \pi \cdot \pi \cdot (1 - \pi) \cdot (1 - \pi)$$

In general, the probability of

$$\underbrace{SS \cdots S}_{x \text{ times}} \ \underbrace{FF \cdots F}_{n - x \text{ times}}$$

is $\pi \cdot \pi \cdots, (1 - \pi) \cdot (1 - \pi) \cdots$

$$= \pi^x (1 - \pi)^{n-x}$$

this multiplication being justified by the independence of the trials. We further note that any outcome in this event has the same probability. For example, the probability of SFSSF is

$$\pi \cdot (1 - \pi) \cdot \pi \cdot \pi \cdot (1 - \pi) = \pi^3 (1 - \pi)^2$$

The same factors appear; they are only ordered differently.

Now we only have to determine how many such sequences (outcomes) are included in this event. This is precisely the number of ways in which the three S's and two F's can be arranged. This number of ways is designated as

$\binom{5}{3}$ or C_3^5,

and is[6]

$$\binom{5}{3} = \frac{5!}{3!\,(5-3)!} = 10$$

or, in general

$$\binom{n}{x} = \frac{n!}{x!\,(n-x)!}$$

To summarize:

Our event

$$(X = 3) \qquad\qquad (X = x)$$

includes

$$\binom{5}{3} = 10 \qquad\qquad \binom{n}{x}$$

outcomes, each with a probability

$$\pi^3 (1 - \pi)^2 = (\tfrac{1}{2})^3 (\tfrac{1}{2})^2 = \tfrac{1}{32} \qquad\qquad \pi^x (1 - \pi)^{n-x}$$

Hence its probability is:

$$p(3) = \Pr(X = 3) = \binom{5}{3} \pi^3 (1 - \pi)^2 \qquad\qquad p(x) = \binom{n}{x} \pi^x (1 - \pi)^{n-x} \quad (4\text{-}7)$$

$$= \frac{5!}{3!\,2!} (\tfrac{1}{2})^3 (\tfrac{1}{2})^2 = \tfrac{10}{32}$$

We summarize with Figure 4-6.

---

[6] This formula is developed as follows. Suppose we wish to fill five spots with five objects, designated S_1, S_2, S_3, F_1, F_2. We have a choice of 5 objects to fill the first spot, 4 the second, and so on; thus the number of options we have is: $5.4.3.2.1 = 5!$ (4-6)

(*cont'd*)

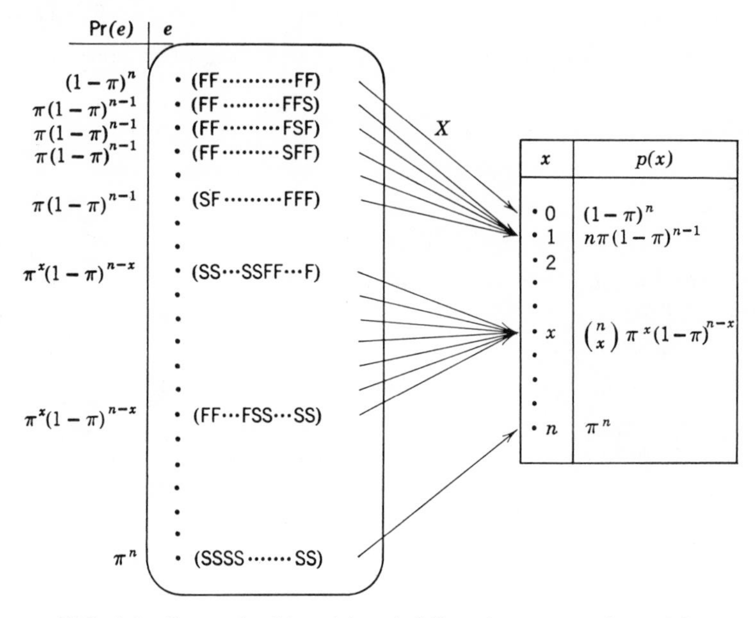

FIG. 4-6 Computing binomial probability of x successes in n trials.

As a final example, we return to our previous experiment of tossing three fair coins. What is the probability of two heads? Each toss is an independent trial in which $\pi = 1/2$. Noting also that $n = 3$ and $x = 2$, we have

$$\Pr(X = 2) = \binom{3}{2}(\tfrac{1}{2})^2(\tfrac{1}{2})^{3-2} = \frac{3!}{2!\,1!}(\tfrac{1}{2})^3 = \tfrac{3}{8} \tag{4-8}$$

a confirmation of a previous result.

But this is not the problem at hand; in our case we cannot distinguish between S_1, S_2, and S_3—all of which appear as S. Thus many of our separate and distinct arrangements in (4-6), (e.g. $S_1S_2S_3F_1F_2$ and $S_2S_1S_3F_1F_2$) cannot be distinguished, and appear as the single arrangement SSSFF. Thus (4-6) involves serious double counting. How much?

We double counted 3.2.1 = 3! times because we assumed in (4-6) that we could distinguish between S_1, S_2, and S_3 when in fact we cannot. (3! is simply the number of distinct ways of arranging S_1, S_2, and S_3.) Similarly, we double counted 2.1 = 2! times because we assumed in (4-6) that we could distinguish between F_1 and F_2, when in fact we cannot. When (4-6) is deflated for double-counting in both these ways, we have

$$\binom{5}{3} = \frac{5!}{3!\,2!} = \frac{5!}{3!\,(5-3)!}$$

PROBLEMS

Note that $\binom{n}{x}$, as well as the complete binomial distribution $p(x)$ are tabulated in Table III of the Appendix, for your optional use.

4-12 (a) Construct a diagram similar to Figure 4-6 to obtain the probability function for the number of heads X when 4 coins are tossed; use general π.
(b) Then set $\pi = \frac{1}{2}$, to obtain the results for a fair coin.
(c) From (b), calculate μ and σ^2.
(d) Graph the probability function of (b), showing μ.

4-13 A ball is drawn from a bowl containing 2 red, 1 blue, and 7 black balls. The ball is replaced, and a second ball is drawn, and so on until 3 balls have been drawn (sampling with replacement).
(a) Let $X =$ the total number of red balls drawn. Tabulate its probability function. Find μ and σ^2. Graph.
(b) Repeat (a), for $Y =$ the total number of blue balls drawn.

4-14 Check the probability function of Problem 4-9 using the formulas of this section.

(4-15) In rolling 3 dice, let X be the number of aces that occur. Tabulate the probability function of X. Find μ and σ^2. Graph.

4-16 On a blind toss of a dart, suppose the probability of hitting the target is 1/5. What is the probability that in 6 tosses you will hit the target exactly 2 times? At most 2 times? At least 3 times?

⇒ 4-17 On the basis of these questions, can you guess the mean of a general binomial variable, in terms of n and π? Can you guess the variance? (This leads into Chapter 6-6.)

*4-18[1] (For calculus students only, leading into section 4-5). Graph the function $f(t) = e^{-t^2/2}$, showing its
(a) Symmetry.
(b) Asymptotes.
(c) Maximum.
(d) Points of inflection.

4-4 CONTINUOUS DISTRIBUTIONS

In Chapter 2 we saw how a continuous quantity such as height was best graphed with a relative frequency histogram. The histogram of heights of

[1] Starred problems are optional, since they are more theoretical and/or difficult than the rest.

Figure 2-3 is reproduced in Figure 4-7*a* below. (For purposes of illustration, we measure height in feet, rather than inches. Furthermore, the y-axis has been shrunk to the same scale as the x-axis.) Note that in Figure 4-7*a* relative frequency is given by the height of each bar; but since its width (or base) is 1/4, its *area* (height times width) is numerically only 1/4 as large. Thus we can't use area in this figure to represent relative frequency, since it would badly understate. In fact, if we wish area to represent relative frequency each

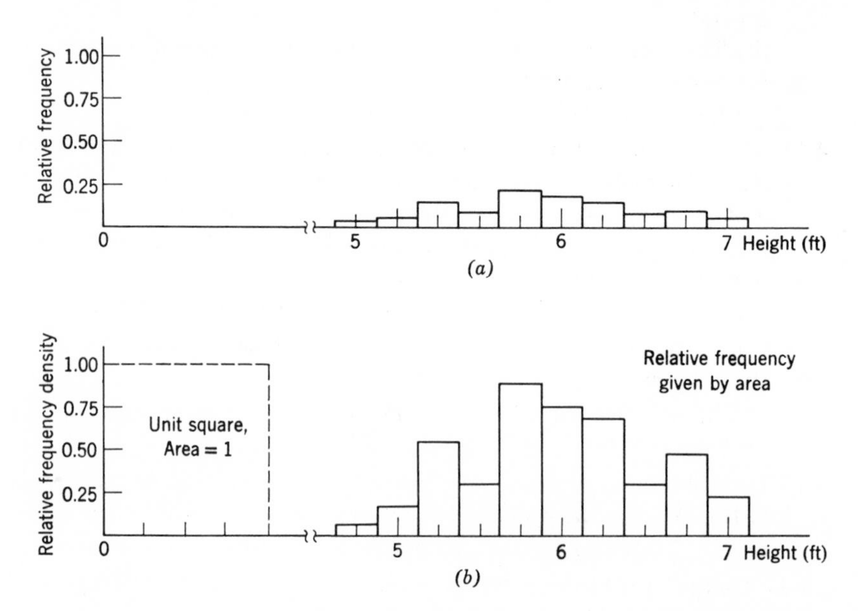

FIG. 4-7 Relative frequency histogram (*a*) transformed into relative frequency density in (*b*) making total area = 1.

height must be increased fourfold. This is done in Figure 4-7*b*, where the area of each bar is relative frequency, and the height of each bar is called *relative frequency density*.

In general

$$(\text{relative frequency density})(\text{cell width}) = (\text{relative frequency})$$

i.e.,

$$\text{area of any bar} = \text{relative frequency}.$$

There is but one more important observation. In Figure 4-7*a*, the heights sum to one (the sum of all relative frequencies must be one). From the

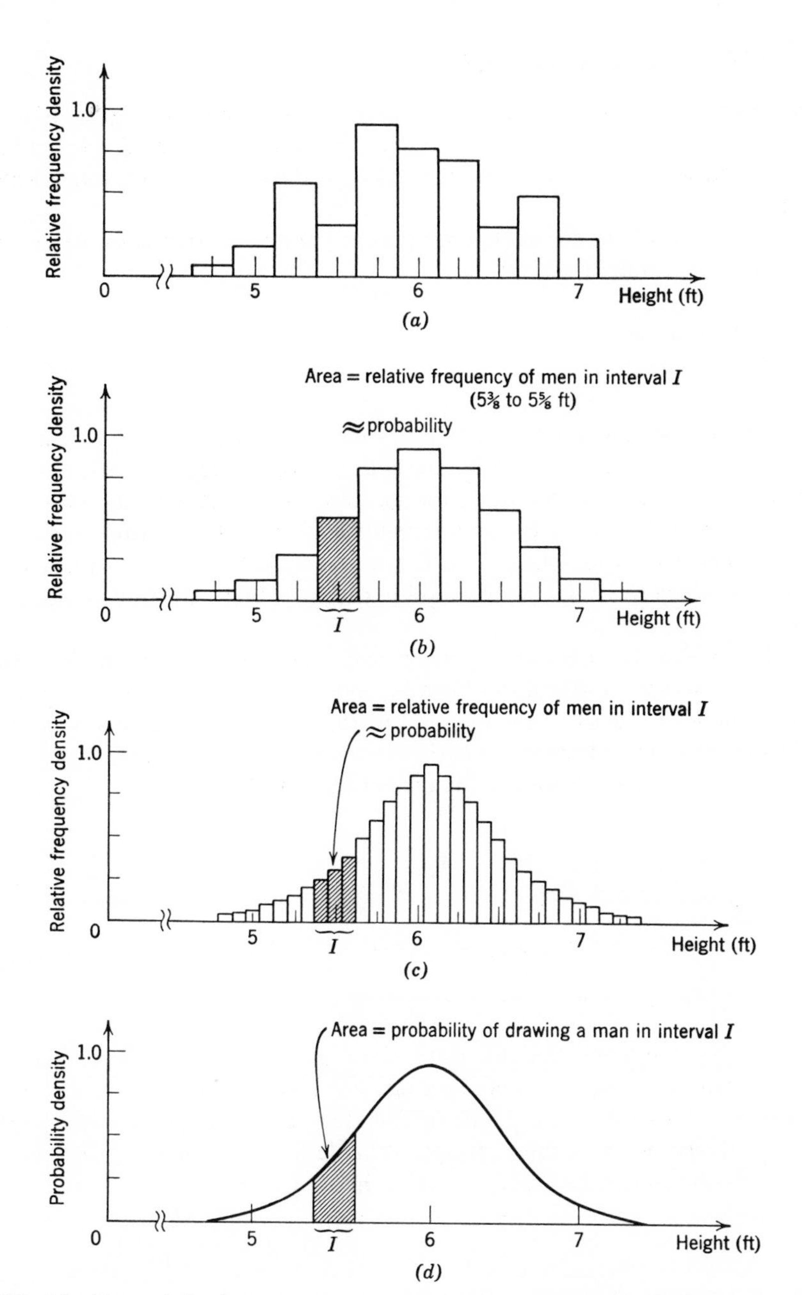

FIG. 4-8 How relative frequency density may be approximated by a probability density function as sample size increases, and cell size decreases. (*a*) Small n, as in Fig. 4-7*b*. (*b*) Large enough n to stabilize relative frequencies. (*c*) Even larger n, to permit finer cells while keeping relative frequencies stable. (*d*) For very large n, this becomes (approximately) a smooth probability density curve.

numerical equivalence of height in Figure 4-7*a* to area in Figure 4-7*b*, it follows that the areas in Figure 4-7*b* must also sum to one. And this is a key characteristic of a density function in statistics: it encloses an area numerically equal to 1.

In Figure 4-8 we show what happens to the relative frequency density of a continuous random variable as

1. Sample size increases.
2. Cell size decreases.

With a small sample, chance fluctuations influence the picture. But as sample size increases, chance is averaged out, and relative frequencies settle down to probabilities. At the same time, the increase in sample size allows a finer definition of cells. While the area remains fixed at 1, the relative frequency density becomes approximately a curve, the so-called probability density function, which we shall refer to simply as the probability function, designated $p(x)$.

If we wish to compute the mean and variance from Figure 4-8*c*, the discrete formulas (4-3) and (4-4) can be applied. But if we are working with the probability density function in Figure 4-8*d*, then integration (which calculus students will recognize is the limiting case of summation) must be used; if a and b are the limits of X, then (4-3) and (4-4) become

$$\text{Mean, } \mu = \int_a^b x\, p(x)\, dx \tag{4-9}$$

$$\text{Variance, } \sigma^2 = \int_a^b (x - \mu)^2\, p(x)\, dx \tag{4-10}$$

All the theorems that we state about discrete random variables are equally valid for continuous random variables, with summations replaced by integrals. Proofs are also very similar. Therefore, to avoid tedious duplication, we give theorems for discrete random variables only, leaving it to the reader to supply the continuous case himself, if he so desires.

4-5 THE NORMAL DISTRIBUTION

For many random variables, the probability density function is a specific bell-shaped curve, called the *normal* curve, or *Gaussian* curve, as shown in

Figures 4-9 to 4-12. It is the single most useful probability function in statistics. Many variables are normally distributed; for example, errors that are made in measuring physical and economic phenomena often are normally distributed. In addition, there are other useful probability functions (such as the binomial) which often can be approximated by the normal curve.

(a) Standard Normal Distribution

The probability function of the standard normal variable Z is

$$p(z) = \frac{1}{\sqrt{2\pi}} e^{-\frac{1}{2}z^2} \tag{4-11}$$

The constant $1/\sqrt{2\pi}$ is a scale factor required to make the total area 1. The symbols π and e denote important mathematical constants, approximately 3.14 and 2.718 respectively. We draw the normal curve in[7] Figure 4-9 to

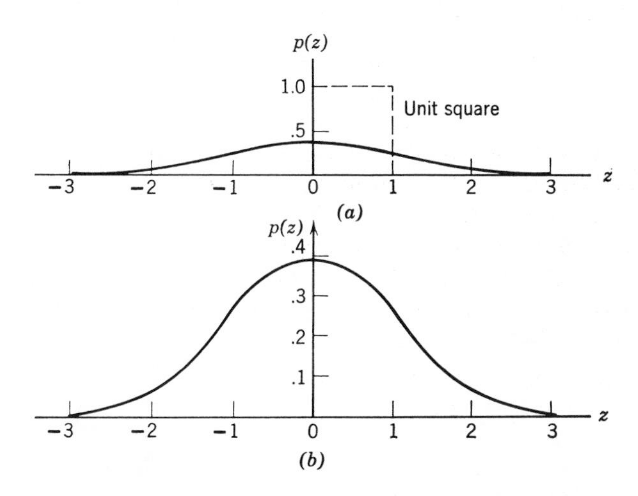

FIG. 4-9 (*a*) Standard normal curve. (*b*) Vertical axis rescaled.

reach a maximum at $z = 0$. We confirm in (4-11) that this is so: as we move to the left or right of 0, z^2 increases; since its negative exponent is increasing

[7] In Problem 4-18 you may have confirmed that the graph of (4-11) is that shown in Figure 4-9.

The mathematical constant $\pi = 3.14$ is not to be confused with the π used in Section 4-3 to designate probability of success.

in size, $p(z)$ decreases. Moreover, the further we move away from zero, the more $p(z)$ decreases; as z takes on very large (positive or negative) values, the negative exponent in (4-11) becomes very large and $p(z)$ approaches zero. Finally, this curve is symmetric. Since z appears only in squared form, $-z$ generates the same probability in (4-11) as $+z$. This confirms the shape of this standard normal curve as we have drawn it in Figure 4-9. The mean and variance of Z can be calculated by integration using (4-9) and (4-10); since

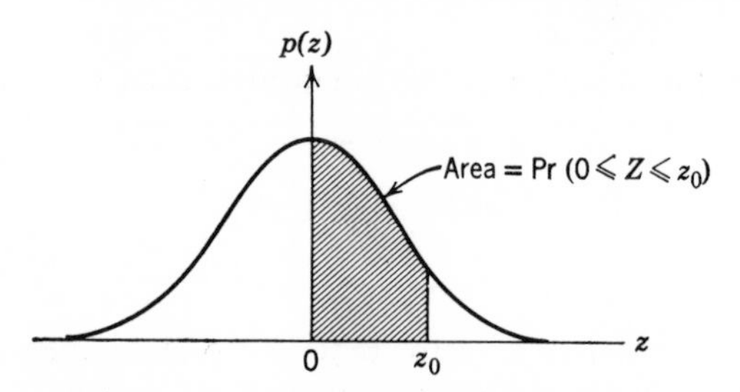

FIG. 4-10 Probability enclosed by the normal curve between 0 and z_0.

this requires calculus, we quote the results without proof:

$$\mu_Z = 0$$
$$\sigma_Z = 1$$

It is for this very reason, in fact, that Z is called a *standard* normal variable. Later when we speak of "standardizing" any variable, this is precisely what we mean: shifting it so that its mean is 0 and shrinking (or stretching) it so that its standard deviation (or variance) is one.

The probability (area) enclosed by the normal curve between the mean (0) and any specified value (say z_0) also requires calculus to evaluate precisely, but may be easily pictured in Figure 4-10.

This evaluation of probability, done once and for all, has been recorded in Table IV of the Appendix. Students without calculus can think of this as accumulating the area of the approximating rectangles, as in Figure 4-8*c*.

To illustrate this table, consider the probability that Z falls between .6 and 1.3, as shown in Figure 4-11*a*. From Table IV in the Appendix we note that the probability that Z falls between 0 and .6 is .2257; similarly the probability that Z falls between 0 and 1.3 is .4032. We require the difference in these two, namely:

$$\Pr(.6 \leq Z \leq 1.3) = .4032 - .2257 = .1775$$

In Figure 4-11*b* we consider the probability that Z falls between -1 and $+2$. Because of the symmetry of the normal curve, the probability that Z falls

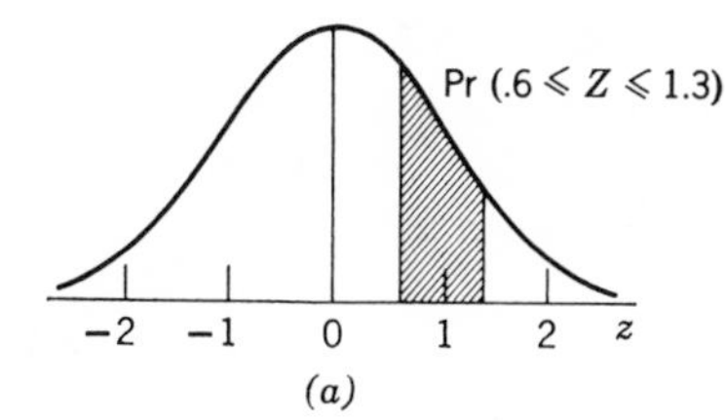

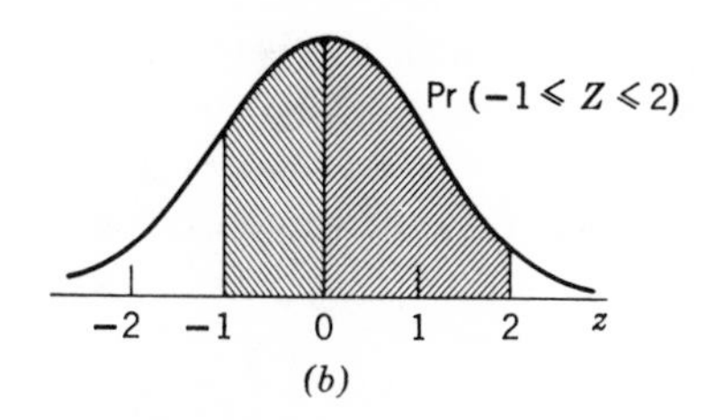

FIG. 4-11 Standard normal probabilities.

between 0 and −1 is identical to the probability between 0 and +1, which is .3413. In this instance we add this to the probability of Z between 0 and 2—namely .4772—which yields

$$\Pr(-1 \leq Z \leq 2) = .3413 + .4772 = .8185$$

Finally, the student may confirm that the probability enclosed between one standard deviation above and below the mean ($-1 \leq Z \leq +1$) is .6826, or just over 2/3 of the area of the normal curve.

PROBLEMS

4-19 If Z is a standard normal variable, use Appendix Table IV to evaluate:
- (a) $\Pr(-2 \leq Z \leq +2)$.
- (b) $\Pr(-\infty \leq Z \leq 1.64)$.
- (c) $\Pr(-2.33 \leq Z \leq \infty)$.
- (d) $\Pr(-2 \leq Z)$.
- (e) $\Pr(Z \leq 2)$.

4-20 (a) If $\Pr(-z_0 \leq Z \leq z_0) = .95$, what is z_0?
(b) If $\Pr(-z_0 \leq Z \leq z_0) = .99$, what is z_0?

(b) General Normal Distribution

If a random variable X has a normal probability curve, with mean μ and standard deviation σ, it probability function is[8] written:

$$p(x) = \frac{1}{\sqrt{2\pi}\sigma} e^{-\frac{1}{2}\left(\frac{x-\mu}{\sigma}\right)^2} \tag{4-12}$$

[8] To prove that (4-12) is centered at μ, we note that the peak of the curve occurs when the negative exponent attains its smallest value 0, i.e., when $x = \mu$. It may also be shown that (4-12) is scaled by the factor σ. Finally, it is bell shaped for the same reasons given in part (a).

We notice that in the very special case in which $\mu = 0$ and $\sigma = 1$, (4-12) reduces to the standard normal distribution (4-11). But more important, regardless of what μ and σ may be, we can translate *any* normal variable X in (4-12) into the standard form (4-11) by defining:

$$\frac{X - \mu}{\sigma} = Z \tag{4-13}$$

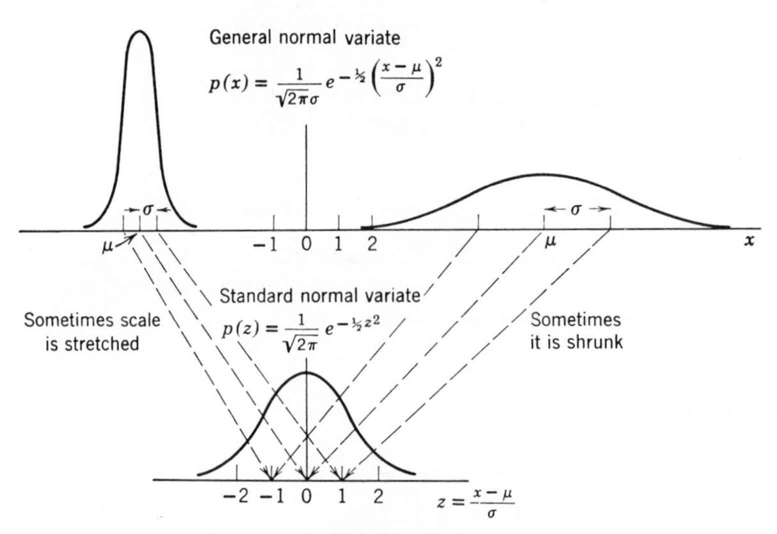

FIG. 4-12 Linear transformation of any normal variable into the standard normal variable.

Z is recognized as just a linear transformation of X, as shown in Figure 4-12. Notice that whereas the mean and standard deviation of a general normal variate X can take on any values, the standard normal variate Z is unique—with mean 0 and standard deviation 1 as proved in Problem 4-10.

To evaluate any normal variate X, we therefore translate X into Z, and then evaluate Z in the standard normal table (Appendix Table IV). For example, suppose that X is normal, with $\mu = 100$ and $\sigma = 5$. What is the probability of getting an X value of 110 or more? That is, we wish to evaluate

$$\Pr(X \geq 110) \tag{4-14}$$

First (4-14) can be written equivalently[9] as

$$\Pr\left(\frac{X - 100}{5} \geq \frac{110 - 100}{5}\right) \tag{4-15}$$

[9] Any inequality is preserved if both sides are diminished by the same amount (100) and divided by the same positive amount (5).

which, noting (4-13), is

$$\Pr(Z \geq 2) \tag{4-16}$$

We see that (4-16) is the standardized form of (4-14), and from Table IV we evaluate this probability to be .0228. Moreover, the standardized form (4-16) allows a clearer interpretation of our original question; in fact, we were asking "What is the probability of getting a normal value at least two standard deviations above the mean?" The answer is: very small—about one in fifty.

As a final example, suppose a bolt picked at random from a production line has a length X which is a normal random variable with mean 10 cm and standard deviation 0.2 cm. What is the probability that its length will be between 9.9 and 10.1 cm? That is

$$\Pr(9.9 \leq X \leq 10.1)$$

This may be written in the standardized form

$$\Pr\left(\frac{9.9-10}{.2} \leq \frac{X-10}{.2} \leq \frac{10.1-10}{2}\right)$$

$$= \Pr(-.50 \leq Z \leq .50)$$

$$= .38$$

These calculations confirm our earlier observation from Figure 4-12: although there is any number of normal curves, there is only one standard normal curve. This is fortunate; instead of requiring a whole book of tables, we only need one (Appendix Table IV).

PROBLEMS

4-21 Draw a diagram similar to Figure 4-12 for both the examples solved in the text directly above. Shade the area being evaluated.

4-22 If X is normal, calculate:

(a) $\Pr(4.5 \leq X \leq 6.5)$ where $\mu_X = 5$ and $\sigma_X = 1$

(b) $\Pr(X \leq 800)$ where $\mu_X = 400$ and $\sigma_X = 200$

(c) $\Pr(800 \leq X)$ where $\mu_X = 400$ and $\sigma_X = 200$

4-23 Suppose that a population of men's heights is normally distributed with a mean of 68 inches, and standard deviation of 3 inches. Find the proportion of the men who

(a) Are over 6 feet

(b) Are under 5 feet 6 inches

(c) Are between 5 feet 6 inches and 6 feet.

To check your 3 answers, see whether they sum to 1.

4-6 A FUNCTION OF A RANDOM VARIABLE

Looking again at the experiment of tossing three coins, let us suppose that there is a reward R depending upon the number X of heads we toss. Formally, we might state that R is a function of X, or

$$R = g(X)$$

Now let us suppose that the specific form of this function is

$$R = (X - 1)^2$$

which is equally well given by Table 4-4.

TABLE 4-4 Tabled Form of the Function $R = (X - 1)^2$

Value of X	Value of R $r = g(x) = (x - 1)^2$
0	$(0 - 1)^2 = 1$
1	$(1 - 1)^2 = 0$
2	$(2 - 1)^2 = 1$
3	$(3 - 1)^2 = 4$

The values of R are customarily rearranged in order as shown in the third column of Table 4-5. Furthermore, the values of R have certain probabilities which may be deduced from the previous probabilities of X, (just as the probabilities of X were deduced from the probabilities in our original sample

TABLE 4-5 Calculation of the Probability of Each R Value from the Probabilities of Various X Values

(1) x	(2) $p(x)$	(3) $r = g(x)$	(4) $p(r)$	(5) $rp(r)$
·0	1/8	·0	3/8	0
·1	3/8	·1	4/8	4/8
·2	3/8	·4	1/8	4/8
·3	1/8			
				$\mu_R = 1$

(Arrows in the table map $x=0 \to r=1$, $x=1 \to r=0$, $x=2 \to r=1$, $x=3 \to r=4$.)

space, in Figure 4-1). Thus we note from Table 4-4 that two values of X, (0 or 2) give rise to an R value of 1. This is indicated with arrows in Table 4-5. The third and fourth column in this table show the probability distribution of R. The last column shows the calculation of the mean of R.

R is a random variable; although it has been "derived" from X, it has all the properties of an ordinary random variable. The mean of R can be computed from its probability distribution, as in Table 4-5, and is found to be 1.0. But if it is more convenient, the answer can be derived from the probability distribution of X, as in Table 4-6.

TABLE 4-6 Mean of $R = (X - 1)^2$, calculated from $p(x)$

x	$g(x)$	$p(x)$	$g(x)\,p(x)$
0	1	1/8	1/8
1	0	3/8	0
2	1	3/8	3/8
3	4	1/8	4/8
			$\mu_R = 8/8 = 1.0$

It is easy to see why this works; in a disguised way we are calculating μ_R in the same way as in Table 4-5. The first and third lines of Table 4-6 appear together as the second line of Table 4-5. Also, the second and fourth lines of Table 4-6 correspond to the first and third lines of Table 4-5. Thus Table 4-6 contains precisely the same information as Table 4-5; it therefore yields the same value for μ_R. The only difference in the two tables is that 4-6 is ordered according to X values, while 4-5 is ordered (and condensed) according to R values.

This example can be generalized, as follows. If X is a random variable, and g is any function, then $R = g(X)$ is a random variable. μ_R may be calculated either from the probability function of R, or alternatively from the probability function of X according to

Theorem

$$\boxed{\mu_R = \sum_x g(x)\,p(x)} \tag{4-17a}$$

4-7 NOTATION

Some new notation will help us better understand the various viewpoints of the mean. For any random variable, X let us say, all the following terms mean exactly the same thing:[10]

$$\begin{aligned}\mu_X &= \text{mean of } X\\ &= \text{average } X\\ &= \text{expectation of } X\\ &= E(X), \text{ the expected value of } X\end{aligned}$$

The term $E(X)$ is introduced because it is useful as a reminder that it represents a weighted sum, i.e.,

$$E(X) \triangleq \sum_x x\, p(x) \tag{4-3}$$

With this new notation, result (4-17a) can be written

$$E(R) = \sum_x g(x)\, p(x) \tag{4-17b}$$

Finally, we recall that R was just an abbreviation for $g(X)$, so that we may equally well write (4-17b) in an easily remembered form:

Theorem

$$\boxed{E[g(X)] = \sum_x g(x)\, p(x)} \tag{4-17c}$$

As an example of this notation, we may write

$$E(X-\mu)^2 = \sum_x (x-\mu)^2\, p(x) \tag{4-18}$$

By (4-4),

$$E(X-\mu)^2 = \sigma^2 \tag{4-19}$$

Thus we see that σ^2 may be regarded as just a kind of expectation—namely, the expectation of the random variable $(X-\mu)^2$.

PROBLEMS

4-24 As in Problem 4-1, let X be the number of heads when 4 coins are fairly flipped.

(a) If $R(X) = X^2 - 3X$, find its probability function, and μ_R and σ_R^2.

(b) Find $E\,|X-2|$ in 2 ways:

(1) Using the probability function of $|X-2|$; and

(2) Using the probability function of X in (4-17)

(c) Find $E(X^2)$

(d) Find $E(X-\mu_X)^2$. Is this related to σ_X in any way?

---

[10] The reason for the plethora of names is historical. For example, gamblers and economists use the term "*expected* gain," meteorologists use the term "*mean* annual rainfall," and teachers use the term "*average* grade."

(4-25) Repeat 4-24, letting X be the number of changes in sequence when 4 coins are tossed.

4-26 The time T, in seconds, required for a rat to run a maze, is a random variable with the following probability function.

t	$p(t)$
22	.1
23	.1
24	.3
25	.2
26	.2
27	.1

(a) Find the average time.
(b) Suppose the rat is rewarded with 1 biscuit for each second faster than 25. (For example, if he takes just 23 seconds, he gets a reward of 2 biscuits. Of course, if he takes 25 seconds or longer, he gets no reward.) What is the rat's average reward?

Review Problems

4-27 In a recent presidential election, 60% of the voters went Democratic, 40% went Republican. If Gallup took a sample of 5 voters at random, find
(a) The probability that the sample would be all Democrats.
(b) The probability that the sample would correctly forecast the election winner, i.e., that a majority of the sample would be Democratic.
(c) In what way is a sample of 5 better than a sample of 1?

4-28 Three coins are independently flipped; let $X =$ number of heads. Make a table of the probability function, and find μ_X and σ_X^2 assuming
(a) The coins are fair.
(b) The last coin is biased, coming "heads up" 3/4 of the time.

4-29 Suppose the amount of cereal in a package cannot be weighed exactly. In fact, it is a normally distributed random variable, with $\mu = 10.10$ oz. and $\sigma = .040$ oz. On the package is claimed, "net weight, 10 oz."
(a) What proportion of the packages are underweight?
(b) To what value must the mean μ be raised in order that only 1/10 of 1% of the packages be underweight?

4-30 Eight volunteers had their breathing capacity measured before and after a certain treatment. The data might have looked like this:

Person	Breathing Capacity Before	After	Improvement
A	2750	2850	+100
B	2360	2380	+20
C	2950	2800	−150
D	.	.	.
E	.	.	.
.	.	.	.
.	.	.	.
.	.	.	.

Let us concentrate on whether a given person improves or deteriorates, i.e., whether the sign of the improvement is + or −. Supposing that treatment has no effect, on average, what is the probability that there will be 6 or more + signs? (Assume that measurements are so precise that a tie is practically impossible.)

(4-31) A person performs a task 3 times in succession. He learns rapidly, so that his chance of error is 1/2 the first time, 1/4 the second time, and 1/6 the third time.

We assume that he learns equally well from his successes and failures, so that the three trials may be considered independent.

(a) Find the probability table and mean of X = the total number of errors.

(b) What is the probability of more than 1 error?

*4-32 (Requires calculus)

A random variable X is continuous, and has a probability function

$$p(x) = \tfrac{3}{8}x^2 \qquad 0 < x < 2$$
$$= 0 \qquad \text{otherwise}$$

(a) Graph $p(x)$.

(b) Find the mean, median, and mode. Are they in the order you expect?

(c) Find σ^2.

chapter 5

Two Random Variables

5-1 DISTRIBUTIONS

The first section is a simple extension of the last two chapters. The main problem will be to recognize the old ideas behind the new names. Therefore we outline this section in Table 5-1, as both an introduction and review.

TABLE 5-1 Review of Section 5-1, Showing the Origins of the Ideas

Old Idea		Application (new terminology)	
		Joint probability function	
$(G \cap H)$	(3-11)		
applied to			
$\Pr(X = 2 \cap Y = 1)$		$p(2, 1)$	(5-2a)
$\Pr(X = x \cap Y = y)$ in general		$p(x, y)$ in general	(5-2b)
$\Pr(H/G) = \dfrac{\Pr(H \cap G)}{\Pr(G)}$	(3-22)	*Conditional probability function*	
applied to			
$\Pr(X = 2/Y = 1)$		$p(2/Y = 1)$	
$\Pr(X = x/Y = y)$ in general		$p(x/Y = y)$ or $p(x/y)$	
Event F is independent of E if		*Variable X is independent of Y if*	
$\Pr(F/E) = \Pr(F)$	(3-24)	$p(x/y) = p(x)$	
or $\Pr(E \cap F) = \Pr(E)\Pr(F)$	(3-25)	or $p(x, y) = p(x)p(y)$	

(a) Joint Probability

In the experiment of tossing a coin three times, let us define (on our single sample space) two random variables:

$$X = \text{number of heads}$$
$$Y = \text{number of changes in sequence}$$

TABLE 5-2 Two Random Variables Defined on the Original Sample Space

(1) Outcomes e	(2) Corresponding X value	(3) Corresponding Y value
· HHH	3	0
· HHT	2	1
· HTH	2	2
· HTT	1	1
· THH	2	1
· THT	1	2
· TTH	1	1
· TTT	0	0

We might be interested in the probability of 2 heads and 1 change of sequence occurring together. As usual, we refer to the sample space of the experiment (in column 1 of Table 5-2), and look for the intersection of these two events, obtaining

$$\Pr(X = 2 \cap Y = 1) = 2/8 \tag{5-1}$$

For convenience $\Pr(X = 2 \cap Y = 1)$ is abbreviated to $p(2, 1)$ (5-2a)

Similarly we could compute $p(0, 0), p(0, 1), p(0, 2), p(1, 2) \ldots$, obtaining in Table 5-3 what is called the *joint (or bivariate) probability function of X and Y.*

The formal definition is

$$p(x, y) \triangleq \Pr(X = x \cap Y = y) \tag{5-2b}$$

The general case is illustrated in Figure 5-1. The events $X = 0$, $X = 1$, $X = 2 \ldots$ form a partition of the sample space, shown schematically as a

TABLE 5-3 $p(x, y)$, The Joint Probability of X and Y in Three Tosses of a Coin.

y = value of $Y \rightarrow$ $\downarrow x$ = value of X	0	1	2	$p(x)\downarrow$
0	1/8	0	0	1/8
1	0	2/8	1/8	3/8
2	0	2/8	1/8	3/8
3	1/8	0	0	1/8
$p(y)$ $\rightarrow$	2/8	4/8	2/8	1 √

horizontal slicing. Similarly, the events $Y = 0$, $Y = 1$ . . . form a partition shown as a vertical slicing of the sample space. The intersection of the horizontal slice $X = x$ and the vertical slice $Y = y$ is the event $(X = x \cap Y = y)$. Its probability is collected into $p(x, y)$ in the table.

This table, or specifically Table 5-3, may be graphed, but we run into some typographical difficulties in trying to represent 3 dimensions on a two-dimensional piece of paper. We shall suggest some possible ways to resolve this difficulty. First, since the outlay of the x and y in Table 5-3 is arbitrary, we shall change it for convenience, running x across and y up as in Figure 5-2*a* (this is the custom in analytic geometry). Then the functional values $p(x, y)$ may be plotted in the direction of an axis which we imagine

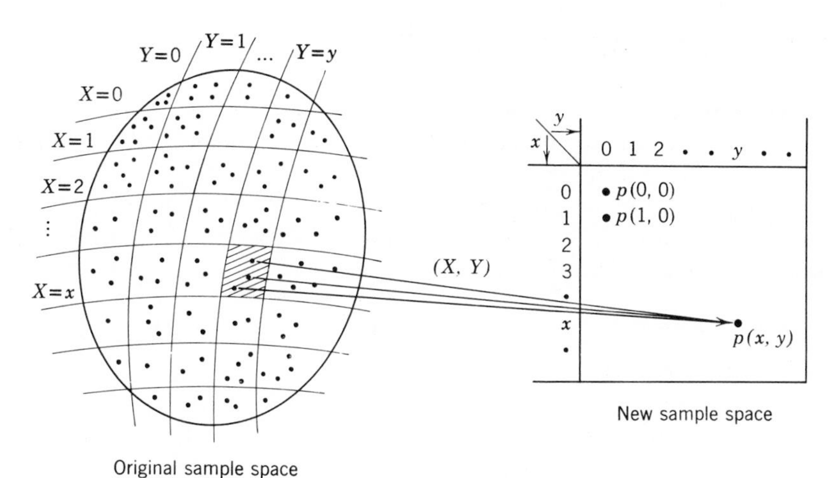

FIG. 5-1 Two random variables (X, Y), showing their sample space and joint probability function.

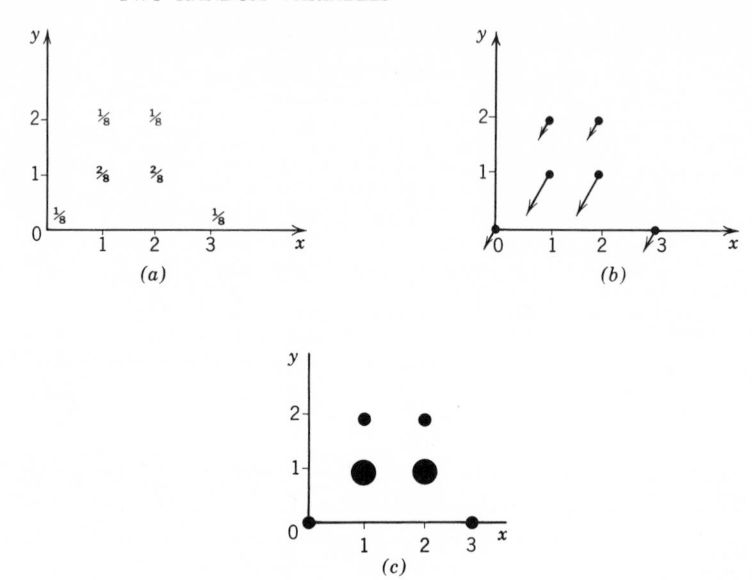

FIG. 5-2 Various graphic presentations of the bivariate probability function of Table 5-3. (*a*) Realignment of the axes. (*b*) $p(x, y)$ is represented by a line segment "coming up out of the paper." (*c*) $p(x, y)$ is represented by the size of the dot.

coming up out of the paper, as in Figure 5-2*b*, or the functional value may be represented by the size of the dot, as in Figure 5-2*c*.

(b) Marginal Probability Function

Suppose we are interested only in X, yet have to work with the joint probability function of X and Y. How can we compute the probability function of X, for example $p(2) = \Pr(X = 2)$?

It appears that the probability of this event (i.e., the horizontal slice $X = 2$ in the schematic sample space of Figure 5-1) is the sum of the probabilities of all those chunks comprising it, i.e.,

$$p(2) = p(2, 0) + p(2, 1) + p(2, 2) + p(2, 3) + \cdots p(2, y) + \cdots \quad (5\text{-}3)$$

$$= \sum_y p(2, y) \quad (5\text{-}4)$$

and in general, for any given x,

$$\boxed{p(x) = \sum_y p(x, y)} \quad (5\text{-}5)$$

For example, this idea may be applied to Table 5-3. We thus find,

$$p(2) = 0 + \tfrac{2}{8} + \tfrac{1}{8} = \tfrac{3}{8}$$

and place this sum in the right-hand margin. Similarly, $p(x)$ is computed for every x, thus providing the whole column in the right-hand margin. This is sometimes called the *marginal* probability distribution of X, to describe how it was obtained. But, of course, it is just the ordinary probability function of X (which could have been found without any reference to Y, as indeed it was in Figure 4-1).

In conclusion, the word "marginal" has no specific technical meaning. It simply describes how the probability distribution of X may be calculated when another variable Y is in play; a row sum is calculated and placed "in the margin."

In an identical way we calculate $p(y)$, the (marginal) probability distribution of Y; this is set out in the marginal *row* of Table 5-3; each element in this row is the sum of the column above. Finally, we note as expected, the exact correspondence of this marginal probability distribution of Y with the probability distribution of Y calculated in Figure 4-4 without any reference whatsoever to X.

(c) Conditional Probability Function

In the example of tossing three coins, we might wish to know the probabilities of various numbers of heads, given one change in sequence. And, in general, it is often of interest to know the probability distribution of X, when Y is given. Thus, let us suppose that Y is known to be 1. The conditional probability distribution of X, given $Y = 1$, is designated as $p(x/Y = 1)$. How is it to be evaluated?

Clearly, we should examine the vertical slice for $Y = 1$, shown in Figure 5-1 generally, or Table 5-3 specifically. The appropriate vertical slice for $Y = 1$ appears as the third column in Table 5-3; it is reproduced as the second column in Table 5-4 below. The problem is that the joint probabilities in this column do not sum to 1, hence they cannot represent a probability distribution. They do, however, give us the *relative* probabilities of various X values. Thus, if we know $Y = 1$, we know that X cannot be 0 or 3, but X values of 1 or 2 are equally probable. Intuitively, therefore, we arrive at the conditional probability distribution of X given $Y = 1$ as shown in the third column. How did we get these numbers? Since all elements in column 2 summed to only 1/2, we simply doubled them all. The result (column 3) must sum to 1; hence it is a bona fide probability distribution.

TABLE 5-4 Derivation of the Conditional Distribution of X, Given $Y = 1$

Values of X	$p(x, 1)$	$p(x/Y = 1)$
0	0	0
1	2/8	1/2
2	2/8	1/2
3	0	0
	Sum $= \Pr(Y = 1)$ $= p(1)$ $= 1/2$	Sum $= 1$ ✓

Formally, doubling all elements in column 2 is justified rigorously by the theory in Chapter 3, where conditional probability was found to be:

$$\Pr(H/G) = \frac{\Pr(H \cap G)}{\Pr(G)} \tag{3-22 repeated}$$

We merely substitute for G and H, events defined in terms of random variables, as follows:

$$\begin{aligned} &\text{For } H, \text{ substitute } (X = x) \\ &\text{For } G, \text{ substitute } (Y = 1) \end{aligned} \tag{5-6}$$

Thus

$$\Pr(X = x/Y = 1) = \frac{\Pr(X = x \cap Y = 1)}{\Pr(Y = 1)}$$

Using new notation

$$p(x/Y = 1) = \frac{p(x, 1)}{p(1)} \tag{5-7}$$

In our example, $p(1) = 1/2$, so that (5-7) becomes

$$p(x/Y = 1) = 2p(x, 1) \tag{5-8}$$

thus justifying the doubling in Table 5-4.

The generalization of (5-7) is clearly

$$p(x/Y = y) = \frac{p(x, y)}{p(y)} \tag{5-9}$$

The conditional probability distribution may be further abbreviated to $p(x/y)$, giving

$$\boxed{p(x/y) = \frac{p(x, y)}{p(y)}} \tag{5-10}$$

Note how similar this is to equation (3-22).

Since the conditional distribution is a bona fide distribution, it can be used for example to obtain the conditional mean

$$E(X/Y = y) \quad \text{or} \quad \mu_{X/y} \triangleq \sum_x x\, p(x/y) \tag{5-11}$$

(d) Independence

We define the independence of 2 random variables by extending the concept of the independence of 2 events developed in Chapter 3.

Definition.

The random variables X and Y are called independent if for every x and y, the events $(X = x)$ and $(Y = y)$ are independent.	(5-12)

The consequences are easily derived. From (3-25) we know that the independence of events $(X = x)$ and $(Y = y)$ means that

$$\Pr(X = x \cap Y = y) = \Pr(X = x)\Pr(Y = y)$$

i.e.,

$$p(x, y) = p(x)\,p(y) \tag{5-13}$$

Returning to our example, we easily show that X and Y are not independent. For independence, (5-13) must hold for every (x, y) combination. We ask whether it holds, for example, when $x = 0$ and $y = 0$? The answer is no; from the probabilities in Table 5-3, (5-13) is shown to be violated since

$$\tfrac{1}{8} \neq \tfrac{1}{8} \cdot \tfrac{2}{8}$$

PROBLEMS

5-1 In 4 tosses of a coin, again let
 $X =$ number of heads
 $Y =$ number of changes of sequence
List the sample space, and then find
- (a) The bivariate probability function; illustrate with a dot graph as in Figure 5-2*c*.
- (b) The (marginal) probability function of X.
- (c) The mean and variance of X.
- (d) The conditional probability function $p(x/Y = 2)$.
- (e) The conditional mean and variance of X, given $Y = 2$.
- (f) Are X and Y independent?

5-2 Suppose X and Y have the following joint probability function

x \ y	2	4	6	
5	.10	.20	.10	
10	.15	.30	.15	

Answer the same questions as in Problem 5-1.

5-3 Suppose X and Y have the following joint distribution

x \ y	1	2	3	
0	.1	.1	0	
1	.1	.4	.1	
2	0	.1	.1	

Answer the same questions as in Problem 5-1.

5-2 FUNCTIONS OF TWO RANDOM VARIABLES

In Section 4-7, we analyzed a derived random variable R which was some function of an (original) random variable X:

$$R = g(X) \tag{5-14}$$

In this chapter we shall analyze a derived variable T which is some function of a *pair* of random variables X, Y:

$$T = g(X, Y) \tag{5-15}$$

The concepts and proofs of this section will therefore run parallel to those of the previous chapter, the main difference being that the joint probability function $p(x, y)$ will replace the probability function $p(x)$.

We shall be particularly interested in the distribution and mean of the new variable T.

Example

Following our normal procedure, we develop the argument in terms of simple examples, and then generalize. To use our example of tossing three coins, shown in Figure 5-3, suppose S is just the sum[1] of X and Y. In this specific case (5-15) becomes:

$$S = X + Y \tag{5-16}$$

We use the symbol S in (5-16) rather than T to emphasize that this function of X and Y is a very special case of (5-15), being a *s*imple *s*um.

In Figure 5-3, we show how $p(s)$, the probability function of S may be derived directly from the original sample space, or indirectly by means of the joint probability function $p(x, y)$. In either case, the result is the same.

FIG. 5-3 Two views of the derivation of the probability function of S = X + Y
(a) Directly

	x	y	$s = x + y$	Reordered s	$p(s)$
· (TTT)	0	0	0	0	1/8
· (TTH)	1	1	2	1	0
· (THT)	1	2	3	2	2/8
· (THH)	2	1	3	3	4/8
· (HTT)	1	1	2	4	1/8
· (HTH)	2	2	4		
· (HHT)	2	1	3		
· (HHH)	3	0	3		

(b) Using the joint probability function of X and Y as an intermediate condensation

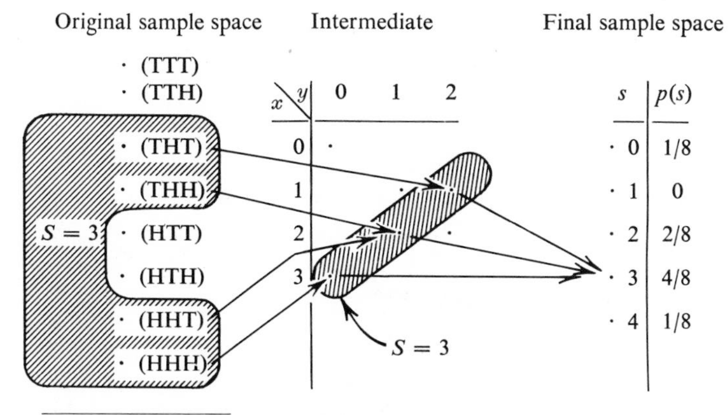

[1] To give some motivation as to why this random variable might be of practical interest, we may reinterpret the tossing of 3 coins as "having 3 children," and then consider X = number of girls and Y = number of sex changes. Since girls are more expensive to clothe than boys, and since sex changes interfere with the convenient passing on of clothing from one child to the next child, we might interpret $S = X + Y$ as a rough index of the clothing costs for the family. Of course, a weighted average of X and Y might be even more appropriate.

On the one hand, consider the direct derivation in Figure 5-3*a*. To illustrate, we note that four of the eight equiprobable outcomes are associated with $S = 3$. Hence $p(3) = \Pr(S = 3)$ is 4/8. Other S probabilities are similarly evaluated.

On the other hand, in Figure 5-3*b*, $p(3)$ may be evaluated indirectly, by first deriving the joint probability function $p(x, y)$. Then the three circled (x, y) combinations all yield $S = 3$, and the sum of their probabilities is 4/8.

The expectation $E(S)$ may similarly be derived in two ways. On the one hand, applying (4-3) directly to the probability distribution of S, we have, by definition:

$$\begin{aligned} E(S) &= \sum_s s\, p(s) \qquad (5\text{-}17) \\ &= 0(\tfrac{1}{8}) + 1(0) + 2(\tfrac{2}{8}) + 3(\tfrac{4}{8}) + 4(\tfrac{1}{8}) \\ &= 2\tfrac{1}{2} \end{aligned}$$

On the other hand, we may arrive at the same result by using the joint distribution of X and Y. Specifically, we wonder if (4-17*c*) can be extended to:

$$\begin{aligned} E(S) = E(X + Y) &= \sum_{x,y} (x + y)\, p(x, y) \qquad (5\text{-}18) \\ &= (0 + 0)(\tfrac{1}{8}) + (0 + 1)(0) \\ &\quad + (1 + 0)(0) + (1 + 1)(\tfrac{2}{8}) \cdots \\ &\quad + (1 + 2)(\tfrac{1}{8}) + (2 + 1)(\tfrac{2}{8}) + (3 + 0)(\tfrac{1}{8}) \\ &= 2\tfrac{1}{2}, \text{ the same result derived in (5-17).} \end{aligned}$$

So (5-18) does, in fact, work—at least in this example. Why? The last 3 terms of (5-18) amount to

$$3(\tfrac{1}{8} + \tfrac{2}{8} + \tfrac{1}{8}) = 3(\tfrac{4}{8})$$

which is the same as the second last term of (5-17). Continuing in this fashion, we see that (5-18) is just a disguised form of the more condensed form (5-17).

In a similar way we could prove generally

Theorem. If $T = g(X, Y)$ is any function of two random variables, then

$$\boxed{E(T) = E[g(X, Y)] = \sum_{x,y} g(x, y)\, p(x, y)} \qquad (5\text{-}19)$$

[compare (4-17c)].

For an example of how this works for a more complicated function of X and Y, we return to the tossing of three coins, and consider

$$T = X^2 - 2Y \qquad (5\text{-}20)$$

Following the method of Figure 5-3(a), we can derive the following probability distribution for T:

Calculation of $E(T)$, using (5-17)

t	$p(t)$	$tp(t)$
-3	$1/8$	$-3/8$
-1	$2/8$	$-2/8$
0	$2/8$	0
2	$2/8$	$4/8$
9	$1/8$	$9/8$
	$\Sigma = 1 \surd$	$E(T) = 1$

from which $E(T)$ is directly calculated to be 1.

Alternatively, we could calculate $E(T)$ from (5-19), using $p(x, y)$ as given in Table 5-3. Thus, noting (5-20):

$$\begin{aligned} E(T) &= \sum_{x,y} (x^2 - 2y)\, p(x, y) \\ &= (0^2 - 2(0))(1/8) + (0^2 - 2(1))(0) + (0^2 - 2(2))(0) \\ &\quad + (1^2 - 2(0))(0) \cdots + (3^2 - 2(2))(0) \\ &= 1. \end{aligned}$$

PROBLEMS

5-4 Let $U = X(X + Y)$

$V = (X - 8)(Y - 4)$

where X and Y have the same joint distribution as in Problem 5-2, namely

x \ y	2	4	6
5	.10	.20	.10
10	.15	.30	.15

(a) Find the distribution of U, and from this its mean.
(b) Find the mean of U using (5-19).
(c) Find $E(V)$.

(5-5) Let $U = XY - 1$

$V = (X - 1)(Y - 2)$

where X and Y have the same joint distribution as in Problem 5-3, namely

$x \backslash y$	1	2	3
0	.1	.1	0
1	.1	.4	.1
2	0	.1	.1

(a) Find the distribution of U, and from this its mean.
(b) Find the mean of U using (5-19).
(c) Find $E(V)$.

5-3 COVARIANCE

This is a measure of the degree to which two variables are linearly related. As an example, consider the joint probability function of Table 5-5, graphed in Figure 5-4*a*. We notice some tendency for these two variables to move together (i.e., a large X tends to be associated with a large Y; and a small X with a small Y).

Our measure of how the variables move together should be independent of our choice of origin. It will, therefore, be convenient in Figure 5-4*b* to translate both axes from the (0, 0) origin to μ_X and μ_Y (which are calculated to be 3 and 3); this means defining two new variables

$$X - \mu_X \qquad \text{and} \qquad Y - \mu_Y$$

Now suppose we multiply the new coordinate values together,

$$(X - \mu_X)(Y - \mu_Y)$$

TABLE 5-5 Joint Probability $p(x, y)$

$x \backslash y$	1	2	3	4	5
1	.1	0	0	0	0
2	0	.2	0	.1	0
3	0	0	.2	0	0
4	0	.1	0	.2	0
5	0	0	0	0	.1

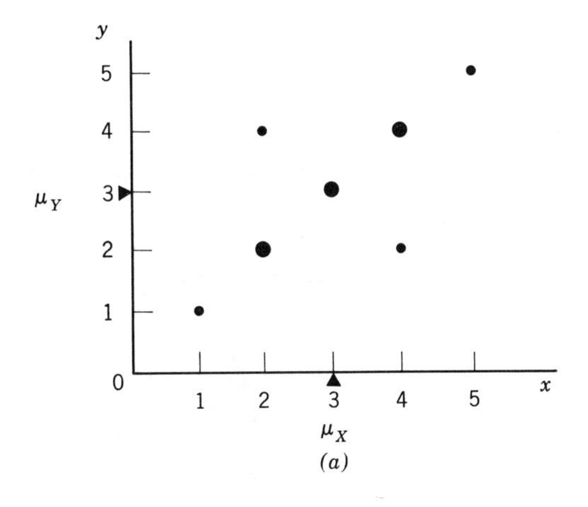

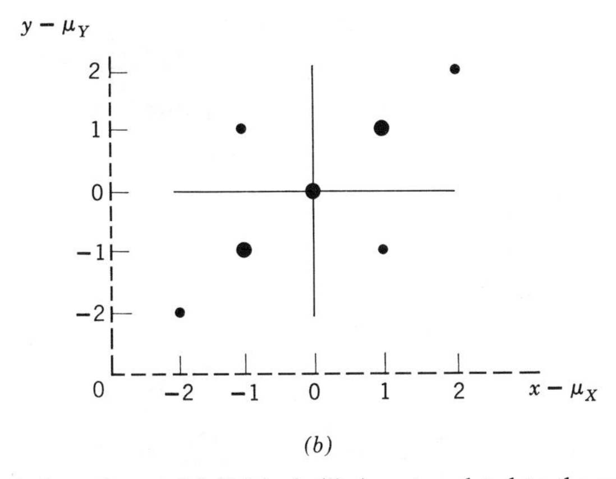

FIG. 5-4 Translation of axes. (*a*) Original. (*b*) Axes translated to the center of the distribution.

For any point in quadrant I in Figure 5-4*b* both its $(X - \mu_X)$ coordinate and $(Y - \mu_Y)$ coordinate will be positive; hence this product will be positive. It will also be positive for any point in the third quadrant, since both factors are negative. But for points in the other two quadrants the product is negative. If we sum all of these, attaching the appropriate probability weights to each, i.e.,

$$\sum_x \sum_y (x - \mu_X)(y - \mu_Y)\, p(x, y) \tag{5-21}$$

this gives us a good measure of how the variables move together, and is in fact σ_{XY}, the "covariance of X and Y."

In our example, the heavier probability weights appear in quadrants I and III; thus the positive terms in this calculation will outweight the negative.

Consequently, covariance will be positive,[2] indicating, as expected, some tendency for the variables to move together. Alternatively, if the larger probabilities had occurred in quadrants II and IV, covariance would be negative, indicating the tendency for X and Y to move in opposite directions. Finally, had the probabilities been evenly distributed in the four quadrants, there would be no discernible tendency for X and Y to move together and, as expected, their covariance would be zero.

We notice that (5-21) is equivalent to the following formal definition.[3]

Definition.

$$\boxed{\begin{array}{l}\text{Covariance of } X \text{ and } Y,\\ \sigma_{XY} \triangleq E(X-\mu_X)(Y-\mu_Y)\end{array}} \tag{5-22}$$

[2] Calculated as follows:

$$\sigma_{XY} = (-2)(-2)(.1) + (-1)(-1)(.2) + (-1)(+1)(.1) + (+1)(-1)(.1) + (+1)(+1)(.2) + (+2)(+2)(.1) = +1.0$$

[3] The computation of σ_{XY} may often be simplified by using

$$\sigma_{XY} = E(XY) - \mu_X\mu_Y \tag{5-23}$$

This formula, with its proof, is analogous to (4-5):

$$\sigma_X^2 = E(X^2) - \mu_X^2 \tag{5-24}$$

Proof of (5-23): beginning with (5-21),

$$\begin{aligned}\sigma_{XY} &= \sum_x\sum_y (x-\mu_X)(y-\mu_Y)\,p(x,y)\\ &= \sum_x\sum_y (xy - x\mu_Y - y\mu_X + \mu_X\mu_Y)\,p(x,y)\\ &= \sum_x\sum_y xy\,p(x,y) - \mu_Y\sum_x\sum_y x\,p(x,y) - \mu_X\sum_x\sum_y y\,p(x,y) + \mu_X\mu_Y\sum_x\sum_y p(x,y)\end{aligned} \tag{5-25}$$

In the second term, we find that

$$\sum_x\sum_y x\,p(x,y) = \sum_x x\left[\sum_y p(x,y)\right]$$

and by (5-5),

$$\begin{aligned}&= \sum_x x\,p(x)\\ &= \mu_X\end{aligned} \tag{5-26}$$

Similarly, in the third term of (5-25),

$$\sum_x\sum_y y\,p(x,y) = \mu_Y \tag{5-27}$$

Finally, in the last term of (5-25),

$$\sum_x\sum_y p(x,y) = 1$$

Thus (5-25) reduces to

$$\begin{aligned}\sigma_{XY} &= \sum_x\sum_y xy\,p(x,y) - \mu_Y(\mu_X) - \mu_X(\mu_Y) + \mu_X\mu_Y\\ &= E(XY) - \mu_X\mu_Y \qquad \text{(5-23) proved}\end{aligned}$$

The variance of X [ref. (4-19)] is recognized as just a special case of this, being the covariance of X with itself.

Since σ_{XY} measures the extent to which the two variables move together, we find it plausible (indeed, it may be proved[4]) that

Theorem.

$$\boxed{\text{If } X \text{ and } Y \text{ are independent, } \sigma_{XY} = 0} \tag{5-28}$$

PROBLEMS

5-6 For the following joint probability table,

x \ y	0	1	2
0	.2	.2	0
1	0	.4	.2

Calculate σ_{XY}

(a) From the definition (5-22).

(b) From the easier formula (5-23).

(5-7) Repeat Problem 5-6, for the following joint probability distribution:

x \ y	0	1	2
0	0	0	.4
1	0	.2	0
2	.2	.2	0

5-8 Suppose X and Y have the following joint distribution:

x \ y	1	2
1	.40	.10
2	.20	.05
3	.20	.05

[4] *Proof.* If X and Y are independent, then

$$p(x, y) = p(x)\,p(y) \qquad \text{(5-13) repeated}$$

Thus (5-21) becomes

$$\begin{aligned}\sigma_{XY} &= \sum\sum (x - \mu_X)(y - \mu_Y)\,p(x)\,p(y)\\ &= [\sum (x - \mu_X)\,p(x)]\,[\sum (y - \mu_Y)\,p(y)]\\ &= 0.0 = 0 \qquad \text{(5-28) proved}\end{aligned}$$

(a) Find $p(x)$ and $p(y)$; then by verifying that $p(x)p(y) = p(x, y)$ confirm that X and Y are independent [ref. equation (5-13)].

(b) What is σ_{XY}?

5-9 (a) Referring to Problems 5-4 and 5-5, is it true that $E(V) = \sigma_{XY}$?

(b) Referring to Problem 5-1, find

(1) σ_{XY}

(2) $E(X + Y)$.

*5-10

x \ y	0	1	2
0	.1	.3	.1
1	.2	.1	.2

(a) Find the probability function of X and the probability function of Y. Compute $E(X)$ and $E(Y)$.

(b) Are X and Y independent?

(c) Calculate σ_{XY}.

(d) Which statements are true, for any X and Y?

(1) If X and Y are independent, then σ_{XY} must be zero.

(2) If $\sigma_{XY} = 0$, then X and Y must be independent.

⇒ 5-11 In a certain gambling game, a pair of honest three-sided dice are thrown. Let

X_1 = number on first die

X_2 = number on the second die

The joint probability distribution of X_1 and X_2 is, of course

x_1 \ x_2	1	2	3
1	1/9	1/9	1/9
2	1/9	.	.
3	.	.	.

The total number of dots S is:

$$S = X_1 + X_2$$

(a) Find the distribution of S, and its mean and variance.

(b) Find the mean and variance of X_1 and X_2.

(c) Do you see the relation between (a) and (b)?

⇒ 5-12 Suppose the gambling game of Problem 5-11 is complicated by using loaded dice, as follows:

x_1	$p(x_1)$
1	.4
2	.3
3	.3

x_2	$p(x_2)$
1	.5
2	.4
3	.1

Assuming that the dice are tossed independently, tabulate the joint distribution of X_1 and X_2, and then answer the same questions as in Problem 5-11.

5-4 LINEAR COMBINATION OF TWO RANDOM VARIABLES

(a) Mean

First, we take leave of more complicated functions, and return to the simple example of Section 5-2 in which S was just the sum of X and Y. When we calculated $E(S)$ the student's suspicions may have been aroused; the mean of S, $(2\frac{1}{2})$ turned out to simply the sum of the mean of X, $(1\frac{1}{2})$ and the mean of Y, (1). Moreover, this was exactly the conclusion in the problems. In fact, for any X and Y, it may be proved[5] that

Theorem.

$$\boxed{E(X + Y) = E(X) + E(Y)} \tag{5-29}$$

Mathematicians often refer to this important property as the "additivity" or "linearity" of the expectation operator. It may be easily generalized to cover the case of a "weighted sum"

$$W = aX + bY \tag{5-30}$$

[5] *Proof* For $S = X + Y$, (5-19) becomes

$$E(X + Y) = \sum_x \sum_y (x + y)\, p(x, y)$$
$$= \sum_x \sum_y x\, p(x, y) + \sum_x \sum_y y\, p(x, y)$$

Considering the first term, we may write it as

$$\sum_x \sum_y x\, p(x, y) = \sum_x x \left[\sum_y p(x, y)\right]$$

by (5-5)

$$= \sum_x x\, p(x)$$
$$= E(X)$$

Similarly the second term reduces to $E(Y)$, so that

$$E(X + Y) = E(X) + E(Y) \qquad \text{(5-29) proved}$$

where a and b are any two constants. W is also known as a "linear combination of X and Y." For example, $S = X + Y$ is just the special case in which $a = b = 1$. As another example, the average of two random numbers X and Y is $(X + Y)/2 = \frac{1}{2} X + \frac{1}{2} Y$, which is just a weighted sum with weights 1/2. Similarly, any weighted average is just a linear combination with a and b satisfying $a + b = 1$.

We might guess that if we know the average of X and the average of Y, we might plug these into (5-30) to find the average of W. Fortunately this simple operation is always justified; thus[6]

Theorem.

$$\boxed{E(W) = E(aX + bY) = aE(X) + bE(Y)} \tag{5-31}$$

As a review, the student should compare (5-19) and (5-31). Both provide a means of calculating the expected value of a function of X and Y. However, (5-19) applies to *any* function of X and Y, whereas (5-31) is restricted to *linear* functions only. When we are dealing with this restricted class of linear functions, (5-31) is generally preferred to (5-19) because it is much simpler. Whereas evaluation of (5-19) involves working through the whole joint probability distribution of X and Y (e.g., Table 5-3), (5-31) requires only the marginal distributions of X and Y (e.g., the last row and column of that table).

(b) Variance

Again, we consider a simple sum first, and any linear combination later. The variance of a sum is a little more complicated than its mean. It may be proved[7] that

Theorem.

$$\boxed{\text{var}\,(X + Y) = \text{var}\,X + \text{var}\,Y + 2\,\text{cov}\,(X, Y)} \tag{5-32}$$

[6] Since the proof parallels the proof of (5-29), it is left as an exercise.

[7] *Proof.* It is time to simplify our proofs by using brief notation such as $E(W)$ rather than the awkward $\sum w\,p(w)$, or the even more awkward $\sum_x \sum_y w(x, y)\,p(x, y)$. First, from (4-19),

$$\text{var}\,S = E(S - \mu_S)^2$$

Substituting for S and μ_S,

$$\begin{aligned} \text{var}\,S &= E[(X + Y) - (\mu_X + \mu_Y)]^2 \\ &= E[(X - \mu_X) + (Y - \mu_Y)]^2 \\ &= E[\underbrace{(X - \mu_X)^2} + \underbrace{2(X - \mu_X)(Y - \mu_Y)} + \underbrace{(Y - \mu_Y)^2}] \end{aligned}$$

each of these is a random variable

Realizing that (5-31) holds for any random variables,

$$\begin{aligned} \text{var}\,S &= E(X - \mu_X)^2 + 2E(X - \mu_X)(Y - \mu_Y) + E(Y - \mu_Y)^2 \\ &= \text{var}\,X + 2\,\text{cov}\,(X, Y) + \text{var}\,Y \end{aligned}$$

(5-32) proved

where var X and cov (X, Y) are alternate notations for σ_X^2 and σ_{XY} respectively. An interesting simplification occurs when X and Y have zero covariance (are "uncorrelated"); this occurs whenever X and Y are independent, for example in the dice Problems 5-11 and 5-12. Then (5-32) simplifies to:

Corollary.

If X and Y are uncorrelated,

$$\text{var}(X + Y) = \text{var } X + \text{var } Y \quad (5\text{-}33)$$

Finally, (5-32) may be generalized to any linear combination.[8]

Theorem.

$$\text{var}(aX + bY) = a^2 \text{ var } X + b^2 \text{ var } Y + 2ab \text{ cov}(X, Y) \quad (5\text{-}34)$$

This and the other theorems of this section are summarized in Table 5-6, a very important table for future reference. The general function $g(X, Y)$ is dealt with in the first row, while the succeeding rows represent increasingly restricted special cases.

TABLE 5-6 Summary of the Mean and Variance of Various Functions of the Random Variables X and Y

Function of X and Y	Mean and Variance Derived by:	Mean	Variance
1. Any function $g(X, Y)$		$E[g(X, Y)] = \sum_{x,y} g(x, y)\, p(x, y)$ (5-19)	
2. Linear combination $aX + bY$	Row 1	$E(aX + bY) = aE(X) + bE(Y)$ (5-31)	$\text{var}(aX + bY) = a^2 \text{ var } X + b^2 \text{ var } Y + 2ab \text{ cov}(X, Y)$ (5-34)
3. Simple sum $X + Y$	Setting $a = b = 1$ in row 2	$E(X + Y) = E(X) + E(Y)$ (5-29)	$\text{var}(X + Y) = \text{var } X + \text{var } Y + 2 \text{ cov}(X, Y)$ (5-32)
4. Function of one variable, aX	Setting $b = 0$ in row 2	$E(aX) = aE(X)$ (ref. Table 4-2)	$\text{var}(aX) = a^2 \text{ var } X$ (Table 4-2)

[8] Since the proof parallels the proof of (5-32), it is left as an exercise. Note also that (5-34) has a corollary similar to (5-33).

Example

Suppose we choose a family at random from a certain population, letting

$$B = \text{number of boys in the family}$$
$$G = \text{number of girls in the family}$$

so that $C \triangleq B + G =$ number of children.

Suppose it is known that

$$E(B) = 1.2 \qquad \text{var } B = 2.0$$
$$E(G) = 1.1 \qquad \text{var } G = 2.2$$
$$\text{cov }(B, G) = 0.3$$

Then we can calculate the average number of children, and the variance:

From (5-29)

$$E(C) = 1.2 + 1.1 = 2.3$$

From (5-32)

$$\text{var }(C) = 2.0 + 2.2 + 2(0.3) = 4.8$$

PROBLEMS

5-13 Continuing Problems 5-11 and 5-12, suppose the pair of 3-sided dice are not only loaded, but dependent, so that the joint probability function of the 2 numbers is

x_1 \ x_2	1	2	3
1	.1	.1	.1
2	.1	.1	.2
3	.1	.1	.1

(a) Find the distribution of S (the total number of dots), and its mean and variance.

(b) Find the mean and variance of X_1 and of X_2.

(c) Find the covariance of X_1 and X_2, and then verify that (5-29) and (5-32) hold true.

(5-14) When a coin is fairly tossed 3 times, let

$X =$ number of heads on the first two coins

$Y =$ number of heads on the last coin

$Z =$ total number of heads

(a) Are X and Y independent? What is their covariance?
(b) For each of X, Y, and Z, find the distribution, mean, and variance.
(c) Verify that (5-29) and (5-32) hold true.

(5-15) Repeat Problem 5-14 for a coin (Problem 3-26) which is not fairly tossed, having in fact the following sample space:

e	$\Pr(e)$
. (H H H)	.15
. (H H T)	.10
. (H T H)	.10
. (H T T)	.15
. (T H H)	.15
. (T H T)	.10
. (T T H)	.10
. (T T T)	.15

5-16 The students of a certain large class wrote 2 exams, each time obtaining a distribution of grades, with the following characteristics:

	Class Mean μ	Standard Deviation σ	Variance σ^2	
1st exam, X_1	50	20	?	covariance
2nd exam, X_2	80	20	?	$\sigma_{12} = 50$
(a) Average, $\bar{X}$	?	?	?	
(b) Weighted average W	?	?	?	

Fill in the blanks in the table, assuming
(a) The instructor calculated a simple average of the two grades, $\bar{X} = (X_1 + X_2)/2$
(b) The instructor thought the second exam was twice as important, so took a weighted average

$$W = \tfrac{1}{3}X_1 + \tfrac{2}{3}X_2$$

5-17 Repeat Problem 5-16, if the covariance is -200. How might you interpret such a negative covariance? What has it done to the variance of the average grade?

(5-18) Repeat Problem 5-16, if the covariance is 0.

Review Problems

5-19 If X and Y have the following joint probability function

x \ y	5	6	7
5	.1	.3	.1
6	.1	.1	.3

Find the probability distribution and mean of
(a) X.
(b) Y.
(c) The sum $S = X + Y$.
(d) Y, given $X = 5$.
(e) Are X and Y independent? Briefly, why?
(f) Find $\Pr(X < Y)$.

5-20 In a small community of ten working couples, yearly income (in thousands of dollars) has the following distribution:

Couple	Man's Income	Wife's Income
1	10	5
2	15	15
3	15	10
4	10	10
5	10	10
6	15	5
7	20	10
8	15	10
9	20	15
10	20	10

A couple is drawn by lot to represent the community at a convention. Let M and W be the (random) income of the man and wife respectively. Find:
(a) The bivariate probability distribution, and its dot graph.
(b) The probability distribution of M; also μ_M and σ_M^2.
(c) The probability distribution of W; also μ_W and σ_W^2.
(d) The covariance σ_{MW}.

(e) $E(W/M = 10)$, $E(W/M = 20)$. Note that as M increases, the conditional mean of W increases too. This is another expression of the "positive relation" between M and W.
(*f*) If C represents the total combined income of the man and wife, what is its mean and variance?
(g) What is Pr $(C \geq 25)$?
(h) If income is taxed a straight 20 percent, what is the mean and variance of the tax on a couple's income?
(i) If the income of a couple is taxed according to the following progressive tax table, what is the mean and variance of the tax?

Combined Income	Tax
10	1
15	2
20	3
25	5
30	7
35	10
40	13

(5-21) Ten people in a room have the following heights and weights

Person	Height (inches)	Weight (pounds)
A	70	150
B	65	140
C	65	150
D	75	160
E	70	150
F	70	140
G	65	140
H	75	150
I	75	160
J	70	160

For a person drawn by lot (with height H and weight W), find:
(a) The bivariate probability distribution, and graph it.
(b) The probability distribution of H, and its mean and variance.
(c) The probability distribution of W, and its mean and variance.
(d) The covariance, σ_{HW}.
(e) $E(W/H = 65)$, $E(W/H = 70)$, $E(W/H = 75)$.

(As height increases, the conditional mean weight increases, which is another view of the positive covariance of H and W.)

(f) Are H and W independent?

(g) If a "size index" I were defined as

$$I = 2H + 3W$$

find the mean, variance and standard deviation of I; then find the distribution of I and verify directly.

5-22 Suppose a game involves dropping 3 coins on the table—a nickel, a dime, and a quarter. Each coin that lands "heads up" you are allowed to keep, so that the possible reward R ranges from 0 to 40¢.

(a) List the sample space.

(b) What is the distribution of R, its mean, and variance?

We shall now work through an alternate way to find the mean and variance of R, *without* going to the trouble of finding its exact distribution. To begin with, let us define

X_1 = the nickel's contribution to the reward

X_2 = the dime's contribution to the reward

X_3 = the quarter's contribution to the reward

Thus

$$R = X_1 + X_2 + X_3 \qquad (5\text{-}35)$$

(c) What is the distribution of X_1, its mean, and variance?

(d) Similarly find the distribution, mean, and variance of X_2 and X_3.

(e) Apply (5-29) and (5-33) to find $E(R)$ and var (R).

(5-23) Continuing, suppose that instead of 3 coins, there were 4 coins dropped on the table—a nickel, a dime, and 2 quarters. Answer the same questions as in Problem 5-22.

5-24 Continuing Problem 5-22, suppose that instead of 3 coins, we dropped 3 nickels, 2 dimes, and 5 quarters. What is the range, mean, and variance of R?

⇒ 5-25 A bowl contains 6 chips numbered from 1 to 6. One chip is selected at random and then a second is selected (random sampling without replacement). Let X_1 and X_2 be the first and second numbers drawn.

(a) Tabulate the joint probability function of X_1 and X_2.

(b) Tabulate the (marginal) probability functions of X_1 and X_2.

(c) Are X_1 and X_2 independent?

(d) What is the covariance of X_1 and X_2?

(e) Find the mean and variance of X_1 and X_2.

(f) Find the mean and variance of $S = X_1 + X_2$ in two different ways.

⇒ 5-26 Repeat Problem 5-25 with the following change. The first chip is drawn and recorded, then replaced in the bowl before the second is drawn (random sampling with replacement). Isn't this sampling problem (with replacement) mathematically identical to tossing a die twice?

⇒ 5-27 Let Y be the total number of dots showing when 10 fair dice are tossed.

(a) What are the mean and variance of Y?

(b) What is the range of possible values of Y?

⇒ 5-28

(a) A bowl contains 50 chips numbered 0, and 50 chips numbered 1. A sample of two chips is drawn with replacement; the sum is denoted by S. Tabulate the probability function of S. What are the mean and variance of S?

(b) Repeat for a sample of three chips.

(c) Repeat for a sample of five chips.

(d) Do you recognize the probability functions in (a), (b), and (c)?

chapter 6

Sampling

6-1 INTRODUCTION

In the last three chapters we have analyzed probability and random variables; we shall now employ this essential theory to answer the basic deductive question in statistics: "What can we expect of a random sample drawn from a known population?"

We have already met several examples of sampling: the poll of voters sampled from the population of all voters; the sample of light bulbs drawn from the whole production of bulbs; a sample of men's heights drawn from the whole population; a sample of 2 chips drawn from a bowl of chips (Problem 5-25). All of these are sampling *without replacement;* an individual once sampled, is out. Since he is no longer part of the population he cannot appear again in the sample. On the other hand, sampling *with replacement* involves returning any sampled individual to the population. The population remains constant; hence any individual may appear more than once in a sample, as in Problems 5-26 and 5-28. Polls of voters are typically samples without replacement; but there is no reason why a poll could not be taken with replacement. Thus no record would be kept of those already selected, and, for example, John Q. Smith of Cincinnati might vote twice in the poll—a privilege he will not enjoy on election day.

As defined earlier, a random sample is one in which each individual in the population is equally likely to be sampled. There are several ways to actually carry out the physical process of random sampling. For example, suppose a random sample is to be drawn from the population of students in the classroom.

1. The most graphic method is to put each person's name on a cardboard chip, mix all these chips in a large bowl and then draw the sample.

2. A more practical method is to assign each person a number, and then draw a random sample of numbers. Thus for a population of less than a

hundred, 2-digit numbers suffice. A random 2-digit number may be obtained by throwing a 10-sided die twice, or by consulting a table of random numbers (Appendix Table II) and reading off a pair of digits for each individual required in the sample.

These two sampling methods are mathematically equivalent. Method 2 is simpler to employ, hence it is used in practical sampling. However, the first method is conceptually easier to deal with and to visualize; consequently in our theoretical development of random sampling, we talk of drawing chips from a bowl. Moreover, if we are studying men's heights, then the height alone is all that is required on the chip and the man's name is irrelevant. Hence we can view the population simply as a collection of numbered chips in a bowl, which is stirred and then sampled.

How can random sampling be *mathematically* specified? If we draw one chip at random, its number can be regarded as a random variable taking on values that range over the whole population of chip values, with probabilities corresponding to the relative frequencies in the population.

As an example, suppose a population of 80 million men's heights has the frequency distribution shown in Table 6-1. For future reference, we also compute μ and σ^2 from Table 6-1, and call them the mean and variance of X, where X represents the *parent population* of men's heights.

TABLE 6-1 A Population of Men's Heights[1]

(1) Height (Midpoint of cell) x	(2) Frequency	(3) Relative Frequency, also $p(x)$
51	825,000	.01
54	791,000	.01
57	2,369,000	.03
60	5,505,000	.07
63	9,483,000	.12
66	16,087,000	.20
69	20,113,000	.25
72	14,480,000	.18
75	7,891,000	.10
78	1,633,000	.02
81	823,000	.01
	$\Sigma = 80{,}000{,}000$	$\Sigma = 1.00$

[1] We approximate each height by the cell midpoint to keep concepts simple. To be more precise, we ought to have used a very fine subdivision of height into many cells, as in Figure 4-8*c*.

From (4-3):

$$\mu = 51(.01) + 54(.01) \cdots 81(.01) = 67.8$$

From (4-4):

$$\sigma^2 = (51 - 67.8)^2(.01) + (54 - 67.8)^2(.01) \cdots + (81 - 67.8)^2(.01) = 28.4$$

$$\sigma = 5.3$$

Random sampling from this population is equivalent mathematically to placing the 80 million chips of column 2 in a bowl with each chip carrying the x value shown in column 1. The first chip selected at random can take on any of these x values, with probabilities shown in column 3. This random variable we designate as X_1; the second draw is designated as the random variable X_2, and so on. But each of these random variables $X_1, X_2, \ldots X_n$ (together representing our sample of n chips) has the same probability distribution $p(x)$, the distribution of the parent population; that is[2]

$$p(x_1) = p(x_2) = p(x_3) = \cdots p(x_n) \tag{6-1}$$

This equality, of course, holds true if we sample with replacement, since the second chip is drawn from exactly the same bowlful as is the first chip, etc. Fortunately, (6-1) also holds true for sampling without replacement, even though $X_1, X_2, \ldots X_n$ are now dependent; since this is not at all obvious, we must show why.

We have already noted that the distribution of X_1 is the same as the distribution of the population. However, the *conditional* distribution of X_2 given X_1 is not the same. Once that first sample value has been taken from the population (and not replaced), the population changes[3], along with relative frequencies (probabilities). Thus X_2 is dependent on X_1; or to restate, the conditional distribution of X_2 will depend on the value of X_1 selected in the first draw. But this is not the issue in (6-1). In that equation $p(x_2)$ is *the* distribution of X_2, which is not the conditional distribution, but rather the *marginal* distribution of X_2—without any condition, i.e., without any knowledge of X_1. And if we have no knowledge of X_1 and consider the distribution of X_2, there is no reason for it to differ from the distribution of X_1.

Our intuition in this case is a good guide. We could formally confirm this result by considering the full table showing the joint probability function of X_1 and X_2. It is symmetric around its main diagonal; hence although conditional distributions (rows or columns) vary in this table, the marginal

[2] Strictly speaking, (6-1) is not precise enough. It would be more accurate to let p_1 denote the probability function of X_1, p_2 of X_2, etc., and then write

$$p_1(x) \equiv p_2(x) \equiv p_3(x) \cdots \equiv p_n(x) \equiv p(x)$$

where $\equiv$ means "identically equal for all x."

[3] In our example, with a population of 80 million heights, this change would be of no practical consequence. But with smaller populations it might.

distributions of X_1 and of X_2 are necessarily identical. (See Problem 5-25b.) Thus equation (6-1) holds true, even in the case of sampling without replacement.

Before leaving Table 6-1, we have one further observation. When the parent population is extremely large, such as 80 million, sampling without replacement is *practically* the same as sampling with replacement. It hardly matters whether the individual sampled is replaced in the population or not—one individual hardly changes the frequencies in column 2 or the relative frequencies in column 3.

Thus the second draw (X_2) will be practically independent of the first (X_1). This leads us to the conclusion that sampling without replacement from an infinite population is equivalent to sampling with replacement; this is important enough that we shall return to it in Section 6-5.

Conclusion. Any population to be sampled may be simulated by a bowl of chips, with the following mathematical characteristics:

1. The number on the first chip drawn is a random variable X_1, with a distribution identical to the distribution of the population random variable X.

2. The sample of n chips gives us n random variables ($X_1, X_2, \ldots X_n$). Each X_i has the same (marginal) distribution—that of the population X. This fundamental characteristic (6-1) holds in all cases—regardless of sample replacement or population size. However, the *independence* of $X_1, X_2, \ldots X_n$ is a more complex issue. If the population is finite and sampling is without replacement, then the X_i are dependent, since the conditional distribution of any X_i depends on the previous X values drawn. In all other cases the X_i are independent; for simplicity, we shall assume this independence in the rest of the book (except Section 6-5).

6-2 SAMPLE SUM

Now we are ready to use the heavy artillery drawn up in Chapter 5. First consider S, the sum of the sample observations, defined as:

$$S \triangleq X_1 + X_2 + X_3 + \cdots + X_n \tag{6-2}$$

The expected value of S is obtained by using[4] Theorem (5-29), as:

$$E(S) = E(X_1) + E(X_2) + \cdots + E(X_n) \tag{6-3}$$

---

[4] $E(S) = E(X_1 + X_2 + \cdots + X_n) = E[(X_1 + X_2 + \cdots + X_{n-1}) + X_n]$

by Theorem (5-29): $\quad = E(X_1 + X_2 + \cdots + X_{n-1}) + E(X_n)$

$\quad = E[(X_1 + X_2 + \cdots + X_{n-2}) + X_{n-1}] + E(X_n)$

Again by Theorem (5-29): $\quad = E(X_1 + X_2 + \cdots + X_{n-2}) + E(X_{n-1}) + E(X_n)$

$\cdot$

$\cdot$

$\quad = E(X_1) + E(X_2) + \cdots + E(X_n)$

This generalization of the special two-variable case in (5-29) is an example of proof by induction.

Noting from (6-1) that each $X_1, X_2, \ldots X_n$ has the same distribution as the population, it follows that each has the same mean as the population (μ). (6-3) can therefore be written:

$$E(S) = \mu + \mu + \cdots + \mu$$

$$E(S) = n\mu \tag{6-4}$$

or

$$\boxed{\mu_S = n\mu} \tag{6-5}$$

Thus, the expected value of a sample sum is simply the mean of the parent population times the sample size.

In the same way, the variance of S is obtained by using Theorem (5-33):

$$\text{var } S \triangleq \text{var } (X_1 + X_2 + \cdots + X_n)$$

$$= \text{var } X_1 + \text{var } X_2 + \cdots + \text{var } X_n \tag{6-6}$$

Note that this depends on the assumed independence of $X_1, X_2, \ldots X_n$. Again, since all the $X_1, X_2, \ldots X_n$ have the same distribution as the population, they also have the variance σ^2 of the population. Thus (6-6) becomes:

$$\text{var } S = \sigma^2 + \sigma^2 + \cdots + \sigma^2$$

$$= n\sigma^2 \tag{6-7}$$

or

$$\boxed{\sigma_S = \sqrt{n}\, \sigma} \tag{6-8}$$

Formulas (6-5) and (6-8) are illustrated in Figure 6-1*a*. As another example, suppose a machine produces a population of bicycle chain links with average length $\mu = .40$ inch and standard deviation $\sigma = .02$ inch. A chain is made by joining together a random sample of 100 of these links. Its length S is a random variable, fluctuating from sample to sample. Its expected length is

$$\mu_S = n\mu = 100(.40) = 40.0 \text{ inches}$$

Moreover, because our sample is drawn from an infinite population, X_1, $X_2, \ldots X_{100}$ are independent. Therefore, we may apply (6-8) to compute the standard deviation of S.

$$\sigma_S = \sqrt{n}\, \sigma = 10(.02) = .20 \text{ inch}$$

The student will notice that this is an example of statistical deduction; characteristics of a sample (μ_S, σ_S) have been deduced from known characteristics (μ, σ) of the parent population.

We pause to interpret (6-5) and (6-8) intuitively. It was no surprise that μ_S was n times μ. But why should σ_S be only $\sqrt{n}$ times σ? Typically, a sample

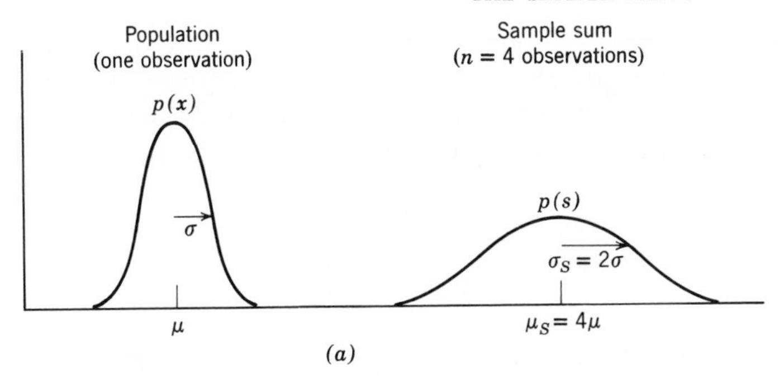

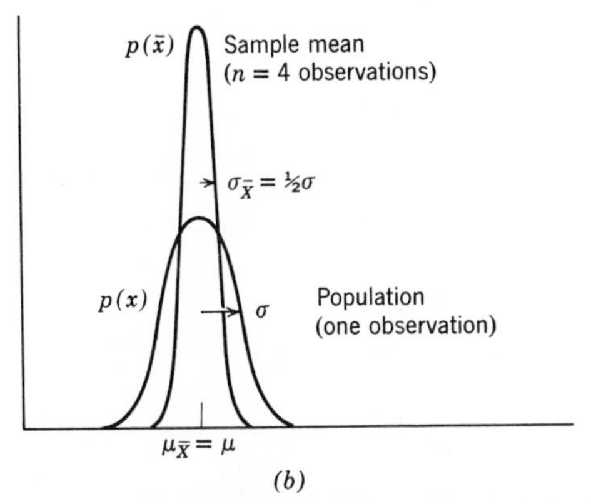

FIG. 6-1 (*a*) Relation of the sample sum S to the parent population. (*b*) Relation of the sample mean $\bar{X}$ to the parent population.

sum (e.g. chain) will include some individuals (links) which are oversized, and some which are undersized so that some cancellation occurs. Thus while the spread in the chain (σ_S) does exceed the spread in an individual link (σ), it is substantially less than it would be if the errors in all the links were accumulated without cancellation ($n\sigma$).

6-3 THE SAMPLE MEAN

Recall the definition of the sample mean,

$$\bar{X} \triangleq \frac{1}{n}(X_1 + X_2 + \cdots + X_n) \qquad \text{(2-1a) repeated}$$

$$= \frac{1}{n} S \qquad (6\text{-}9)$$

We recognize that $\bar{X}$ is just a linear transformation of S, and hence $\bar{X}$ can easily be analyzed in terms of S.

It is important to remember that $\overline{X}$, as well as S, is a random variable that fluctuates from sample to sample. It seems intuitively clear that $\overline{X}$ will fluctuate about the same central value as an individual observation, but with less deviation because of "averaging out." We thus find plausible the formulas

$$\mu_{\overline{X}} = \mu \tag{6-10}$$

$$\sigma_{\overline{X}} = \frac{\sigma}{\sqrt{n}} \tag{6-11}$$

Proof. First, for the mean, we apply the last row of Table 5-6 to (6-9).

$$\mu_{\overline{X}} = \frac{1}{n}\,\mu_S$$

and from (6-5)

$$= \frac{1}{n}\,(n\mu)$$

$$\mu_{\overline{X}} = \mu \qquad \text{(6-10) proved}$$

Now, for the variance, we apply the last row of Table 5-6 to (6-9) again.

$$\sigma^2_{\overline{X}} = \left(\frac{1}{n}\right)^2 \sigma^2_S$$

and from (6-7)

$$= \frac{1}{n^2}\,(n\sigma^2)$$

$$\sigma^2_{\overline{X}} = \frac{\sigma^2}{n} \tag{6-12}$$

$$\sigma_{\overline{X}} = \frac{\sigma}{\sqrt{n}} \qquad \text{(6-11) proved}$$

Formulas (6-10) and (6-11) are illustrated in Figure 6-1*b*. A graph of the distribution of the sample mean for $n = 9$ and $n = 25$ is left as an exercise; this will confirm how this distribution concentrates about μ as sample size increases.

We review this section by reconsidering a familiar problem—the rolling of a die. Two rolls (X_1, X_2) can be regarded as a sample of 2 taken from the infinite population of all possible rolls of the die. This is also equivalent to a sampling of 2 chips from a bowl, as discussed in Problem 5-26. The probability distribution of the parent population is shown in Table 6-2a, along with its mean (μ) and standard deviation (σ).

Because this experiment has such simple probability characteristics, we can also compute the probability distribution of S and of $\overline{X}$ for a sample of 2 rolls of the die as shown in Table 6-2b; the moments of both S and $\overline{X}$ are also calculated in this table.

TABLE 6-2 (a) Probability Distribution of the Roll of a Die (Population)

x	$p(x)$	$x\,p(x)$
1	1/6	1/6
2	1/6	2/6
3	1/6	3/6
4	1/6	4/6
5	1/6	5/6
6	1/6	6/6

$\mu = 21/6 = 3.5$

similarly

$$\sigma = \sqrt{\frac{35}{12}} = 1.71$$

TABLE 6-2 (b) Probability Distribution of the Sample S and $\overline{X}$, with $n = 2$

(1) Outcome Set or Sample Space: First Die Second Die	(2) Sum s	Mean $\bar{x}$	(3) Probability	(4) $s\,p(s)$	(5) $\bar{x}\,p(\bar{x})$
.(1, 1)	2	1	1/36	2/36	1/36
.(1, 2) .(2, 1)	3	1.5	2/36	6/36	3/36
.(1, 3) .(2, 2) .(3, 1)	4	2	3/36	12/36	6/36
.	5	2.5	4/36	.	
.	6	3	5/36	.	
.	7	3.5	6/36	.	
36 equiprobable	8	4	5/36		
outcomes	9	4.5	4/36		
.	10	5	3/36		
.	11	5.5	2/36		
.(6, 6)	12	6	1/36		
				$\mu_S = 252/36 = 7.0$ similarly $\sigma_S = 2.4$	$\mu_{\bar{X}} = 126/36 = 3.5$ similarly $\sigma_{\bar{X}} = 1.2$

TABLE 6-2 (c) Alternative Calculation of Mean and Variance

Moment	Direct Calculation from Table 6-2b	On the Other Hand, This Relevant Formula Gives the Short-cut Calculation (using population μ and σ from Table 6-2a)		
μ_S	7.0	(6-5)	$\mu_S = n\mu$	$= 2(3.5) = 7$
σ_S	2.4	(6-8)	$\sigma_S = \sqrt{n}\,\sigma$	$= \sqrt{2}(1.71) = 2.4$
$\mu_{\bar{X}}$	3.5	(6-10)	$\mu_{\bar{X}} = \mu$	$= 3.5$
$\sigma_{\bar{X}}$	1.2	(6-11)	$\sigma_{\bar{X}} = \dfrac{\sigma}{\sqrt{n}}$	$= \dfrac{1.71}{\sqrt{2}} = 1.2$

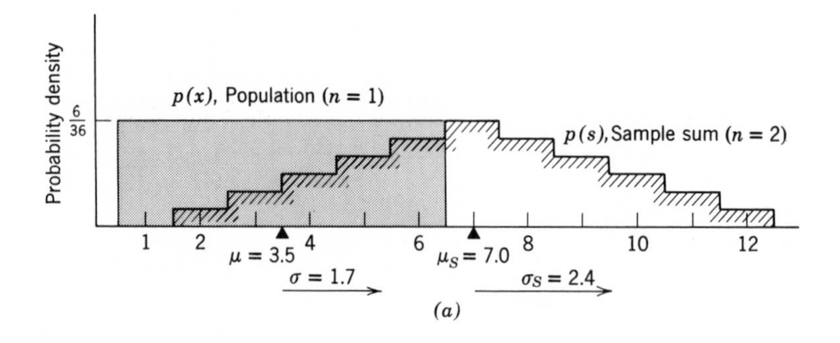

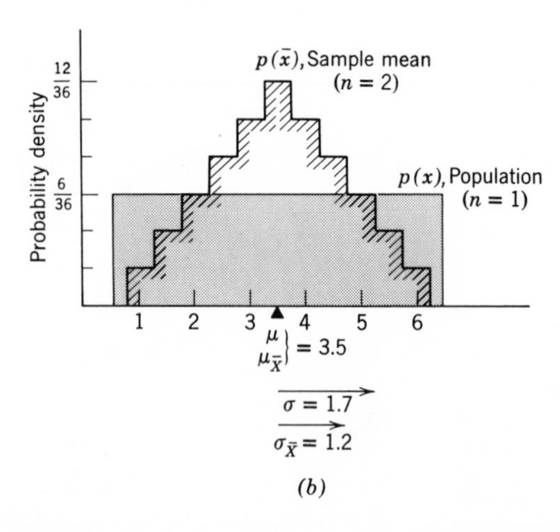

FIG. 6-2 Throwing a die twice (a specific illustration of Fig. 6-1). (*a*) Relation of the sample sum S to the parent population. (*b*) Relation of the sample mean $\bar{X}$ to the parent population. (*Note.* In order to facilitate graphing, the probabilities were converted to probability *densities*, so that they would all have the same comparable area = 1.)

In Table 6-2c we show how these moments could have been obtained more simply using the formulas of this section. Finally, this die-tossing example is summarized in Figure 6-2.

PROBLEMS

6-1 True or false? If false, correct the errors:

When a die is rolled twice, the average of the 2 numbers ($\bar{X}$) is a random variable having an expectation of $3\frac{1}{2}$, the same expectation as for a single roll X. This illustrates $\mu_{\bar{X}} = \mu$.

The range of $\bar{X}$ is from 1 to 6, also the same as for a single roll X. However, $\bar{X}$ does not take on all values equally likely—the extreme values are rare. Thus $\bar{X}$ has a smaller standard deviation than X, illustrating $\sigma_{\bar{X}} = \sigma/\sqrt{n}$.

Incidentally, this illustrates why the range of a random variable is a better measure of spread than the standard deviation.

6-2 True or false? If false, correct the errors:

If 10 men were randomly sampled from the population of Table 6-1, and then laid end to end, the expectation of the total length would be $n\mu = 678$ inches. The total length would vary (from sample to sample) with a standard deviation of $n\sigma = 53$ inches.

On the other hand, if the 10 men in the random sample were averaged, the expectation of the average would be $\mu = 67.8$ inches, and its standard deviation would be $\sigma = 5.3$ inches. This is how the long and short men in the sample tend to "average out," making $\bar{X}$ fluctuate less than a single observation.

6-3 (Classroom Exercise)

(a) Make a relative frequency (probability) graph of the population of heights of the men in the class.

(b) Take a few random samples of size 4 (with replacement), showing how in each sample the tall students tend to be offset by short students.

(c) For each sample, calculate $\bar{X}$. Plot the values of $\bar{X}$ and compare to (a).

6-4 The population of employees in a certain large office building has weights distributed around a mean of 150 pounds, with a standard deviation of 20 pounds. A random group of 25 employees get in the elevator each morning. Find the mean and variance of:

(a) The total weight S.

(b) The average weight $\bar{X}$.

⇒ 6-5 A bowl is full of many chips, one-third marked 2, one-third marked 4, and one-third marked 6.

(a) When one chip is drawn, let X be its number. Find μ and σ, (the population mean and standard deviation.)
(b) When a sample of 2 chips is drawn, let $\bar{X}$ be the sample mean. Find
(1) The probability table of $\bar{X}$.
(2) From this calculate $\mu_{\bar{X}}$ and $\sigma_{\bar{X}}$; check your answers using (6-10) and (6-11).
(c) Repeat (b) for a sample of 3 chips.
(d) Graph $p(\bar{x})$ for each case above, i.e., for sample size $n = 1, 2, 3$. Comparison is facilitated by using probability density, i.e., by using a bar graph with probability = area = (height) (width).

As n increases, notice that $p(\bar{x})$ becomes more concentrated around μ. What else is happening to the *shape* of $p(\bar{x})$?

6-4 THE CENTRAL LIMIT THEOREM

In the preceding section we found the mean and standard deviation of $\bar{X}$. The one question we have not yet addressed is the *shape* of its distribution. We consider two cases.

(a) The Distribution of the Sample Mean When the Population is Normal

In this case $\bar{X}$ is exactly normal. This follows from a theorem on linear combinations, which we quote without proof:

> If X and Y are normal, then any linear combination $Z = aX + bY$ is also a normal random variable. (6-13)

With a normal population, each observation in a sample $X_1, X_2, \ldots X_n$ is normal. The sample mean $\bar{X}$ can be written as a linear combination of these n normal variables,

$$\bar{X} = \frac{1}{n} X_1 + \frac{1}{n} X_2 \cdots + \frac{1}{n} X_n \tag{6-14}$$

so that (6-13) can be used to establish that $\bar{X}$ is normal. Finally, we re-emphasize that its distribution concentrates about μ as sample size n increases (ref. 6-11).

(b) The Distribution of $\bar{X}$ When the Population is *Not* Normal

It is surprising that, even in this case, most of the same conclusions follow. As an example, consider the bowl of 3 kinds of chips in Problem 6-5.

This is obviously a nonnormal population; in fact, it is a rectangular distribution. As a larger and larger sample is taken, the distribution of $\overline{X}$ is graphed[5] in Figure 6-3a. As well as the increasing concentration of this distribution, we notice the tendency to the normal bell shape.

This same tendency to the normal shape occurs for the sample of dice throws ($n = 1, 2, 3, \ldots$ throws from a population of all possible throws), as shown in Figure 6-3b.

Finally, in Figure 6-3*c* a third population is shown, having chips numbered 2, 4, and 6, with proportions 1/4, 1/4, and 1/2. Sample means from this population also show the same tendency to normality.

These three examples display an astonishing pattern—the sample mean becomes normally distributed as n grows, no matter what the parent population is. This pattern is of such central importance that mathematicians have formulated it as

> *The Central Limit Theorem.* As the sample size n increases, the distribution of the mean, $\overline{X}$, of a sample taken from practically[6] any population approaches a *normal* distribution, (with mean μ and standard deviation $\sigma/\sqrt{n}$). (6-15)

The central limit theorem is not only remarkable, but very practical as well. For it completely specifies the distribution of $\overline{X}$ in large samples, and is therefore the key to large-sample statistical inference. In fact, as a rule of thumb it has been found that usually when the size n reaches about 10 or 20, the distribution of $\overline{X}$ is practically normal. This is certainly the case in the 3 examples of Figure 6-3.

In conclusion, we can assume that $\overline{X}$ is normal for any sample taken from a normal population, and for large samples taken from practically any population. With our previous conclusions on the mean and standard deviation of $\overline{X}$, we can now be very specific in our deduction about a sample mean taken from a known population.

Example

Consider the marks of all students on a statistics test. If the marks have a normal distribution with a mean of 72 and standard deviation of 9, compare

[5] The student has already done the first 3 graphs of Figure 6-3*a* (in Problem 6-5), and the first 2 graphs of Figure 6-3*b* (in Table 6-2). The rest of the graphs may be similarly calculated.

[6] The one qualification is that the population have finite variance. For a proof of this theorem, see, for example, P. Hoel, *Introduction to Mathematical Statistics*, 3rd ed., pp. 143–5, John Wiley & Sons, 1962.

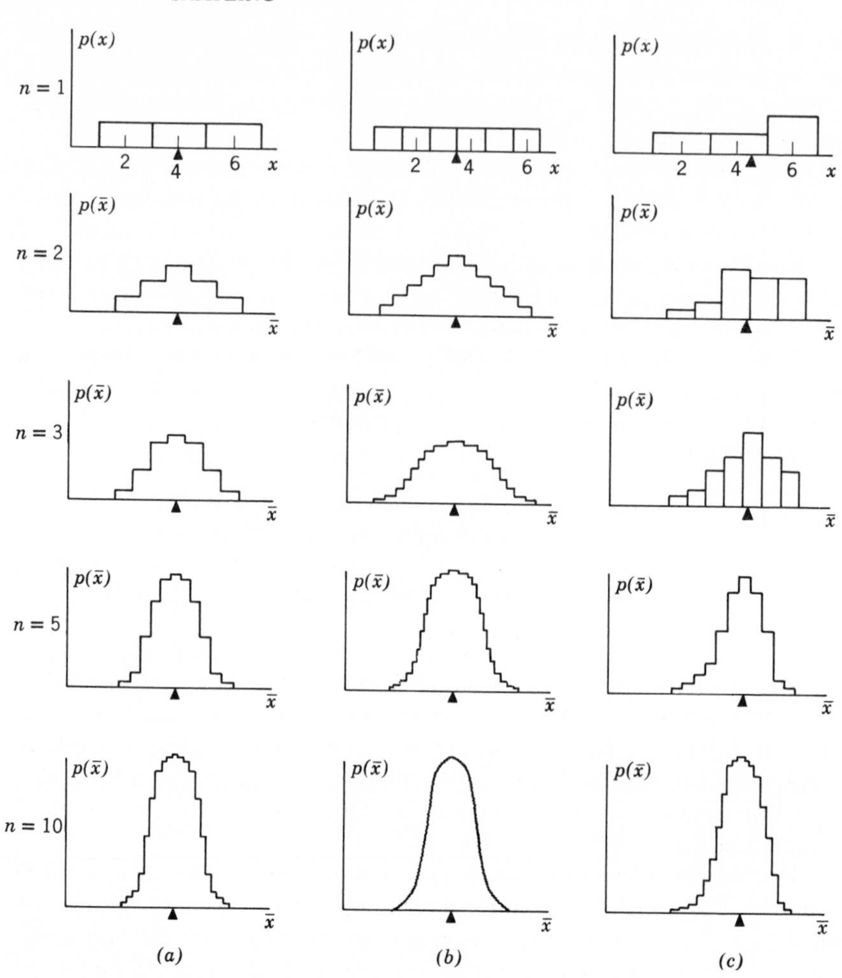

FIG. 6-3 The limiting normal shape for $p(\bar{x})$. (*a*) Bowl of three kinds of chips. (*b*) Bowl of six kinds of chips (or die). (*c*) Bowl of three kinds of chips of different frequency.

(1) the probability that any one student will have a mark over 78 with (2) the probability that a sample of 10 students will have an average mark over 78.

1. The probability that a single student will have a mark over 78 is found by standardizing the normal population

$$\Pr(X > 78) = \Pr\left(\frac{X-\mu}{\sigma} > \frac{78-72}{9}\right)$$

$$= \Pr(Z > .67) = .50 - .2486 = .2514.$$

2. Now consider the distribution of the sample mean. From the theorems above we know it is normal, with a mean of 72 and a standard deviation $\sigma/\sqrt{n} = 9/\sqrt{10}$. From this we calculate the probability of a sample mean exceeding 78 to be:

$$\begin{aligned} \Pr(\bar{X} > 78) &= \Pr\left(\frac{\bar{X} - \mu}{\sigma/\sqrt{n}} > \frac{78 - 72}{9/\sqrt{10}}\right) \\ &= \Pr(Z > 2.11) \\ &= .0174 \end{aligned} \tag{6-16}$$

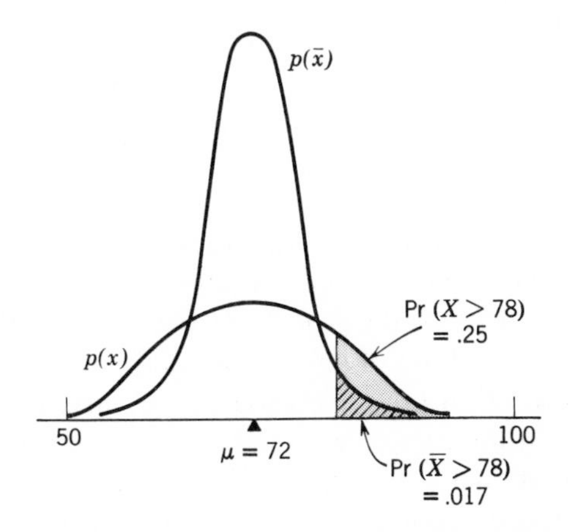

FIG. 6-4 Comparison of probabilities for the population and for the sample mean.

Hence, although there is a reasonable chance (about 1/4) that a single student will get over 78, there is very little chance (about 1/60) that a sample average of ten students will perform this well. This is shown in Figure 6-4.

PROBLEMS

6-6 The weights of packages filled by a machine are normally distributed about a mean of 25 ounces, with a standard deviation of one ounce. What is the probability that n packages from the machine will have an average weight of less than 24 ounces if $n = 1, 4, 16, 64$?

6-7 Suppose that the education level among adults in a certain country has a mean of 11.1 years, and a variance of 9. What is the probability that in a random survey of 100 adults you will find an average level of schooling between 10 and 12?

6-8 Does the central limit theorem (6-15) also hold true for the sample sum? Justify briefly.

6-9 An elevator is designed with a load limit of 2000 lb. It claims a capacity of 10 persons. If the weights of all the people using the elevator are normally distributed with a mean of 185 lb and a standard deviation of 22 lb, what is the probability that a group of 10 persons will exceed the load limit of the elevator?

6-10 Suppose that bicycle chain links have lengths distributed around a mean $\mu = .50$ cm, with a standard deviation $\sigma = .04$ cm. The manufacturer's standards require the chain to be between 49 and 50 cm long.
(a) If chains are made of 100 links, what proportion of them meets the standards?
(b) If chains are made of only 99 links, what proportion now meets the standards? How many links should be put in a chain?
(c) Using 99 links, to what value must σ be reduced (how much must the quality control on the links be improved) in order to have 90 percent of the chains meet the standards?

(6-11) The amount of pocket money that persons in a certain city carry has a nonnormal distribution with a mean of \$9.00 and a standard deviation of \$2.50. What is the probability that a group of 225 individuals will be carrying a total of more than \$2100?

6-12 In Problems 6-6 to 6-11, the variance formulas required that the n individuals in the sample were independently drawn. Do you think this is a questionable assumption? Why?

*6-13 A farmer has 9 wheatfields planted. The distribution of yield from each field has a mean of 1000 bushels and variance 20,000. Furthermore, the yields of any 2 fields are correlated, because they share the same weather conditions, weed control, etc; in fact the covariance is 10,000. Letting S denote the total yield from all 9 fields, find
(a) The mean and variance of S. [*Hint.* How must the footnote proof to (5-32) be adjusted?]
(b) Pr $(S < 8{,}000)$, assuming S is normal.

*6-5 SAMPLING FROM A FINITE POPULATION, WITHOUT REPLACEMENT

In the preceding analysis, we have assumed either sampling with replacement, or alternatively, sampling from an infinite population—in which case it doesn't matter whether we replace or not. This leaves one remaining possibility—sampling from a finite population, without replacement.

* This is a starred section, and like a starred problem, it is optional; the student may skip it without loss of continuity.

We have already argued in Section 6-1 that all the X_i in a sample of n observations $(X_1, X_2, \ldots X_n)$ will have the same (marginal) distribution whether or not we replace; i.e., equation (6-1) holds regardless, so that (6-5) still follows from (6-3):

$$\mu_S = n\mu \qquad \text{(6-5) repeated}$$

And similarly

$$\mu_{\bar{X}} = \mu \qquad \text{(6-10) repeated}$$

On the other hand, the *variance* of the sample mean *does* depend on whether or not we replace; it is easy to see why. Suppose we sample 10 of the heights of the male students on a college campus; suppose further that the first student we sample is the star of the basketball team (say Lew Alcindor, at 7 feet 1 inch). Clearly, we now face the problem of a sample average that is "off target"—specifically, too high. If we replace, then in the next 9 men chosen, Alcindor *could* turn up again, throwing our sample mean even further off target on the high side. But if we don't replace, then we don't have to worry about Alcindor again. In summary, sampling without replacement yields a more reliable sample mean (i.e., $\bar{X}$ has less variance), because extreme values once sampled, cannot return to haunt us again.

Formally, the argument runs as follows. If we sample without replacement, then $X_1, X_2 \ldots X_n$ are *not* independent. Hence all our theorems on the variance of S and $\bar{X}$ above, based on the independence assumption, do not hold true. Specifically, (6-7)—which assumed replacement—now must be modified to:

$$\text{var } S = \sigma_S^2 = n\sigma^2 \left[\frac{N-n}{N-1}\right] \qquad \text{(6-17)}$$

(sampling without replacement)

where N = population size, and n = sample size. Furthermore (6-12)—which also assumed replacement—must be similarly modified to:

$$\text{var } \bar{X} = \sigma_{\bar{X}}^2 = \frac{\sigma^2}{n}\left[\frac{N-n}{N-1}\right] \qquad \text{(6-18)}$$

(sampling without replacement)

Although we do not prove these two formulas, we interpret them:

1. The variance of $\bar{X}$ without replacement (6-18) is less than the variance of $\bar{X}$ with replacement (6-12); (this is the formal confirmation of our intuitive example of heights of college students). This occurs because the

"reduction factor,"

$$\left(\frac{N-n}{N-1}\right) \tag{6-19}$$

appearing in (6-18) is less than one. [Unless of course, the sample size n is only one. In this case, no distinction can be made between replacement and nonreplacement, and (6-12) and (6-18) must necessarily coincide. If you have wondered where the 1 came from in the denominator, you can see that it is necessary, in order to logically make (6-12) and (6-18) equivalent, as they must be, for a sample size of one.]

2. When $n = N$, the sample coincides with the whole population, every time. Hence every sample mean must be the same—the population mean. The variance of the sample mean, being a measure of its spread, must be zero. This is reflected in (6-19) having a zero numerator; and var $\bar{X}$ in (6-18) becomes zero. (Note that with replacement this is not the case—in this instance, $n = N$ does not guarantee that the sample and the population are identical).

3. On the other hand, when n is much smaller than N, (e.g., when 200 men are sampled from 80 million), then (6-19) is practically 1, so that var $\bar{X}$ is practically the same as with replacement. This, of course, coincides with common sense; if the population is very large, it makes very little difference whether or not the observations are thrown back in again before continuing sampling.

PROBLEMS

*6-14 In the game of bridge, cards are allotted points as follows:

Cards	Points
All cards below jack	0
Jack	1
Queen	2
King	3
Ace	4

(a) For the population of 52 cards, find the mean number of points, and the variance.

(b) In a randomly dealt hand of 13 cards, the number of points Y is a random variable. What are the mean and variance of Y? (Bridge players beware: no points counted for distribution).

(c) What is Pr $(Y \geq 13)$? (*Hint.* The distribution shape is approximately normal, as we might hope from the central limit theorem).

*6-15 Rework Problem 6-9, assuming the population of people using the elevator is no longer very large, but rather
(a) $N = 500$.
(b) $N = 50$.

6-6 SAMPLING FROM BERNOULLI POPULATIONS

We have examined the distribution of a sample mean and a sample sum; the final statistic that we study is the one referred to in our poll of U.S. voters in Chapter 1, the sample proportion P.

(a) The Bernoulli Population

First, we must be clear on the population from which the sample is drawn. We conceive of this as being made up of a large number of individuals,

TABLE 6-3 A Bernoulli Variable

	x	Frequency	$p(x)$	$x\,p(x)$
Republican	0	66,000,000	$\frac{66{,}000{,}000}{150{,}000{,}000} = .44$	0
Democrat	1	84,000,000	$\frac{84{,}000{,}000}{150{,}000{,}000} = .56$	.56
		$\sum = 150{,}000{,}000$		$\mu = .56$

all marked D or R (Democrat or Republican). We can make this look like the familiar bowl of chips by relabelling each D with a 1 and each R with a 0. Thus, if the voting population of 150 million is comprised of 84 million Democrats and 66 million Republicans, the population probability distribution would be as shown in Table 6-3.

The population proportion π of Democrats is .56, which is also the probability, in sampling one individual at random, that a Democrat will be chosen. This is called a "Bernoulli" population and its distribution is graphed later in Figure 6-6*a*. This is the simplest kind of probability distribution, lumped at only two values—0 and 1. (Note that this population is as far from being normal as any that we will encounter). Its mean and variance are easily computed in Table 6-4. In our example, $\mu = .56$, and $\sigma = .5$

The reason that the arbitrary values of 0 and 1 were assigned to the population is now clear. This ensures that μ and π coincide.

TABLE 6-4 Calculation of μ and σ^2 for a Bernoulli Population

x	$p(x)$	$x\,p(x)$	$(x - \mu)$	$(x - \mu)^2\,p(x)$
0	$(1 - \pi)$	0	$-\pi$	$(-\pi)^2(1 - \pi)$
1	π	π	$1 - \pi$	$(1 - \pi)^2\pi$

$$\mu = \pi \tag{6-20}$$

$$\sigma^2 = \pi(1 - \pi) \tag{6-21}$$

$$\sigma = \sqrt{\pi(1 - \pi)} \tag{6-22}$$

(b) Bernoulli Sampling

We now ask, "What can we expect of a sample drawn from this sort of population?" The population is so large that even without replacement, the n observations are practically independent; the probability of choosing a Democrat remains practically .56 regardless of whether or not we replace.

If we take a sample of $n = 50$ let us say, we might obtain, for example, the following 50 numbers:

$$0\ 1\ 1\ 0\ 1\ 0\ 0\ 1\ 0\ 1\ 1\ 1 \ldots 0\ 1\ 1 \tag{6-23}$$

The sample sum, of course, will be just the number of Democrats in the sample. We recall encountering this before as a binomial random variable in Table 4-3; thus a binomial random variable is simply a sample sum in disguise.

Why is this interesting coincidence of any practical value? Suppose we wish to calculate the binomial probability of *at least* 30 Democrats in 50 trials. We could evaluate the probability of exactly 30 Democrats, of 31, 32, and so on. This would require a major computational effort: not only are some twenty odd probabilities involved, but in addition, each is extremely difficult to calculate.[7] But we recognize that this is equivalent to calculating

[7] As an exercise, the student should consider whether it is feasible to evaluate the probability of getting 30 Democrats in a sample of 50, which is:

$$\binom{50}{30}(.56)^{30}(.44)^{20}$$

the probability that S, the sample sum taken from a Bernoulli population, is at least 30 in a sample of 50. This is very easy to calculate, because in the previous section we have completely described the distribution of any sample sum.

S in fact is approximately normally distributed,[8] with the following mean and variance:

From (6-5),

$$\mu_S = n\mu$$

from (6-20),

$$\boxed{\mu_S = n\pi} \qquad \text{Binomial mean} \tag{6-24}$$

From (6-7),

$$\sigma_S^2 = n\sigma^2$$

and using (6-21),

$$\sigma_S^2 = n\pi(1 - \pi)$$

$$\boxed{\sigma_S = \sqrt{n\pi(1 - \pi)}} \qquad \text{Binomial standard deviation} \tag{6-25}$$

Hence the probability of at least 30 Democrats in a sample of 50 is:

$$\Pr(S \geq 30)$$

which, in standardized form is

$$\Pr\left(\frac{S - u_S}{\sigma_S} \geq \frac{30 - 28}{\sqrt{12.3}}\right) \simeq \Pr(Z \geq .58) \simeq .28 \tag{6-26}$$

To confirm the usefulness of this normal approximation to the binomial, the student should compare this simple solution with the calculations involved in evaluating some twenty-odd expressions, each like the one in the footnote on p. 120. The normal approximation to the binomial is graphed[9] in Figure 6-5.

[8] For large n, by the central limit theorem. A useful rule of thumb is that n should be large enough to make $n\pi > 5$ and $n(1 - \pi) > 5$. If n is large, yet π is so small that $n\pi \leq 5$, then there is a better approximation than the normal, called the Poisson distribution.

[9] This graph clearly indicates that a better approximation to the binomial histogram would be the area under the curve above 29.5, not 30. This peculiarity arises from trying to approximate a discrete variable with a continuous one, and is therefore called the *continuity correction*. Our better approximation is

$$\Pr\left(\frac{S - \mu_S}{\sigma_S} \geq \frac{29.5 - 28}{\sqrt{12.3}}\right) \simeq \Pr(Z \geq .43) \simeq .334 \tag{6-27}$$

To keep the analysis uncluttered, this continuity correction is ignored in the rest of the book.

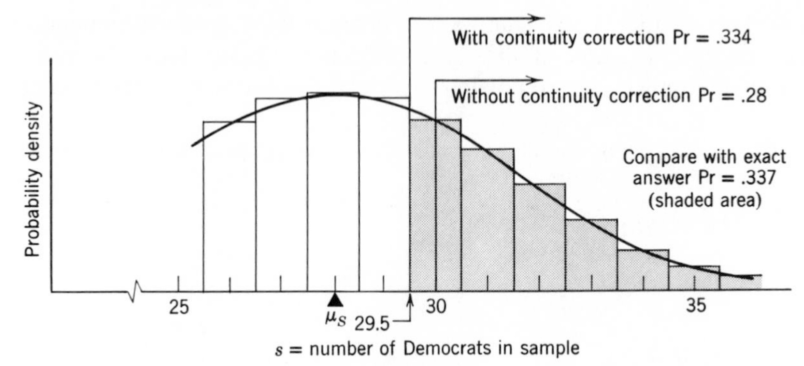

FIG. 6-5 Normal approximation to the binomial. (Compare Fig. 6-1*a*.)

We now turn to the second major issue of this section: with π known, what is the distribution of the sample proportion P?

Just as the total number of successes is merely the sample sum in disguise, so the sample proportion is merely the sample mean in disguise:

$$P = \frac{S}{n} = \bar{X} \tag{6-28}$$

All our theory developed for $\bar{X}$, can now be applied to determine the distribution of the sample statistic P. Thus, from (6-10) and (6-20) the mean of P is:

$$\boxed{\mu_P = \pi} \tag{6-29}$$

From this we note that, on the average, the sample proportion P is on target, i.e., its average value is equal to the population proportion—which (we shall see in Chapter 8) it will be used to estimate. But any specific sample P will be subject to sampling variation and will typically fall above or below π. From (6-11) and (6-22) we discover that its standard deviation is

$$\sigma_P = \sqrt{\frac{\pi(1-\pi)}{n}} \tag{6-30}$$

Finally, since P is a sample mean, its distribution is normal for large samples (central limit theorem).

As an example, consider the population of voters shown in Figure 6-6*a*. What is the probability that in a random sample of 50 voters between 50 and 60 percent will be Democrats? From (6-29) and (6-30)

$$\mu_P = \pi = .56$$

$$\sigma_P = \sqrt{\frac{\pi(1-\pi)}{n}} = \sqrt{\frac{.56(1-.56)}{50}} = .070$$

These two values, along with our knowledge that P is normal, completely define the distribution of P shown in Figure 6-6*b*. Even though our population was nowhere near normal, our sample statistic P *is* approximately normal.

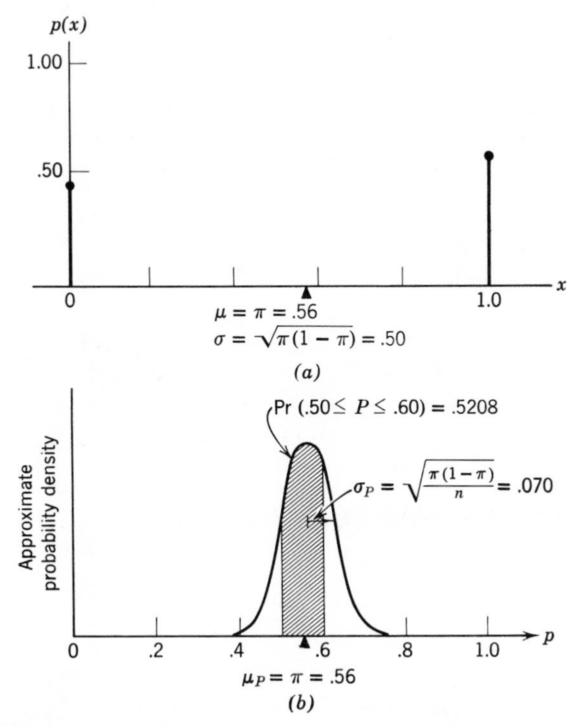

FIG. 6-6 Relation of the sample proportion to the population proportion (compare Fig. 6-1*b*). (*a*) Population of voters. (*b*) In a sample of 50 voters, distribution of P.

The evaluation of the area of this normal distribution between .50 and .60 is now a straightforward matter:

$$\begin{aligned}\Pr(.50 \leq P \leq .60) &= \Pr\left(\frac{.50 - .56}{.070} \leq \frac{P - \mu_P}{\sigma_P} \leq \frac{.60 - .56}{.070}\right) \\ &= \Pr(-.857 \leq Z \leq .572) \\ &= .5208\end{aligned}$$

PROBLEMS

(Note that if you want high accuracy in your answers, you should make continuity corrections.)

6-16 Suppose Gallup takes a poll of 1000 voters from a population which is 56 percent Democratic. Letting P be the sample proportion, find
(a) Pr $(.52 < P < .60)$.
(b) Pr $(P > .5)$, i.e., the probability that the sample will correctly predict the election. Note how we are beginning to answer the problems raised in Chapter 1.

6-17 In tossing a fair coin 50 times, what is the probability that the proportion of heads will exceed .55?

6-18 If a fair die is rolled 100 times, what is the probability that at least one quarter of these are aces? Answer two ways:
(a) Pr $(P \geq \frac{1}{4})$
(b) Pr (total number of aces ≥ 25)

(6-19) What is the chance that of the first 100 babies born in the New Year, more than 60 will be boys?

6-20 What is the chance that of the first 8 babies born in the New Year, more than 6 will be boys? Answer two ways:
(a) Exactly, using the binomial distribution.
(b) Approximately, using the normal distribution.

6-7 SUMMARY OF SAMPLING THEORY

(a) General Sampling

1. The distribution of the sample mean $\bar{X}$ is approximately normal for large samples—say $n \geq 10$ or 20 as a rule of thumb. (Moreover, if the population is near normal, then a much smaller sample will be approximately normal.)

2. $\bar{X}$ will have an expectation equal to μ, the population expectation.

3. If we sample without replacement, $\bar{X}$ will have a variance equal to:

$$\frac{\sigma^2}{n}\left[\frac{N-n}{N-1}\right]$$

If the population (N) is very large, this reduces to, approximately:

$$\frac{\sigma^2}{n}$$

which is also the formula for variance when we sample with replacement. Thus we may write:

$$\boxed{\bar{X} \sim N\left(\mu, \frac{\sigma^2}{n}\right)} \tag{6-31}$$

which is a useful abbreviation for "$\bar{X}$ is normally distributed, with mean μ and variance σ^2/n."

(b) Bernoulli Sampling

If we apply this sampling theory to a special population—chips coded 0 and 1—then we have the solution to the proportion problem. The sample proportion P is just a disguised $\bar{X}$, and the population proportion π is just a disguised μ, so that

$$\boxed{P \sim N\left(\pi, \frac{\pi(1-\pi)}{n}\right)} \tag{6-32}$$

again assuming n is sufficiently large.

Review Problems

6-21 Five men, selected at random from a normal population with mean weight $\mu = 160$ lb and $\sigma = 20$ lb, get on an elevator. What is the probability that
(a) All five men weigh more than 170?
(b) The average weight is more than 170?
(c) The total weight is more than 850?
(d) Give an intuitive reason why your answers are related.

6-22 A man at a carnival pays \$1 to play a game (roulette) with the following payoff:

Y = Gross Winning	Net Winning = $Y - 1$	Probability
0	-1	20/38
\$2	$+1$	18/38

(a) What is the average net winning in a game?
(b) What is his approximate chance of ending up a loser (net loss) if he plays the game:
 (1) 5 times?
 (2) 25 times?
 (3) 125 times?
(c) How could you get an exact answer for (b)1?
(d) How many times should he play is he wants to be 99% certain of losing?

6-23 Fill in the blanks.

(a) Suppose that in a certain election, the U.S. and California are alike in their proportion of Democrats, π, the only difference being that the U.S. is about 10 times as large a population. In order to get an equally reliable estimate of π, the U.S. sample should be ——— as large as the California sample.

(b) A certain length is measured with an error, which we suppose for simplicity to be $+2''$ or $-2''$, equally likely. A sample of n independent measurements is taken. The sample sum S *could possibly be* in error as much as ———. However, S is *likely* (95%) to be in error by no more than ———. For example, for $n = 100$, these two errors are

1. Worst possible error = ———
2. Likely error $\leq$ ———

⇒ 6-24 Let $\bar{X}$ be the sample mean when a die is thrown 1000 times. Intuitively we feel "fairly certain" that $\bar{X}$ is "quite close" to μ. More precisely, calculate

$$\Pr(\mu - .1 < \bar{X} < \mu + .1)$$

(6-25) In making up a budget, a housewife rounds out to the nearest 10¢.

(a) If the budget consists of 200 items, what is the chance that the rounding error will exceed \$1.00?

(b) Briefly state the assumptions necessary in answer (a).

(6-26) Suppose there are five men in a room, whose heights in inches are 62, 65, 68, 65, 65. One man is drawn at random with his height denoted X.

(a) Graph the probability function of X, i.e., the population distribution. Find its mean μ, and variance σ^2.

Suppose a sample of two men is drawn, with replacement, and the sample mean $\bar{X}$ is calculated.

(b) Construct a table of the probability function of $\bar{X}$. (*Hint*. List the possible samples, i.e., the sample space. Are the outcomes equally likely? For each outcome, calculate $\bar{X}$.)

(c) Graph the probability function of $\bar{X}$.

(d) Find the mean and variance of $\bar{X}$ from its probability distribution.

(e) Check your answers to (d) using the equations of this chapter.

(f) Is the following a valid interpretation of these formulas? If not, correct it:

> $\bar{X}$ fluctuates around μ; sometimes larger, sometimes smaller, but exactly equal to μ on the average. ($\mu_{\bar{X}} = \mu$). $\bar{X}$, the average of n observations, does not fluctuate as much as a single observation, however ($\sigma_{\bar{X}}^2 = \sigma^2/n < \sigma^2$). This is to be expected, because in a sample, a large observation will often be "cancelled out" by a

small observation, or at least swamped by the rest of the observations which will be more typical.

(6-27) Repeat Problem 6-26 for a sample of 2 men drawn *without* replacement. Why is this sampling without replacement preferable?

*6-28 In Chapter 3 it was stated that relative frequency in the long run is "very likely" to be "close to" probability. To make this statement precise, for the rolling of a die for example, let P denote the proportion of aces in 10,000 throws, and calculate

$$\Pr\left(\tfrac{1}{6} - .01 < P < \tfrac{1}{6} + .01\right)$$

chapter 7

Estimation I

7-1 INTRODUCTION

Before beginning statistical induction, we pause in Table 7-1 to review the concepts of sample and population.

It is essential to remember that the population is fixed, so that its mean μ and variance σ^2 are constants (though generally unknown). These are called population *parameters*.

By contrast, the sample mean $\bar{X}$ and sample variance s^2 are random variables, varying from sample to sample, with a certain probability distribution. For example, the distribution of $\bar{X}$ was found to be approximately $\bar{X} \sim N(\mu, \sigma^2/n)$ in Chapter 6. A random variable such as $\bar{X}$ or s^2 which is calculated from the observations in a sample is given the technical name *sample statistic*.

As a specific example of statistical inference, suppose we wish to estimate the average height of American men on a large Midwestern campus. This population mean μ is a fixed, but unknown parameter. We estimate it by taking a sample of 36 students, and compute the sample mean $\bar{X}$; let us suppose this turns out to be 68 inches. We shall see in the next section that this is our best single estimate or "point estimate" of μ. But we also know, from

TABLE 7-1 Review of Sample versus Population

Random Sample is a Random Subset of the	*Population*
1. Relative frequencies f_i/n are used to compute	Probabilities $p(x)$ are used to compute
2. $\bar{X}$ and s^2 which are examples of	μ and σ^2 which are examples of
3. Random statistics, or	Fixed parameters, or
4. Estimators	Targets

the theory of our previous chapter that unless we are extremely lucky in our sample this estimate $\bar{X}$ will not be exactly on target, but rather a bit high or a bit low. Technically, $\bar{X}$ is distributed around μ—above and below it—as shown in Figure 6-1*b*. If we want to be reasonably confident that our inference is correct, we cannot estimate μ to be precisely equal to our observed $\bar{X}$; instead we must estimate that μ is bracketed by some interval—known as a *confidence interval*—of the following form.

$$\mu = \bar{X} \pm \text{an error allowance} \tag{7-1}$$

As an example, we might estimate

$$\mu = 68 \pm 3 \text{ inches} \tag{7-2}$$

We observe that in evaluating the right-hand side of (7-1), there is no problem with the estimator $\bar{X}$; this is a simple calculation of the average of the sample values. The problem is the evaluation of the error allowance.

In this section we will show that we can be very specific in our interval estimate of μ, because we could be specific about the distribution of $\bar{X}$ around μ in the previous chapter. To keep inferences simple, we assume $\bar{X}$ is normally distributed according to the assumptions of Section 6-3, so that its distribution is that of Figure 7-1.

First we must decide: "How confident do we wish to be that our interval estimate is right—that it does bracket μ?" It is common to choose 95% confidence; in other words, we will be using a technique that will give us, in the long run, a correct interval estimate 19 times out of 20.

To get a confidence level of 95%, we select the smallest range under the normal distribution of $\bar{X}$ that will just enclose a 95% probability. Obviously, this is the middle chunk, leaving $2\frac{1}{2}$% probability excluded in each tail, as shown in Figure 7-1. From our normal tables, we note that this involves going above and below the mean by 1.96 standard deviations of $\bar{X}$. We therefore write

$$\Pr\left(\mu - 1.96\frac{\sigma}{\sqrt{n}} < \bar{X} < \mu + 1.96\frac{\sigma}{\sqrt{n}}\right) = 95\% \tag{7-3}$$

The bracketed inequalities may be solved for μ, "turned around" so to speak, obtaining the equivalent statement:[1]

$$\Pr\left(\bar{X} - 1.96\frac{\sigma}{\sqrt{n}} < \mu < \bar{X} + 1.96\frac{\sigma}{\sqrt{n}}\right) = 95\% \tag{7-4}$$

[1] To prove (7-4) more directly, we could begin by standardizing $\bar{X}$, which then has the standard normal distribution. Thus from the standard normal tables:

$$\Pr\left(-1.96 < \frac{\bar{X} - \mu}{\sigma/\sqrt{n}} < 1.96\right) = 95\% \tag{7-5}$$

In (7-5) the bracketed inequalities may be solved for μ, obtaining the equivalent inequalities

$$\Pr\left(\bar{X} - 1.96\frac{\sigma}{\sqrt{n}} < \mu < \bar{X} + 1.96\frac{\sigma}{\sqrt{n}}\right) = 95\% \qquad \begin{matrix}(7\text{-}6)\\ (7\text{-}4) \text{ proved}\end{matrix}$$

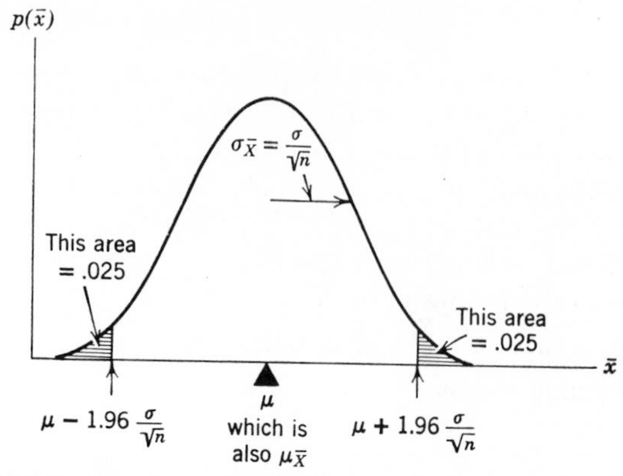

FIG. 7-1 Distribution of sample mean $\bar{X} \sim N[\mu, (\sigma^2/n)]$. (*Note.* μ is an *unknown constant;* we don't know what its value is; all we know is that, whatever μ may be, the *variable* $\bar{X}$ is distributed around it as shown in this diagram.)

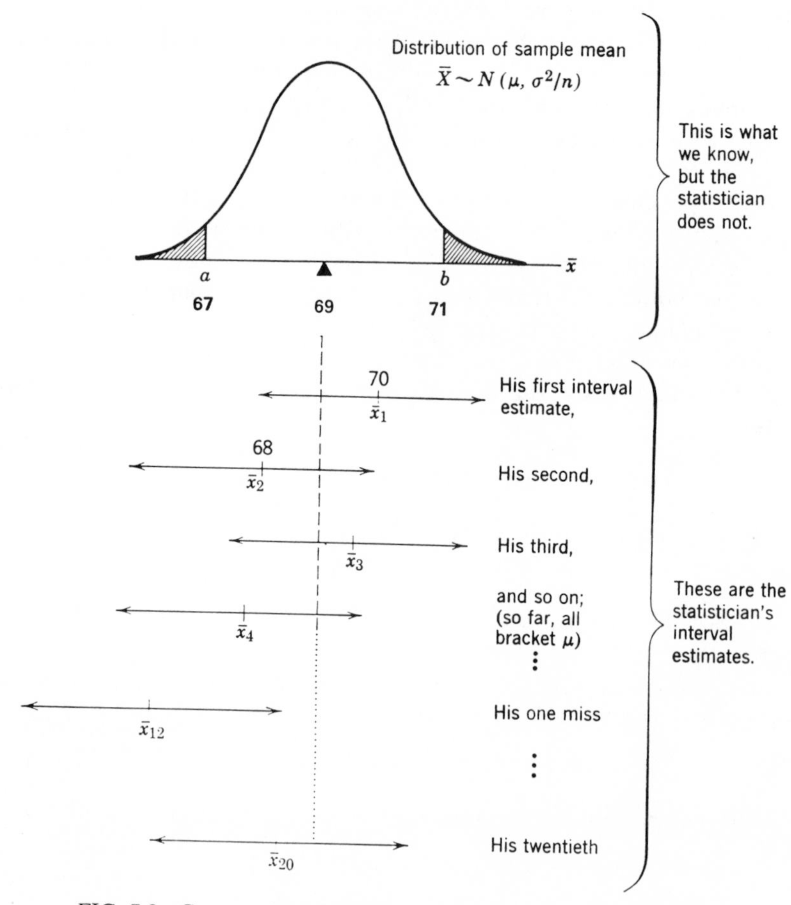

FIG. 7-2 Construction of twenty interval estimates: a typical result.

We must be exceedingly careful not to misinterpret (7-4). μ has *not* changed its character in the course of this algebraic manipulation. It has not become a variable; it remains a population constant. Equation (7-4), like (7-3), is a probability statement about the random variable $\bar{X}$, or more precisely, the "random interval" $\bar{X} - 1.96(\sigma/\sqrt{n})$ to $\bar{X} + 1.96(\sigma/\sqrt{n})$. *It is this interval that varies, not* μ.

To appreciate this fundamental point, let's return to our problem of constructing an interval estimate of average men's heights on our large campus. Moreover, to clearly illustrate what is going on, suppose we have some supernatural knowledge of the population μ (which we know to be 69 inches) and σ (which we know to be 6 inches). Now let's just observe what happens when the statistician (poor mortal that he is) tries to estimate μ using (7-4) above. Just for the sake of illustration, let's suppose he makes 20 such interval estimates, each time from a different random sample of 36. Figure 7-2 illustrates his typical experience.

In that diagram we show the distribution of the sample mean. We know that $\bar{X}$ is normal with mean equal to the population μ (69) and standard deviation equal to $\sigma/\sqrt{n} = 6/\sqrt{36} = 1$ inch. Thus from (7-3) we know that there is a 95% probability that any $\bar{X}$ will fall in the range 67 to 71 inches.

But the statistician doesn't know this; he blindly takes his first random sample, from which he computes the first mean $\bar{x}_1$ to be 70. From (7-4) he calculates[2] the appropriate 95% confidence interval for μ:

$$70 \pm 1.96 \frac{6}{\sqrt{36}} \tag{7-7}$$

$$= 68 \text{ to } 72 \tag{7-8}$$

This interval estimate for μ is the first one shown in Figure 7-2. We note that in his first effort, the statistician is right; μ is enclosed in this interval.

In his second sample, the statistician happens to draw a shorter group of individuals, and duly computes $\bar{x}_2$ to be 68 inches. From a similar evaluation of (7-4) he comes up with his second interval estimate shown in the diagram, and so on. We observe that nineteen of these twenty estimates bracket the constant μ. Only one—the twelfth—does not; in this case he missed the mark, and was wrong.

---

[2] We gloss over one difficulty here. In evaluating (7-4) the statistician has an observed value for $\bar{X}$ and knows that sample size n is 36. But there is one value he does *not* know: σ, the population standard deviation. All he can do is guess at it, and his best guess is, the sample standard deviation, s, which we suppose he computes to be 6 inches. We deal at length with this problem later; but for now we can rest assured that s will be a reasonable approximation for σ in this problem.

We can easily see why he was right most of the time. For each interval estimate he is simply adding and subtracting 2 inches to his sample mean; but this is the same ± 2 inches that defines the range *ab* around μ. Thus, if and only if he observes a sample mean within the range *ab*, will his interval estimate bracket μ. Nineteen of his twenty sample means do fall in the range *ab*, and in all these instances his interval estimate was right. He was wrong only in the one instance when he observed a sample mean outside *ab* (i.e., $\bar{x}_{12}$).

In practice, of course, a statistician does not take many samples—he would only take one (e.g., $\bar{x}_1$). And once this interval estimate is made, he is either right or wrong; this interval brackets μ or it does not. But the important point to recognize is that the statistician is using a method with a 95% probability of success; this follows because there is a 95% probability that his observed $\bar{X}$ will fall within the range *ab*, and as a consequence his interval estimate will bracket μ. This is what is meant by a 95% confidence interval: the statistician knows that in the long run, 95% of the intervals he constructs in this way will bracket μ.

To review, we briefly emphasize the main points:

1. The population parameter is constant, and remains constant. It is the interval estimate that is a random variable, because $\bar{X}$ is a random variable. As long as $\bar{X}$ is a random variable that can take on a whole range of values, it is referred to as an "estimator" of μ.

2. But once the sample has been observed and $\bar{X}$ takes on one specific value (e.g., $\bar{x}_1 = 70$ inches) it is then called an "estimate"[3] of μ. Since it is no longer a random variable, probability statements are no longer strictly valid. For this reason when the estimate $\bar{x}$ is substituted into (7-4), it is no longer called a 95% probability statement, but rather a 95% confidence interval:

$$\boxed{\bar{x} - 1.96\frac{\sigma}{\sqrt{n}} < \mu < \bar{x} + 1.96\frac{\sigma}{\sqrt{n}}} \tag{7-9}$$

Thus, our deduction in (7-4) that $\bar{X}$ is within $1.96\sigma/\sqrt{n}$ of μ is "turned around" into the induction that μ is within $1.96\sigma/\sqrt{n}$ of the observed $\bar{x}$. (7-9) is sometimes abbreviated to

$$\boxed{\begin{array}{l}\text{95\% confidence interval:}\\ \mu = \bar{x} \pm z_{.025}\dfrac{\sigma}{\sqrt{n}}\end{array}} \tag{7-10}$$

[3] For emphasis, the estimate is denoted by the lower case letter $\bar{x}$, while the random estimator is denoted by the capital letter $\bar{X}$. We might call $\bar{x}$ the realized value, and $\bar{X}$ the potential value.

where $z_{.025}$ is the critical value leaving $2\frac{1}{2}\%$ probability in the upper tail of the standard normal distribution.

To recapitulate, once $\bar{X}$ is observed to be $\bar{x}$, then the "die is cast," and the interval estimate (7-9) will be either right for certain, or wrong for certain.

3. Because of our omniscience, we know that the statistician erred only in his twelfth try. But *he* has no idea which estimates, if any, are wrong. All he knows is that he will be right 95% of the time, in the long run.

4. As sample size is increased, the distribution of $\bar{X}$ becomes more concentrated around μ ($\sigma/\sqrt{n}$ decreases as n increases), and the confidence interval narrows (becomes more precise).

5. If we wish to be more confident—e.g., 99% confident—of our conclusions, then we must leave less of the probability in each tail in Figure 7-2; thus the range *ab* increases. Hence our interval estimate becomes less precise. Note how this point and the one preceding verify our casual observations in Chapter 1.

6. An inference about the population parameter μ was feasible because we knew the distribution of its estimator $\bar{X}$. This raises an interesting question, "It is not possible that there are other statistics (for example, the sample median) that could be used to estimate μ? Why did we use the sample mean?" Intuitively, it seems preferable to estimate a mean with a mean. But there are stronger reasons, given in the next section.

PROBLEMS

7-1 An anthropologist measured the heights (in inches) of a random sample of 100 men from a certain population, and found the sample mean and variance to be 71 and 9 respectively.
(a) Find a 95% confidence interval for the mean height μ of the whole population.
(b) Find a 99% confidence interval.

7-2 A research study examines the consumption expenditures (in thousands of dollars) of a random sample of 50 American families (all at the same income and asset level). The sample mean is 5.2 and the standard deviation is .72. Construct a 95% confidence interval for the mean consumption of all American families (at this income and asset level).

7-3 The reaction times of 150 randomly selected drivers were found to have a mean of .83 sec and standard deviation of .20 sec. Find a 95% confidence interval for the mean reaction time of the whole population of drivers.

(7-4) From a very large class in statistics, the following 40 marks were randomly selected:

71	74	65	72	64	42	62	62	58	82
49	83	58	65	68	60	76	86	74	53
78	64	55	87	56	50	71	58	57	75
58	86	64	56	45	73	54	86	70	73

Construct a 95% confidence interval for the average mark of the whole class. (*Hint*. Reduce your work to manageable proportions by grouping into cells of width 5.)

7-5 What is the probability that a statistician who constructs 20 independent 95% confidence intervals will err:
(a) Once (as in our example in Section 7-1)?
(b) Not at all?
(c) More than once?

7-2 DESIRABLE PROPERTIES OF ESTIMATORS

To be perfectly general, we consider any population parameter θ, and denote an estimator for it by $\hat{\theta}$. (In our special example in the preceding section, μ is the population parameter θ, and $\bar{X}$ is its estimator $\hat{\theta}$). We would like the random variable $\hat{\theta}$ to vary within only a narrow range around its fixed target θ; (thus in our example in Figure 7-2, we should like the distribution of $\bar{X}$ to be concentrated around μ, as close to μ as possible). We develop this notion of closeness in several ways.

(a) No Bias

An unbiased estimator is one that is, *on the average*, right on target, as shown in Figure 7-3*a*. Formally, we state

Definition.

$$\boxed{\hat{\theta} \text{ is an unbiased estimator of } \theta \text{ if } E(\hat{\theta}) = \theta} \tag{7-11}$$

For example, $\bar{X}$ is an unbiased estimator of μ, because

$$E(\bar{X}) = \mu \qquad \text{(6-10) repeated}$$

Of course, an estimator $\hat{\theta}$ is called biased if $E(\hat{\theta})$ is different from θ; in fact, bias is defined as this difference:

Definition.

$$\boxed{\text{bias } B \triangleq E(\hat{\theta}) - \theta} \tag{7-12}$$

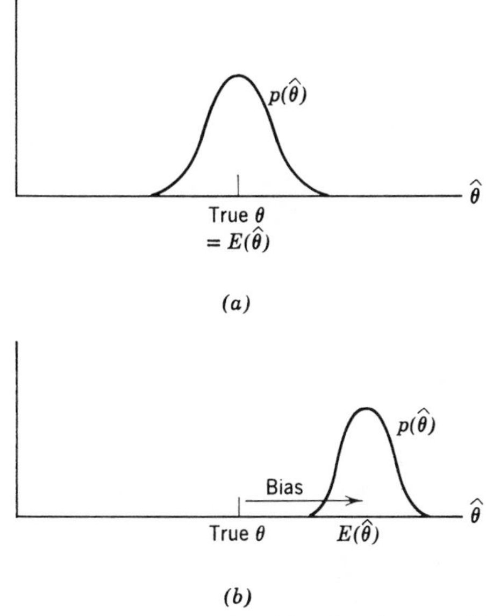

FIG. 7-3 Comparison of a biased and unbiased estimator. (*a*) Unbiased estimator. (*b*) Biased estimator.

Bias is illustrated in Figure 7-3*b*. The distribution of $\hat{\theta}$ is "off target"; since $E(\hat{\theta})$ exceeds θ, there will be a tendency for $\hat{\theta}$ to over-estimate θ.

As an example of a biased estimator, the sample mean squared deviation

$$\text{MSD} = \frac{1}{n} \sum (X_i - \bar{X})^2 \qquad (7\text{-}13)$$

(2-5a) repeated

will on the average underestimate σ^2, the population variance.[4] But if we inflate it just a little, by dividing by $n - 1$ instead of n, we obtain the sample variance

$$s^2 = \frac{1}{n-1} \sum (X_i - \bar{X})^2 \qquad (7\text{-}14)$$

(2-6) repeated

which has been proved an unbiased estimator of σ^2. (When we say "has been proved," we mean that it has been proved in advanced texts. If it has been proved in *this* text, we shall usually say "we have proved.") The student who

[4] This underestimation can be seen very easily in the case of $n = 1$. Then $\bar{X}$ coincides with X_i, so that Eq. (7-13) gives MSD $= 0$, which is an obvious underestimate of σ^2.

On the other hand, Eq. (7-14), when $n = 1$, gives $s^2 = 0/0$, which is undefined. But this is not a drawback; in fact, it is a good way to warn the unwary that since a sample of just one observation has no "spread," it cannot estimate the population variance σ^2 (assuming μ is unknown, of course).

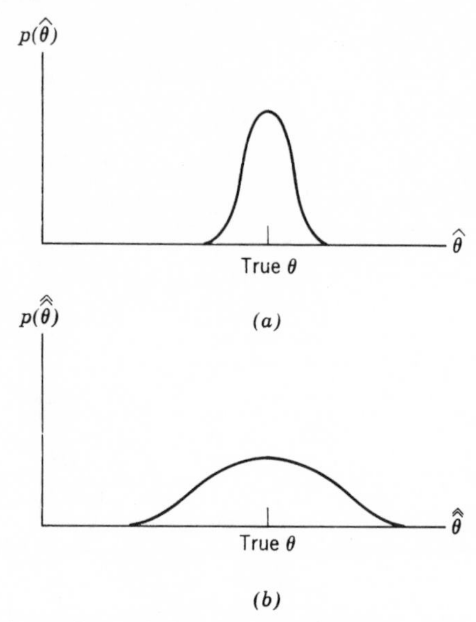

FIG. 7-4 A comparison of an efficient and inefficient estimator (both are unbiased). (*a*) Efficient. (*b*) Inefficient.

was puzzled by our division by $n - 1$ in defining s^2 in Chapter 2 can now see why: we want to use this sample variance as an unbiased estimator of the population variance.

Both the sample mean and median are unbiased estimators of μ in a normal population; thus, in judging which is to be preferred, we must examine their other characteristics.

(b) Efficiency

As well as being on target on the average, we should also like the distribution of an estimator $\hat{\theta}$ to be concentrated, that is, to have a small variance. This is the notion of efficiency, shown in Figure 7-4. We describe $\hat{\theta}$ as more efficient because it has smaller variance. A useful relative measure of the efficiency of two unbiased[5] estimators is:

Definition.

$$\text{Relative efficiency of } \hat{\theta} \text{ compared to } \hat{\hat{\theta}} \triangleq \frac{\operatorname{var} \hat{\hat{\theta}}}{\operatorname{var} \hat{\theta}} \tag{7-15}$$

[5] For biased estimators, the definition of efficiency is

$$\frac{E(\hat{\hat{\theta}} - \theta)^2}{E(\hat{\theta} - \theta)^2}$$

which of course is (7-15) if both estimators have 0 bias.

An estimator which is more efficient than any other is called absolutely efficient, or simply "efficient."

Finally, we are in a position to pass judgement on the merits of the sample mean and median as estimators of μ. In sampling from a normal population, $\bar{X}$ has been proved to be the efficient estimator of μ. We have already established that its variance is σ^2/n. On the other hand, the sample median has been shown to have, for large n, a variance of

$$(\pi/2)(\sigma^2/n) \tag{7-16}$$

Hence in a large sample, the relative efficiency of the sample mean compared to the median is derived from (7-15) as:

$$\frac{(\pi/2)(\sigma^2/n)}{\sigma^2/n} = \pi/2 \simeq 1.5 \tag{7-17}$$

Because it is half again more efficient, $\bar{X}$ is preferred. It will give us a point estimate that will tend to be closer to the target μ; or, it will give us a more precise (i.e., smaller range) interval estimate. Of course, by increasing sample size (n) we can reduce the variance of either estimator. Therefore, an alternative way of looking at the greater efficiency of the sample mean is to recognize that the sample median will yield as accurate a point or interval estimate only if we take a larger sample. Hence, using the sample mean is more efficient, because it costs less to sample; note how the economic and statistical definitions of efficiency coincide.

(c) Consistency

Roughly speaking, a consistent estimator is one that concentrates completely on its target as sample size increases indefinitely, as sketched in Figure 7-5. In the limiting case, as the sample size becomes infinite, a consistent estimator $\hat{\theta}$ will provide a perfect point estimate of the target θ.

We now state consistency more precisely. Just as the variance was a good measure of the spread of a distribution about its mean, so the

$$\text{mean squared error} \triangleq E(\hat{\theta} - \theta)^2 \tag{7-18}$$

is a good measure of how the distribution of $\hat{\theta}$ is spread about its target value θ. Consistency requires this to be zero in the limit:

Definition.

$$\boxed{\hat{\theta} \text{ is consistent}^6 \text{ if } E(\hat{\theta} - \theta)^2 \to 0 \text{ as } n \to \infty} \tag{7-19}$$

[6] This definition is sometimes called "consistency in mean-square." It implies a condition called "consistency in probability": for any positive δ (no matter how small),

$$\Pr(|\hat{\theta} - \theta| < \delta) \to 1 \quad \text{as } n \to \infty \tag{7-20}$$

This is often taken as the definition of consistency.

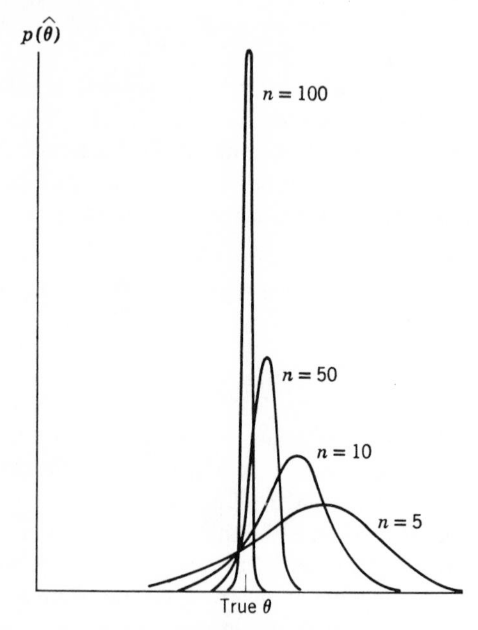

FIG. 7-5 A consistent estimator, showing how the distribution of $\hat{\theta}$ concentrates on its target θ as n increases.

Mean squared error is related to bias and variance by the following theorem.[7]

Theorem.

$$\boxed{E(\hat{\theta} - \theta)^2 = B_{\hat{\theta}}^2 + \sigma_{\hat{\theta}}^2} \tag{7-21}$$

Corollary.

$$\boxed{\begin{array}{l}\hat{\theta} \text{ is a consistent estimator iff}^8 \text{ its variance} \\ \text{and bias } both \text{ approach zero, as } n \to \infty.\end{array}} \tag{7-22}$$

If only the bias approaches zero, the estimator is called "asymptotically unbiased"—a condition that is clearly weaker than[9] consistency.

Consistency does not guarantee that an estimator is a good one. For example, as an estimator of μ in a normal population, the sample median is

---

[7] *Proof.*
$$\begin{aligned} E(\hat{\theta} - \theta)^2 &= E[(\hat{\theta} - \mu_{\hat{\theta}}) + (\mu_{\hat{\theta}} - \theta)]^2 \\ &= E(\hat{\theta} - \mu_{\hat{\theta}})^2 + 2(\mu_{\hat{\theta}} - \theta)E(\hat{\theta} - \mu_{\hat{\theta}}) + (\mu_{\hat{\theta}} - \theta)^2 \\ &= \sigma_{\hat{\theta}}^2 + 0 + (\mu_{\hat{\theta}} - \theta)^2 \\ &= \sigma_{\hat{\theta}}^2 + B_{\hat{\theta}}^2 \end{aligned}$$

[8] Iff is an abbreviation for "if and only if."

[9] Asymptotic unbiasedness is also a weaker condition than unbiasedness—since the latter applies for all n, not just $n \to \infty$.

consistent.[10] But it is not a good estimator; the sample mean is preferred because it is both consistent *and* efficient.

As a final example, the sample MSD is a consistent estimator of σ^2. It is true that it is a biased estimator; but as $n \to \infty$, this bias disappears, i.e., it is asymptotically unbiased.[11] Since it can also be proven that its variance tends to zero, the conditions of corollary (7-22) are satisfied. This concept of a biased, yet consistent estimator is a very important one—for example, in econometrics.

PROBLEMS

7-6 True or false? If false, correct it.

(a) The sample proportion P is an unbiased estimator of the population proportion π.

(b) μ is a random variable (varying from sample to sample), and is used to estimate the parameter $\bar{X}$.

7-7 Based on a sample of 2 observations, consider the two estimators of μ:

$$\bar{X} = (\tfrac{1}{2})X_1 + (\tfrac{1}{2})X_2$$

and

$$W \triangleq (\tfrac{1}{3})X_1 + (\tfrac{2}{3})X_2$$

(a) Prove they are unbiased.

(b) What is the efficiency of W relative to $\bar{X}$? Which estimator is preferable?

7-8 A farmer has a square field, whose area he wants to estimate. When he measures the length of the field, he makes a random error, so that his *observed length* 0_1 is a normal variate centered at 200 (the true but unknown value) with $\sigma = 20$. Worried about his possible error, he decides to take a second observation 0_2 and average. But he is in a dilemma as to how to proceed:

(1) Should he average 0_1 and 0_2, and then square? or

(2) Should he square first, and then average?

Mathematically, it's a question whether

$$\left(\frac{0_1 + 0_2}{2}\right)^2 \quad \text{or} \quad \left(\frac{0_1{}^2 + 0_2{}^2}{2}\right) \text{ is best}$$

---

[10] To prove consistency, we use corollary (7-22), noting that the sample median has zero bias, and a variance given by (7-16) which approaches zero.

[11] To establish this, we note that

$$\text{MSD} = \left(\frac{n-1}{n}\right)s^2$$

Thus MSD $\to s^2$ as $n \to \infty$. Since s^2 is unbiased (for any n), it follows that MSD is unbiased as $n \to \infty$.

(a) Are methods (1) and (2) really different, or are they just 2 different ways of saying the same thing?
(*Hint*. Try a couple of actual values, like $0_1 = 230$ and $0_2 = 200$, and work out (1) and (2).)
(b) If they are different, which has less bias?
(*Hints*. This problem will actually be easier if you avoid arithmetic by generalizing from a length of 200 feet to a length of μ, and also use general σ. Furthermore, the *normality* is irrelevant to questions of expectation. Finally, try using equation (4-5): $E(X^2) = \mu^2 + \sigma^2$.)
(c) Generalize answer (b) to a sample of n measurements.

7-9 As in Problem 6-5b, consider a bowl full of many chips—one-third marked 2, one-third marked 4, and one-third marked 6. When a sample of 2 chips is drawn, construct the probability table of $\bar{X}$, and hence
(a) Show (once more) that $\bar{X}$ is an unbiased estimator of μ.
(b) Is $(2\bar{X} + 1)$ an unbiased estimator of $(2\mu + 1)$?
(c) Is $(\bar{X})^2$ an unbiased estimator of μ^2? (Compare Problem 7-8.)
(d) Is $1/\bar{X}$ an unbiased estimator of $1/\mu$?
How could you have answered parts (a), (b), and (c) theoretically, without going through all the computations?

7-10 To illustrate bias very concretely, consider a sample of $n = 2$ tosses from the population of all tosses of a fair die. The population moments are easily computed:

$$\mu = 3.5, \qquad \sigma^2 = 35/12$$

We shall study sample estimators in 2 ways.
(a) *Empirical approach* (Monte-Carlo technique). Repeat the experiment many times. (You can simulate the roll of 2 dice with the random digits of Appendix Table II. If each student does it, say, 5 times, and the results from the class are pooled, this would save work.) The result will be a table like:

Result of 2 Tosses	$\bar{X}$	MSD	s^2
(3, 1)	2	1	2
(2, 5)	3.5	$2\frac{1}{4}$	$4\frac{1}{2}$
.			
.			
Averages	?	?	?

It may be convenient to array the data in a relative frequency table. Then answer

(1) Does $\bar{X}$ average close to μ?

(2) Does s^2 average close to σ^2?

(3) Does MSD average close to σ^2?

(b) *Theoretical approach.* In (a), if the experiment were repeated endlessly, the relative frequencies would settle down to probabilities. But these probabilities can be calculated very easily, by exploiting the symmetry of the dice. After calculating the relevant probability table, find

(1) $E(\bar{X}) \stackrel{?}{=} \mu$

(2) $E(s^2) \stackrel{?}{=} \sigma^2$

(3) $E(\text{MSD}) \stackrel{?}{=} \sigma^2$

*7-3 MAXIMUM-LIKELIHOOD ESTIMATION (MLE)

(a) Introduction

The next question is, "Does some technique exist for finding estimators with these attractive characteristics?" The maximum-likelihood method is the technique that statisticians most often use. We introduce it with an example of sampling from a Bernoulli population; to be concrete, suppose we flip a biased coin 10 times in order to estimate π, the population proportion[12] of heads, and get 4 heads. We shall temporarily forget the common-sense solution (estimate π with the sample proportion $P = 4/10$) in order to develop some general ideas.

With 4 out of 10 heads before us, we ask, "Is .1 a reasonable estimate of π?" If π were .1, then the probability of four heads (successes) in our ten tosses (trials) would be, according to the binomial formula

$$\binom{n}{x}\pi^x(1-\pi)^{n-x} = \binom{10}{4}(.1)^4(.9)^6 = .011$$

In other words, if $\pi = .1$, there is only about one chance in a hundred that we would get the sample we observed.

Similarly, we might ask ourselves how likely our result of four heads would be if π were .8. The student can verify that the probability of getting 4 heads from this sort of population is only .006; again it seems implausible that a population with $\pi = .8$ would yield the sample result we observed.

[12] This is also, of course, the population probability of heads. But, for simplicity, we refer hereafter only to the "proportion."

TABLE 7-2 Outline of Maximum Likelihood Estimation (MLE)

	Binomial: Special Case in Text	Binomial: General Case	MLE of μ from a Sample Drawn from a Normal Population, $p(x; \mu)$	MLE of any Parameter θ from any Population $p(x; \theta)$
Given:	4 successes in 10 trials.	x successes in n trials.	Sample values: x_1, x_2, x_3.	Sample values: $x_1, x_2, \ldots x_n$.
Find:	MLE of π.	MLE of π.	MLE of μ.	MLE of θ.
As follows:	The probability of 4 heads in 10 trials is $p(4; \pi) = \binom{10}{4}\pi^4(1-\pi)^6$ But in this estimation problem the sample values 10 and 4 have already been observed, hence are given. This therefore must be written as a likelihood function of π only: π : $L(\pi) = \binom{10}{4}\pi^4(1-\pi)^6$ 0 : 0 .1 : .011 .2 : .088 .3 : .200 .4 : .251 $\Leftarrow$ max .5 : .205 .6 : .111 .7 : .037 .8 : .006 1.0 : 0	Our x successes in n trials were generated by: $p(x; \pi) = \binom{n}{x}\pi^x(1-\pi)^{n-x}$ Since n and x are fixed at their observed values we can write this as a likelihood function of π only: $L(\pi) = \binom{n}{x}\pi^x(1-\pi)^{n-x}$ At what value of π is this function a maximum? Calculus shows that this occurs when $\pi = \frac{x}{n} = P$, the sample proportion.	Probability of our sample resulting from any μ is $p(x_1, x_2, x_3; \mu) = \prod_{i=1}^{3}\left[\frac{1}{\sqrt{2\pi\sigma^2}} e^{-(1/2\sigma^2)(x_i-\mu)^2}\right]$ But with (x_1, x_2, x_3) fixed at their observed values, the above becomes the likelihood function $L(\mu)$, an expression in μ: $L(\mu) = \prod_{i=1}^{3}\left[\frac{1}{\sqrt{2\pi\sigma^2}} e^{-(1/2\sigma^2)(x_i-\mu)^2}\right]$ Try out all possible values of μ, selecting that one that maximizes this function. Calculus shows that this is: $\mu = \frac{1}{3}(x_1 + x_2 + x_3)$ $= \bar{x}$, the sample mean.	Probability of our sample, for any θ, is: $p(x_1, x_2, \ldots x_n; \theta)$ $= p(x_1; \theta)p(x_2; \theta) \cdots p(x_n; \theta)$ $= \prod_{i=1}^{n} p(x_i; \theta)$ But with $(x_1, x_2, \ldots x_n)$ fixed at their observed values, the above function becomes $L(\theta)$, an expression in θ: $L(\theta) = \prod_{i=1}^{n} p(x_i; \theta)$ Select the value of θ that maximizes this likelihood function.
Conclude:	MLE of π is .4 which is the sample proportion P.	Hence, the MLE of π is P.	Thus $\bar{X}$ is the MLE of μ.	

Similarly, we consider all the other possible values for π, in each case asking how likely it is that this π would yield the sample that we in fact observed. The results are shown in the first column of Table 7-2, and graphed fully in Figure 7-6. We refer to this as the likelihood function, when the sample values of 4 and 10 are fixed, and the only variable in the function is the hypothetical value of π. For emphasis, we often write this as a function of π alone:

$$L(\pi) = \binom{10}{4} \pi^4 (1 - \pi)^6$$

The maximum likelihood estimate ($\pi = .4$) is the value maximizing this likelihood function. In general:

Definition.

The MLE is the hypothetical population value which maximizes the likelihood of the observed sample.	(7-23)

We note:

(a) The sample proportion P is our MLE of the population proportion π; it is often, but not always the case that the corresponding sample value is the MLE of the population parameter.

(b) Figure 7-6 is the likelihood function for the particular sample we observed, (i.e., 4 heads in 10 tosses). A different sample result would call for a different likelihood function, and hence a different MLE.

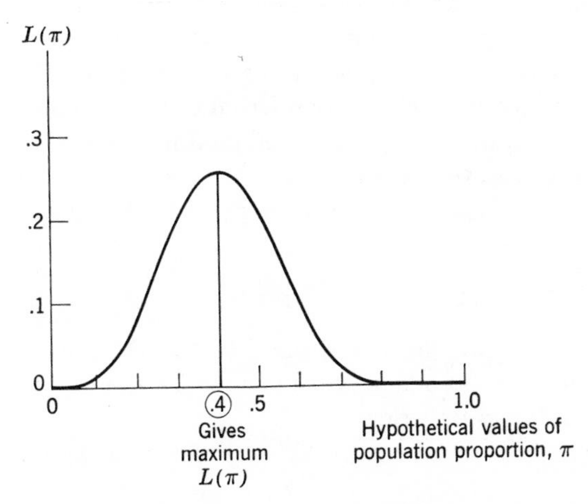

FIG. 7-6 An example of a likelihood function. $L(\pi)$, the likelihood function that various hypothetical population proportions would yield the sample we observed.

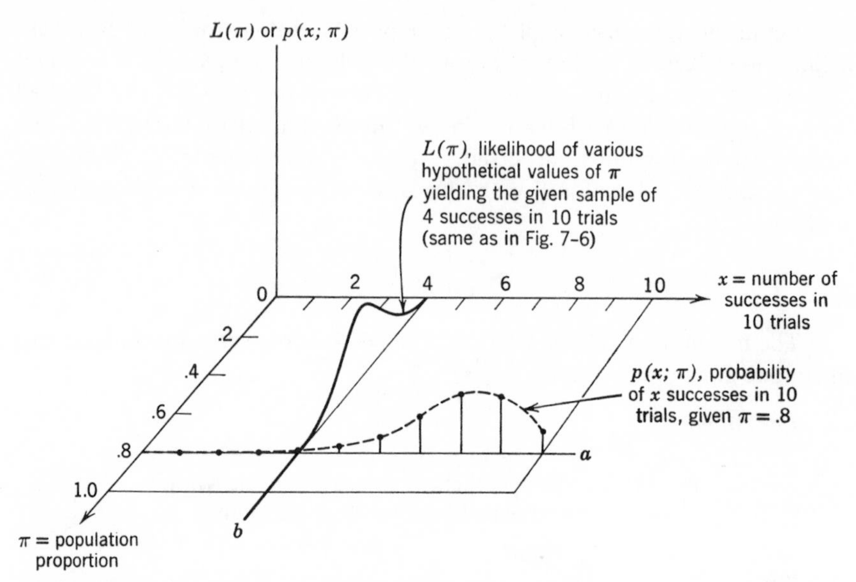

FIG. 7-7 The binomial probability function $p(x; \pi)$ plotted against both x and π.

In Figure 7-7 this discussion is related to our previous deductions about the binomial in Chapters 3, 4, and 6. In this figure we graph the binomial probabilities $p(x; n, \pi)$. [Since n is set at 10 regardless, this function is referred to simply as $p(x; \pi)$]. In earlier chapters we regarded π fixed and x variable, as in slice a; thus the dotted function shows the probability of various numbers of heads if the population proportion is given as .8. In this chapter we regard x—the observed sample result—as fixed, while the population π is thought of as taking on a whole set of hypothetical values; thus slice b shows the likelihood that various possible population proportions would yield 4 heads. Slices in the a direction are referred to as probability functions, while slices in the b direction are called likelihood functions.

We now generalize maximum likelihood estimation. (A summary of our results is shown in the last three columns of Table 7-2 for reference.)

(b) General Binomial

It is very easy to show that our result in the previous section was no accident, and that the maximum likelihood estimate of the binomial π is *always* the sample proportion P.

Given any observed sample of x successes in n trials, the likelihood function is

$$L(\pi) = \binom{n}{x} \pi^x (1 - \pi)^{n-x} \tag{7-24}$$

With calculus it can easily be shown[13] that the maximum value of this likelihood function occurs when $\pi = x/n = P$. Thus

MLE of $\pi = P$, the sample proportion.

We argued in Chapter 1 that it is reasonable to use the sample proportion to estimate the population proportion; but in addition to its intuitive appeal, we now add the more rigorous justification of maximum likelihood: a population with $\pi = P$ would generate with the greatest likelihood the sample we observed.

(c) MLE of the Mean (μ) of any Normal Population

Suppose we have drawn a sample (x_1, x_2, x_3) from a parent population which is $N(\mu, \sigma^2)$; our problem is to find the MLE of the unknown μ for this sample. Because the population is normal, the probability of getting any value x, given a population mean μ is:

$$p(x;\mu) = \frac{1}{\sqrt{2\pi\sigma^2}} e^{-(1/2\sigma^2)(x-\mu)^2} \tag{7-26}$$

Specifically, the probability that we would get the value x_1 that we observed in our first sample draw is

$$p(x_1;\mu) = \frac{1}{\sqrt{2\pi\sigma^2}} e^{-(1/2\sigma^2)(x_1-\mu)^2} \tag{7-27}$$

while the probabilities of drawing the values x_2 and x_3 are, respectively

$$p(x_2;\mu) = \frac{1}{\sqrt{2\pi\sigma^2}} e^{-(1/2\sigma^2)(x_2-\mu)^2} \tag{7-28}$$

---

[13] To find where $L(\pi)$ is a maximum, set the derivative equal to zero.

$$\frac{dL(\pi)}{d\pi} = \binom{n}{x}[\pi^x(n-x)(1-\pi)^{n-x-1}(-1) + x\pi^{x-1}(1-\pi)^{n-x}] = 0 \tag{7-25}$$

Dividing by $\binom{n}{x}\pi^{x-1}(1-\pi)^{n-x-1}$, (7-25) becomes:

$$-\pi(n-x) + x(1-\pi) = 0$$

$$-n\pi + x = 0$$

$$\pi = \frac{x}{n}$$

You can easily confirm that this is a maximum (rather than a minimum or inflection point).

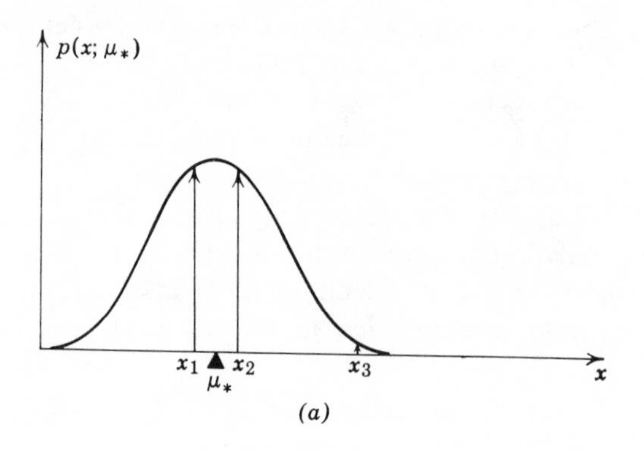

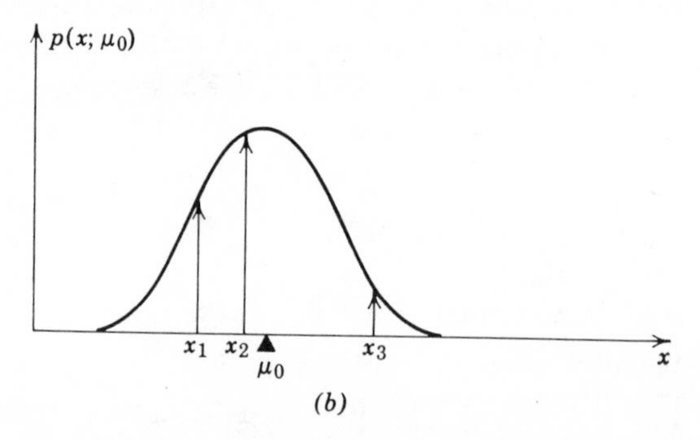

FIG. 7-8 Maximum likelihood estimation of the mean (μ) of a normal population, based on three sample observations (x_1, x_2, x_3). (a) Small likelihood $L(\mu_*)$, the product of the three ordinates. (b) Large likelihood $L(\mu_0)$.

and

$$p(x_3; \mu) = \frac{1}{\sqrt{2\pi\sigma^2}} e^{-(1/2\sigma^2)(x_3-\mu)^2} \tag{7-29}$$

We assume as usual that X_1, X_2, and X_3 are independent so that the joint probability function is the product of (7-27), (7-28), and (7-29):

$$p(x_1, x_2, x_3; \mu) = \prod_{i=1}^{3} \left[\frac{1}{\sqrt{2\pi\sigma^2}} e^{-(1/2\sigma^2)(x_i-\mu)^2} \right] \tag{7-30}$$

where $\prod$ means "the product of," just as $\sum$ means "the sum of." But in our estimation problem the sample values x_i are fixed and only μ is thought of as varying over hypothetical values; we shall speculate on these various possible values of μ, with a view to selecting the most plausible. Thus (7-30) can be written as a likelihood function of μ:

$$L(\mu) = \prod_{n=1}^{3} \left[\frac{1}{\sqrt{2\pi\sigma^2}} e^{-(1/2\sigma^2)(x_i-\mu)^2} \right] \tag{7-31}$$

The MLE of μ is defined as the hypothetical value of μ which maximizes the likelihood function (7-31). Its value may be derived with calculus, but we consider only a geometric interpretation in Figure 7-8.

We "try out" two hypothetical values of μ. We note that a population with mean μ_* as in Figure 7-8*a* is not very likely to yield the sample we observed. Although the probabilities of x_1 and x_2 are large, the probability of x_3 (i.e., the ordinate above x_3) is very small because it is so far distant from μ_*. The product of all three probabilities [i.e., the likelihood of a population with mean μ_* generating the sample (x_1, x_2, x_3)] is therefore quite small. On the other hand a population with mean μ_0 as in Figure 7-8*b* is more likely to generate the sample values. Since the x values are collectively closer to μ_0, they have a greater joint probability. Thus the likelihood is greater for μ_0 than for μ_*; indeed, very little additional shift in μ_0 is apparently required to maximize the likelihood of the sample. It seems that the MLE of μ might be the sample mean—i.e., the average value of x_1, x_2, and x_3; this can, in fact, be proved, as in Problem 7-12.

Finally, the reader who has carefully learned that μ is a fixed population parameter may wonder how it can appear in the likelihood function (7-31) as a variable. This is simply a mathematical convenience. The true value of μ is, in fact, fixed. But since it is unknown, in MLE we must consider all of its possible, or hypothetical values; the way to do this mathematically is to treat it as a variable.

(d) MLE of any Parameter from any Population

We now state MLE in its full generality. A sample $(x_1, x_2 \cdots x_n)$ is drawn from a population with probability function $p(x; \theta)$, where θ is any unknown population parameter that we wish to estimate. From our definition of random sampling (with replacement, or from an infinite population), the X_i are independent, each with the probability function $p(x_i; \theta)$; hence the joint probability of the whole sample is obtained by multiplying.

$$p(x_1, x_2 \cdots x_n; \theta) = p(x_1; \theta)\, p(x_2; \theta) \cdots p(x_n; \theta)$$

$$= \prod_{i=1}^{n} p(x_i; \theta) \tag{7-32}$$

But we regard the observed sample values as fixed, and ask, "Which of all the hypothetical values of θ maximizes this probability?" This is emphasized by renaming (7-32) the likelihood function:

$$L(\theta) = \prod_{i=1}^{n} p(x_i; \theta) \tag{7-33}$$

The MLE is that hypothetical value of θ that maximizes this likelihood function.

(e) Maximum Likelihood vs Method of Moments Estimation (MLE versus MME)

In the analysis above, we have estimated a population proportion with a sample proportion, and a population mean with a sample mean. Why not always use this technique, and estimate any population parameter with the corresponding sample value? This is known as method of moments estimation (MME). Its great advantage is that is it plausible and easy to understand. Moreover, MLE and MME often coincide.

But suppose the two methods do differ (as in Problem 7-14)? In such a circumstance MLE is usually superior. The intuitive appeal of MME is more than offset by the following impressive advantages of MLE. Since MLE is the population value most likely to generate the sample values observed, it is in some sense the population value that "best matches" the observed sample. In addition, under broad conditions MLE has the following asymptotic properties:

1. *Efficient*, with smaller variance than any other estimators.
2. *Consistent*, that is, asymptotically unbiased, with variance tending to zero.
3. *Normally distributed*, with easily computed mean and variance; hence it may be readily used to make inferences.

For example, we have already seen that these three properties are true for $\bar{X}$, the MLE of μ in a normal population. [Property 3 follows from Theorem (6-13); Property 2 follows from (6-10) and (6-11); Property 1 is proved in advanced texts, and has been alluded to in (7-17).]

We emphasize that these properties are *asymptotic*, that is, true for large samples as $n \to \infty$. But for the small samples often used by economists for example, MLE is not necessarily best.

PROBLEMS

*7-11 Following Figure 7-6, graph the likelihood function for a sample of 6 heads in 8 tosses of a coin; show the MLE.

*7-12 Derive the MLE of μ for a normal population, using calculus.

*7-13 (a) Derive the MLE of σ^2 for the normal distribution, assuming μ is known.
(b) Is it unbiased?

*7-14 As $N + 1$ delegates arrived at a convention, they were given successive tags numbered $0, 1, 2, 3, \ldots N$. In order to estimate the unknown number N, a brief walk in the corridor provided a sample of 5 tags, numbered 37, 16, 44, 43, 22.
(a) What is the MME of N? Is it biased?
(b) What is the MLE of N? Is it biased?

FURTHER READING

For a detailed description of the virtues of MLE, see for example

1. Wilks, S. S. *Mathematical Statistics*, New York: John Wiley & Sons (1962).
2. Lindgren, B. W. *Statistical Theory*, New York: Macmillan (1959).

chapter 8

Estimation II

8-1 DIFFERENCE IN TWO MEANS

In the previous chapter, we used a sample mean to estimate a population mean. In this chapter we will develop several other similar examples of how a sample statistic is used to estimate a population parameter.

Whenever two population means are to be compared, it is usually their *difference* that is important, rather than their absolute values. Thus we often wish to estimate

$$\mu_1 - \mu_2 \tag{8-1}$$

A reasonable estimate of this difference in population means is the difference in sample means

$$\bar{X}_1 - \bar{X}_2 \tag{8-2}$$

(Assuming normality of the parent populations, this is the maximum likelihood estimator, with many attractive properties.)

Again, because of the error in point estimates, we are typically interested in an interval estimate. Its development is comparable to the argument in Section 7-1, and involves two steps: the distribution of our estimator $(\bar{X}_1 - \bar{X}_2)$ must be deduced; then this can be "turned around" to make an inference about the population parameter $(\mu_1 - \mu_2)$.

First, how is the estimator $(\bar{X}_1 - \bar{X}_2)$ distributed? From (6-31) we know that the first sample mean $\bar{X}_1$ is approximately normally distributed around the population mean μ_1 as follows.

$$\bar{X}_1 \sim N(\mu_1, \sigma_1^2/n_1) \tag{8-3}$$

where σ_1^2 represents the variance of the first population, and n_1 the size of the sample drawn. Similarly

$$\bar{X}_2 \sim N(\mu_2, \sigma_2^2/n_2) \tag{8-4}$$

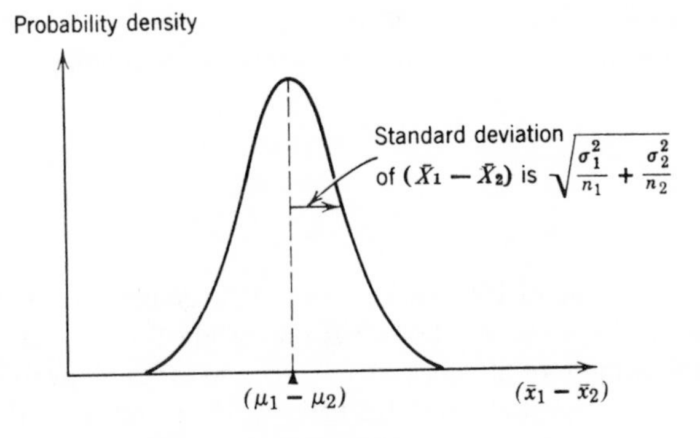

FIG. 8-1 Distribution of $(\bar{X}_1 - \bar{X}_2)$.

Independence of the two sampling procedures will ensure that the two random variables $\bar{X}_1$ and $\bar{X}_2$ are independent; hence (5-31), (5-34), and (6-13) can be applied directly:

$$(\bar{X}_1 - \bar{X}_2) = (1)\bar{X}_1 + (-1)\bar{X}_2 \sim N(\mu_1 - \mu_2, \sigma_1^2/n_1 + \sigma_2^2/n_2) \tag{8-5}$$

This distribution of $(\bar{X}_1 - \bar{X}_2)$ is shown in Figure 8-1. Equation (8-5) is exactly true, assuming that both populations are normal; it still remains approximately true (by the central limit theorem) for large samples from practically any populations.

Under these conditions, our knowledge in (8-5) of how the estimator $(\bar{X}_1 - \bar{X}_2)$ behaves can now be turned around to construct the confidence interval:

95% confidence interval for the difference in means $(\mu_1 - \mu_2)$

$$(\bar{x}_1 - \bar{x}_2) \pm 1.96\sqrt{\frac{\sigma_1^2}{n_1} + \frac{\sigma_2^2}{n_2}} \tag{8-6}$$

When σ_1 and σ_2 have a common value, say σ, the 95% confidence interval for $(\mu_1 - \mu_2)$ becomes:

$$(\bar{x}_1 - \bar{x}_2) \pm 1.96\sigma\sqrt{\frac{1}{n_1} + \frac{1}{n_2}} \tag{8-7}$$

The variances of the two populations, σ_1^2 and σ_2^2 in (8-6) are usually not known; the best the statistician can do is guess at them, with the variances s_1^2 and s_2^2 he observed in his two samples. Provided his sample is large, this is an accurate enough approximation; but with a small sample, this introduces a new source of error. The student will recall that this same problem was encountered in estimating a single population mean in Section 7-1. In the next section we shall give a solution for these problems of small-sample estimation.

PROBLEMS

8-1 A random sample of 100 workers in one large plant took an average of 12 minutes to complete a task, with a standard deviation of 2 minutes. A random sample of 50 workers in a second large plant took an average of 11 minutes to complete the task, with a standard deviation of 3 minutes. Construct a 95% confidence interval for the difference between the two population averages.

8-2 Two samples of 100 seedlings were grown with two different fertilizers. One sample had an average height of 10 inches and a standard deviation of 1 inch. The second sample had an average height of 10.5 inches and a standard deviation of 3 inches. Construct a confidence interval for the difference between the average population heights $(\mu_1 - \mu_2)$

(a) At the 95% level of confidence.

(b) At the 90% level of confidence.

8-3 A random sample of 60 students was taken in two different universities. The first sample had an average mark of 77 and a standard deviation of 6. The second sample had an average mark of 68 and a standard deviation of 10.

(a) Find a 95% confidence interval for the difference between the mean marks in the two universities.

(b) What increase in the sample size would be necessary to cut the error allowance by 1/2?

(c) What increase in the sample size would be necessary to reduce the error allowance to 1.0?

8-2 SMALL SAMPLE ESTIMATION: THE t DISTRIBUTION

We shall assume in this section that the populations are normal.

(a) One Mean, μ

In estimating a population mean μ from a sample mean $\bar{X}$, the statistician generally has no information on the population standard deviation σ; hence he uses the estimator s, the sample standard deviation. Substituting this into

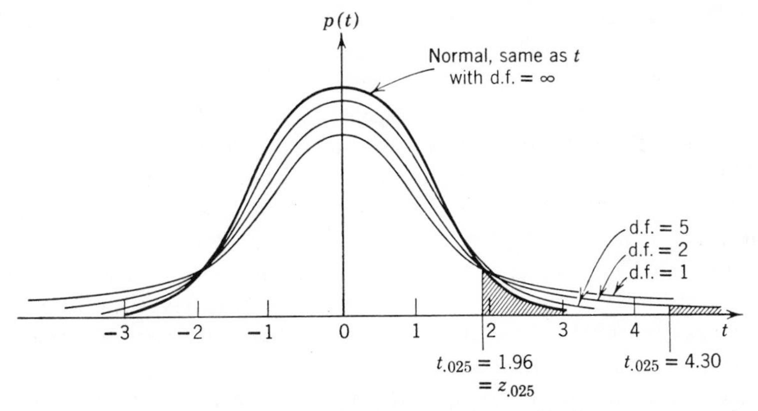

FIG. 8-2 The standard normal distribution and the t distribution compared.

(7-10), he estimates the 95% confidence interval for μ as,

$$\mu = \bar{x} \pm z_{.025} \frac{s}{\sqrt{n}} \tag{8-8}$$

Provided his sample is large (at least 25–50, depending on the precision required), this will be a reasonably accurate approximation. But with a smaller sample size, this substitution introduces an appreciable source of error. Hence if he wishes to remain 95% confident, his interval estimate must be broadened. How much?

Recall that $\bar{X}$ has a normal distribution; when σ is known, we may standardize, obtaining

$$Z = \frac{\bar{X} - \mu}{\sigma/\sqrt{n}} \tag{8-9}$$

where Z is the standard normal variable. By analogy, we introduce a new "Student[1] t" variable, defined as

$$t = \frac{\bar{X} - \mu}{s/\sqrt{n}} \tag{8-10}$$

The similarity of these two variables is immediately evident. The only difference is that Z involves σ, which is generally unknown; but t involves s, which can always be calculated from an observed sample. The precise distribution of t, like Z, has been derived by mathematicians and is shown in Table V of the Appendix. The distribution of t is compared to Z in Figure 8-2.

[1] This t variable was first introduced by Gosset writing under the pseudonym "Student," and later proved valid by R. A. Fisher. We make no attempt to develop the entire proof, because it is not very instructive. It can be found in almost any mathematical statistics text.

(We must emphasize a break in notation. Until now, capital letters denoted random variables, while small letters denoted their realized values. But from now on, in order to conform to common usage, we shall entirely forget this convention; to represent either random variables or realized values, we shall use small letters t and s, and capital letters X, $\bar{X}$, Z, P, etc.)

As expected, the t distribution is more spread out than the normal, since the use of s rather than σ introduces a degree of uncertainty. Moreover, while there is one standard normal distribution, there is a whole family of t distributions. With small sample size, this distribution is considerably more spread out than the normal; but as sample size increases, the t distribution approaches the normal, and for samples of about 50 or more, the normal becomes a very accurate approximation.

The distribution of t is not tabled according to sample size (n), but rather according to "degrees of freedom," the divisor in s^2. Thus, in calculating s^2 we may write[2]:

$$\text{d.f.} = \text{degrees of freedom} = n - 1 \tag{8-11}$$

For example, for a sample with $n = 3$, then d.f. $= 2$, and we find from Appendix Table V that the critical t value which leaves $2\frac{1}{2}\%$ probability in the upper tail is

$$t_{.025} = 4.30$$

This is shown in Figure 8-2. By symmetry, it follows that for any observed t

$$\Pr(-4.30 < t < 4.30) = 95\% \tag{8-12}$$

Substituting for t according to (8-10):

$$\Pr\left(-4.30 < \frac{\bar{X} - \mu}{s/\sqrt{n}} < 4.30\right) = 95\% \tag{8-13}$$

This deduction can now be "turned around" into the following inference: for a sample of size 3, the 95% confidence interval for μ is

$$\mu = \bar{X} \pm 4.30 \frac{s}{\sqrt{n}} \tag{8-14}$$

---

[2] The phrase "degrees of freedom" is explained in the following intuitive way:

Originally there are n degrees of freedom in a sample of n observations. But one degree of freedom is used up in calculating $\bar{X}$, leaving only $n - 1$ degrees of freedom for the residuals $(X_i - \bar{X})$ to calculate s^2.

For example, consider a sample of two observations, 21 and 15, say. Since $\bar{X} = 18$, the residuals are $+3$ and -3, the second residual necessarily being just the negative of the first. While the first residual is "free," the second is strictly determined; hence there is only 1 degree of freedom in the residuals.

Generally, for a sample of size n, it may be shown that if the first $n - 1$ residuals are specified, then the last residual is automatically determined by the requirement that the sum of all residuals be zero, i.e., $\Sigma(X_i - \bar{X}) = 0$.

For a general sample size n, the 95% confidence interval for μ is

$$\boxed{\mu = \bar{X} \pm t_{.025} \frac{s}{\sqrt{n}}} \tag{8-15}$$

where $t_{.025}$ is the critical t value leaving $2\frac{1}{2}\%$ of the probability in the upper tail, with $n - 1$ degrees of freedom.

To sum up, we note the similarity of t estimation in (8-15) and normal estimation in (7-10). The only difference is that an observed sample value (s) is substituted for σ, and as a consequence a critical t value must be substituted for the normal value.

An important practical question is: "When do we use the t distribution and when do we use the normal?" If σ is known, the normal distribution is used; but if σ is unknown, then the t distribution is theoretically appropriate—regardless of sample size. However, if the sample size is large, the normal is an accurate enough approximation[3] of the t. *So in practice the t distribution is used only for small samples when σ is unknown—and the normal is used otherwise. The t distribution has one additional requirement: the parent population from which the sample is drawn is assumed normal.* (But normality is a requirement for all our small-sample estimation, even if σ is known. Recall from Chapter 7 that inference about a nonnormal population was validated by the central limit theorem only if the sample size was large.)

As sample size (n) decreases, estimation becomes less precise (i.e., interval estimates become wider). The two reasons for this are clearly distinguished in (8-15). First, the divisor $\sqrt{n}$ becomes smaller. This appears in (7-10) as well as in (8-15); thus even if σ is known and inference is based on the normal distribution, the error allowance increases and the interval estimate becomes wider as a consequence. The secondary reason for loss of precision occurs if s must be substituted for an unknown σ. The smaller the sample, the more the appropriate t distribution will depart from the normal; and the more spread the t distribution, the broader the interval estimate.

(b) Difference in Two Population Means ($\mu_1 - \mu_2$)

We shall assume, as often occurs in practice, that even though the two populations may have different means, they have a common variance σ^2.

[3] This may be verified from Table V. For example, a 95% confidence interval constructed from a sample of size 61 should use a critical t value of 2.00; but the use of the normal value of 1.96 as an approximation involves very little (2%) error.

As we scan down the $t_{.025}$ column in Table V, these critical values approach $z_{.025} = 1.96$. this verifies Figure 8-2, where the t distributions approach the normal.

When σ^2 is known, (8-7) is appropriate. When unknown, σ^2 must be estimated. The appropriate estimate is to add up all the squared deviations from both samples, and then divide by the degrees of freedom $(n_1 - 1) + (n_2 - 1)$, to obtain an unbiased estimator called the pooled sample variance:

$$s_p^2 = \frac{1}{(n_1 + n_2 - 2)}\left[\sum_{i=1}^{n_1}(X_{1i} - \bar{X}_1)^2 + \sum_{i=1}^{n_2}(X_{2i} - \bar{X}_2)^2\right] \qquad (8\text{-}16)$$

where X_{1i} represents the ith observation in the first sample. Substitution of s_p for σ in (8-7) requires that the t distribution also be used, obtaining the 95% confidence interval:

$$\boxed{(\mu_1 - \mu_2) = (\bar{X}_1 - \bar{X}_2) \pm t_{.025}\, s_p \sqrt{\frac{1}{n_1} + \frac{1}{n_2}}} \qquad (8\text{-}17)$$

where $t_{.025}$ is the critical t value with d.f. $= n_1 + n_2 - 2$.

PROBLEMS

8-4 Sixteen weather stations at random locations in a state measure rainfall. In 1967, they recorded an average of 10 inches and standard deviation of 1.5 inch. For the mean rainfall for the state,
(a) Construct a 95% confidence interval.
(b) Construct a 99% confidence interval.

8-5 100 cars on a thruway were clocked at an average speed of 69 m.p.h., with a standard deviation of 4 m.p.h. Construct a 95% confidence interval for the mean speed of all cars on this thruway.

(8-6) A random sample of 4 students in a large statistics course received the following marks: 56, 70, 55, 59. Construct a 95% confidence interval for the average mark of all students in the course.

8-7 From a sample of five random normal numbers from Table IIb find a 95% confidence interval for the mean of the population.

8-8 Five people selected at random had their breathing capacity measured before and after a certain treatment, obtaining the following data:

	Breathing Capacity		
Person	Before (X)	After (Y)	Improvement
A	2750	2850	+100
B	2360	2380	+20
C	2950	2800	−150
D	2830	2860	+30
E	2250	2300	+50

Let μ_X (and μ_Y) be the mean capacity of the whole population before (and after) treatment.

(a) What is the (point) estimate of the mean improvement $(\mu_Y - \mu_X)$?

(b) Construct a 95% confidence interval for $(\mu_Y - \mu_X)$.

8-9 In a random sample of 10 football players, the average age was 27 and the sum of squared deviations was 300. In a random sample of 20 hockey players, the average age was 25 and the sum of squared deviations was 450. Estimate, with a 95% confidence interval, the difference in the population means, assuming $\sigma_1 = \sigma_2$.

8-10 Given the following random samples from 2 populations:

$$n_1 = 25 \qquad \bar{X}_1 = 60.0 \qquad s_1 = 12$$
$$n_2 = 15 \qquad \bar{X}_2 = 68.0 \qquad s_2 = 10$$
$$\text{and assume} \qquad \sigma_1 = \sigma_2$$

Find a 95% confidence interval for $(\mu_1 - \mu_2)$.

8-11 Derive the confidence interval (8-14) from (8-13). (*Hint*. Use practically the same method as in the footnote to equation (7-4).)

8-12 Derive the confidence interval (8-6) from (8-5).

8-3 ESTIMATING POPULATION PROPORTIONS: THE ELECTION PROBLEM ONCE AGAIN

In Section 6-6, we saw that a sample proportion P is just a sample mean $\bar{X}$ in disguise. For example, if we observe 4 Democrats in a sample of 10, then

$$P = \bar{X} = \tfrac{1}{10}(1 + 1 + 0 + 0 + 0 + 1 + 0 + 1 + 0 + 0) = \tfrac{4}{10}$$

Similarly, the population proportion π is just the population mean μ in disguise. The simplest method of deriving an interval estimate for a proportion is therefore to modify (7-10), the interval estimate for a mean. Thus the 95% confidence interval for π is

$$\pi = P \pm 1.96\sqrt{\frac{\pi(1-\pi)}{n}} \tag{8-18}$$

We confirm that (8-18) is just a recasting of (7-10). $\bar{X}$ is replaced by P, and $\sqrt{\sigma^2/n}$ (the standard deviation of $\bar{X}$) is replaced by $\sqrt{\pi(1-\pi)/n}$ [the standard deviation of P, as given in (6-30)].

But we seem to have reached an impasse; the unknown π appears in the right-hand side of (8-18). Fortunately, the situation has a remedy: substitute the sample P for π in the right side of (8-18). This is a strategy we have used before, when we substituted s for σ in the confidence interval for μ. Again, this approximation introduces another source of error; but with a large sample size, this is no great problem. Thus:

> For large samples, the 95% confidence interval for π is:
>
> $$\pi = P \pm z_{.025}\sqrt{\frac{P(1-P)}{n}} \tag{8-19}$$

where $z_{.025}$ is the critical value leaving $2\frac{1}{2}\%$ of the probability in the upper tail. As an example, the voter poll of Section 1-1 used this formula.

For small samples, there are several options. The simplest is to read the interval estimate for π from Figure 8-4, a table which is constructed in the following manner. The first step is the mathematical deduction of how the variable estimator P is distributed, for any population π. This is shown for a sample size 20 in Figure 8-3. Thus, for example, if $\pi = .4$, then the sample P has the dotted distribution shown in this diagram, and there is a 95% probability that any P calculated from a random sample of 20 will lie in the interval *ab*. For each possible value of π, such a probability function of P defines two critical points like a and b. When all such points are joined, the result is the two curves enclosing a 95% probability band.

This description of how the statistic P is related to the population π can of course be "turned around" to draw a statistical inference about π from a given sample P. For example, if we have observed a sample proportion $P_1 = 11/20 = .55$, then the 95% confidence interval for π is defined by *fg*, the

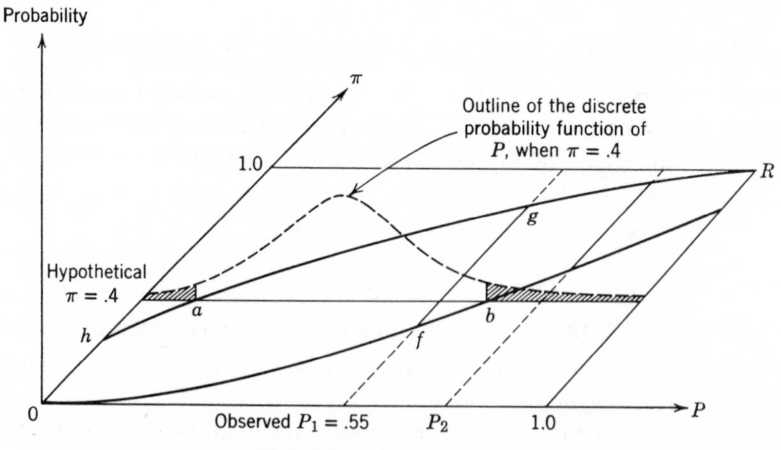

FIG. 8-3 Distribution of P.

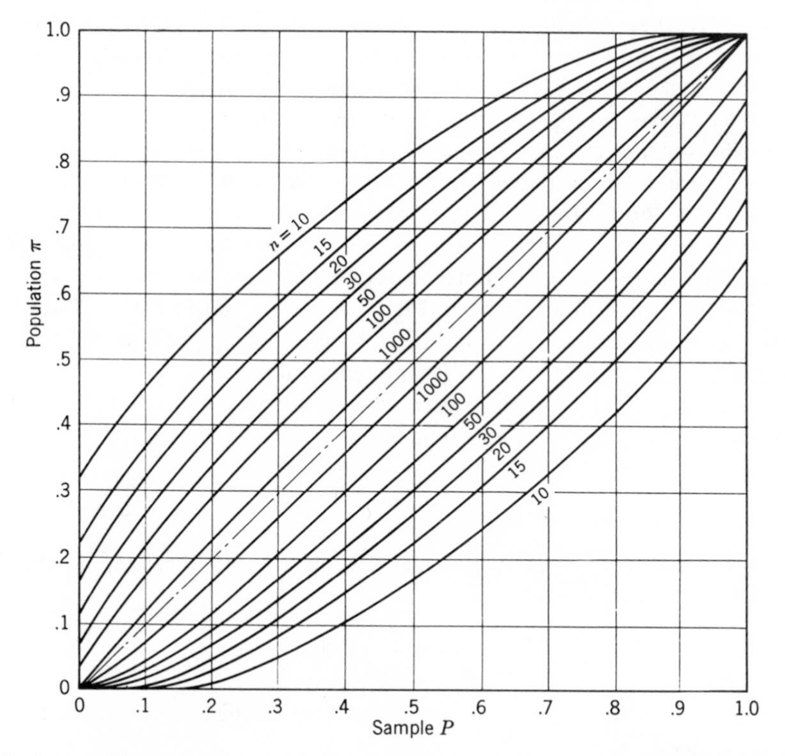

FIG. 8-4 95% confidence intervals for population proportions (π). [Reproduced with the permission of Professor E. S. Pearson from C. J. Clopper and E. S. Pearson, "The Use of Confidence or Fiducial Limits Illustrated in the Case of the Binomial," *Biometrika*, **26** (1954), p. 404.]

width of this probability band above P_1, i.e.,

$$.31 < \pi < .77 \tag{8-20}$$

Whereas the (deduced) probability interval is defined in the horizontal direction of the P axis, the (induced) confidence interval is defined in the vertical direction of the π axis.

This is the same logic we have used in deriving confidence intervals before. We will nevertheless pause briefly to review, because this is a more generalized argument than we have previously encountered. Suppose the true value of π is .4; then there is a probability of 95% that a sample P will fall between a and b. *If and only if it does* (e.g., P_1) will the confidence interval we construct bracket the true π of .4. We are therefore using an estimation procedure which is 95% probable to bracket the true value of π, and thus yield a correct statement. But we must recognize the 5% probability that the sample P will fall beyond a or b (e.g., P_2); in this case our interval estimate will not bracket $\pi = .4$, and our conclusion will be wrong.

Why is this a more general theory of confidence interval estimation? In previous instances (e.g., estimating a population mean, μ) we constructed a confidence interval *symmetrically* about our point estimate $\bar{X}$. But in estimating π, no such symmetry is generally involved.[4] For example, with our observed sample proportion $P_1 = .55$, the confidence interval (8-20) we constructed for π was *not* symmetric about our point estimate .55.

The 95% probability band in Figure 8-3 is set out in Figure 8-4, along with the similar bands appropriate for other sample sizes. This neater diagram is used to construct 95% confidence intervals for π.

As an example, if we have observed a $P = .6$ in a sample of 15, the 95% confidence interval for π is approximately

$$.32 < \pi < .84$$

For the same $P = .6$ in a larger sample of 100, the 95% confidence interval for π is narrower:

$$.50 < \pi < .70$$

Alternatively, with such a large sample, (8-19) could have been used, with the same result, i.e.,

$$\pi = .60 \pm 1.96\sqrt{\frac{(.6)(.4)}{100}}$$

$$= .60 \pm .10$$

Finally, there is a third method of estimating π that we introduce, not so much for its practical value as for its illustration of this useful principle: with a little imagination, several alternative methods of solving a problem can often be developed, and the most appropriate one to use in a given set of circumstances is a matter of judgment.

Let us be conservative, and ask: what is the maximum width of the interval estimate in (8-18), i.e., what is the maximum value that the error

[4] The student may have wondered why the 95% probability band does not converge on the two end points O and R. It is true that one half of this band (made up of all points similar to b) does intersect the P axis at 0; this means that if π is zero (e.g., no Socialists in the U.S.), then any sample P must also be zero (no Socialists in the sample). But the other half of this band does not intersect the π axis at 0; instead it intersects at h. This means that an observed P of zero (e.g., no Socialists in a sample) does *not* necessarily imply that π is zero (no Socialists in the U.S.).

allowance $1.96\sqrt{\frac{\pi(1-\pi)}{n}}$ can have? It is easily shown[5] that the maximum value of $\pi(1-\pi)$ is 1/4.

Then (8-18), the 95% confidence interval for π (with a large sample), can be written

$$\pi = P \pm 1.96\frac{\sqrt{1/4}}{\sqrt{n}}$$

or simply:

$$\pi = P \pm \frac{.98}{\sqrt{n}} \tag{8-21}$$

But this is assuming the worst; if, in fact, π is not 1/2, then $\pi(1-\pi)$ is less than 1/4, and our interval estimate need not be this wide; or to restate, (8-21) is an interval estimate for π with *at least* a 95% level of confidence. For example, this very simple formula is sometimes used in political polls where it is known on the basis of historical experience that the proportion of Democrats is close to 1/2. In these circumstances (8-21) becomes a very accurate approximation.

For completeness, we write the 95% confidence interval for the *difference* in 2 proportions:

For large n,

$$(\pi_1 - \pi_2) = (P_1 - P_2) \pm 1.96\sqrt{\frac{P_1(1-P_1)}{n_1} + \frac{P_2(1-P_2)}{n_2}} \tag{8-22}$$

This is derived in essentially the same way as (8-6).

[5] The simplest way is with calculus, setting the derivative of $\pi(1-\pi)$ equal to zero.
To prove it without calculus, we may simply graph $f(\pi) = \pi(1-\pi)$, as follows:

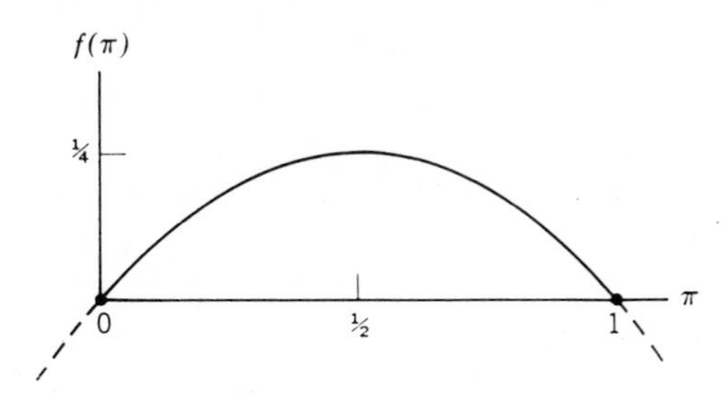

Note that for either extreme value of π (1 or 0) the value of $f(\pi)$ is zero; and if $\pi = 1/2$, then $\pi(1-\pi)$ reaches its maximum value because of symmetry of the parabola.

PROBLEMS

8-13 Construct a 95% confidence interval for π, the proportion of Republican voters in the U.S., if there were 4820 Republicans in a random sample of 10,000.

8-14 In a random sample of tires produced by a certain company, 20% did not meet the company's standards. Construct a 95% confidence interval for the proportion π (in the whole population of tires) which do not meet the standards,
(a) If the sample size $n = 10$.
(b) If $n = 25$.
(c) If $n = 2500$.

(8-15) By talking to 15 voters you discover that only 3 favor a certain candidate. Construct a 95% confidence interval for the proportion of all voters favoring this candidate.

(8-16) In a random sample of 100 British smokers, 28 preferred brand X. Construct a 95% confidence interval to estimate the proportion of all British smokers who prefer X.

8-17 In a survey of U.S. consumer intentions, 498 families in a random sample of 2500 indicated that they intended to buy a new car within a year. Construct a 95% confidence interval for the proportion of all U.S. families intending a new car purchase.
(a) Answer two ways:
  (1) Using the usual formula (8-19).
  (2) Using the simplified formula (8-21).
(b) If the sample P had been .40, would the error in (a2) have been greater?

8-18 If $\pi = 3/4$, what is the precise percentage error introduced by using (8-21) rather than (8-19)? Does this suggest that (8-21) is a reasonable approximation, provided $1/4 < \pi < 3/4$?

8-19 A sample of 100 cars was taken in each of 2 cities. In one city 72 of the cars passed the safety test; in the second only 66 passed. Construct a 95% confidence interval for the difference between the proportions of safe cars in the two cities.

8-20 A sample of 3182 voters yielded the following frequency table,* relating their attitudes to Senator Joseph McCarthy and their vote in 1948. Construct a 95% confidence interval for $(\pi_1 - \pi_2)$, where π_1

* From S. M. Lipset, "The Radical Right" in Bell, Daniel, ed. *The New American Right*, New York Criterion Books [1955].

is the proportion of all Democrat voters who were pro-McCarthy, and π_2 is the proportion of Republican voters who were pro-McCarthy.

↓1958 Vote \ Attitude to McCarthy →	Pro	Anti
Democrat	506	1381
Republican	563	732

(8-21) In an urban survey of 1000, 790 favored certain legislation. In a rural survey of 300, 180 opposed the same legislation.

Construct a 99% confidence interval for the difference between the proportions of city and country voters who favor the legislation.

*8-4 ESTIMATING THE VARIANCE OF A NORMAL POPULATION: THE CHI-SQUARE DISTRIBUTION

There is one further example of a confidence interval, interesting not so much for its practical value[6] as for the insight it provides.

Consider a normal population $N(\mu, \sigma^2)$ with both μ and σ^2 unknown. So far we have estimated σ^2 with s^2 only as a means of finding a confidence interval for μ. Now suppose, on the other hand, that our primary interest is in σ^2, rather than μ. For example, we may wish to ask "How much variance is there in Japan's balance of payments?" in order to get some indication of the country's requirement of foreign exchange reserves. Or, we may ask "What is the variance of farm income?" in order to evaluate whether a policy aimed at stabilizing farm income is necessary.[7] To estimate variance we shall assume that the random variable (e.g., farm income) is normally distributed; if so, how do we proceed?

We have already seen, in Section 7-2, that s^2 is an unbiased estimator of σ^2; but to construct an interval estimate for σ^2 we must ask: "how is the estimator s^2 distributed around σ^2?" To answer this, it is customary to

[6] One reason that the confidence interval for σ^2 is of limited practical use is that it depends crucially on the assumption that the parent population is normal. By contrast, most of the confidence intervals for means remain approximately true even if the parent population is nonnormal; such confidence intervals are called *robust*.

[7] Income stabilization policies are almost always designed to stabilize income around a reasonably high level. Thus they aim both at reducing variance (σ^2) *and* raising average income (μ). Here we concentrate only on the variance problem.

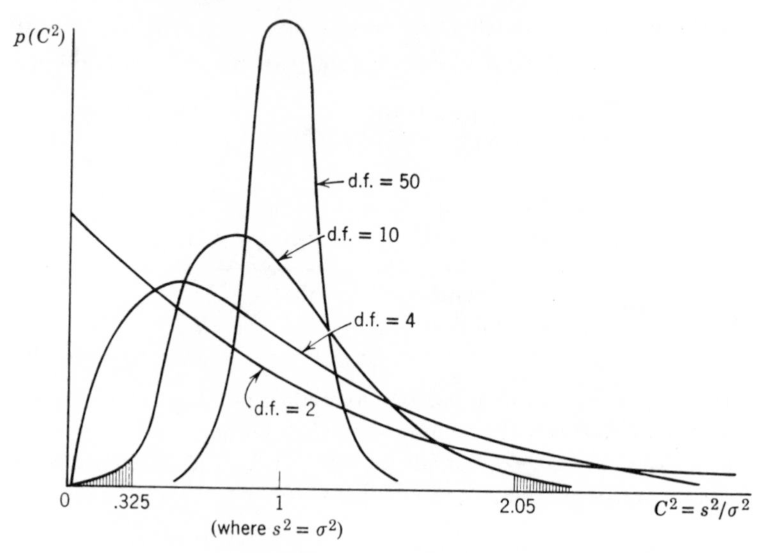

FIG. 8-5 Distribution of the modified chi-square, C^2.

define a new variable:

$$C^2 \triangleq \frac{s^2}{\sigma^2} \tag{8-23}$$

Of course, when $s^2 = \sigma^2$, this ratio is 1; thus our question can be rephrased: "how is C^2 distributed around 1?"

C^2 is called a modified Chi-square variable, with $n - 1$ degrees of freedom.[8] It has been proved by advanced calculus that the distribution of C^2 is that of Figure 8-5; critical values are given in Appendix Table VI.

Since its numerator s^2 and denominator σ^2 are both positive, the variable C^2 is also always positive, with its distribution falling to the right of zero in Figure 8-5. For small sample values we note that it is also skewed to the right; but as n gets large, this skewness disappears and the C^2 distribution approaches normality. Since s^2 is an unbiased estimator of σ^2, this implies that the expected value of each of these C^2 distributions is 1. Moreover as sample size increases C^2 becomes more and more heavily concentrated around 1, indicating that s^2 is becoming an increasingly accurate estimator of σ^2.

With this deduction of how the estimator s^2 is distributed around its target σ^2, we may now infer a 95% confidence interval for σ^2 using our now-familiar technique. We illustrate with sample size $n = 11$ (d.f. = 10). From

[8] C^2 is comprised of the constant parameter σ^2, and the variable s^2. Thus it has the same degrees of freedom as s^2 [explained in the footnote to equation (8-11)].

Figure 8-5, or more precisely from Table VI, we find the critical points cutting off $2\frac{1}{2}\%$ of the distribution in each tail; thus

$$\Pr\left(.325 < \frac{s^2}{\sigma^2} < 2.05\right) = 95\% \tag{8-24}$$

Solving for σ^2, we obtain the equivalent statement

$$\Pr\left(\frac{s^2}{2.05} < \sigma^2 < \frac{s^2}{.325}\right) = 95\% \tag{8-25}$$

If the observed value of s^2 turns out to be 3.6, then the 95% confidence interval for σ^2 is

$$1.76 < \sigma^2 < 11.1 \tag{8-26}$$

We note that this is another example of an asymmetrical confidence interval.

In general, the upper and lower critical values of C^2 are denoted $C^2_{.025}$ and $C^2_{.975}$ and the 95% confidence interval is written

$$\boxed{\frac{s^2}{C^2_{.025}} < \sigma^2 < \frac{s^2}{C^2_{.975}}} \tag{8-27}$$

PROBLEMS

*8-22 If a sample of 25 IQ scores from a certain population has $s^2 = 120$, construct a 95% confidence interval for the population σ^2.

*8-23 From the sample of Problem 8-6, construct a 95% confidence interval for σ^2.

Review Problems

8-24 Two machines are used to produce the same good. In 400 articles produced by machine A, 16 were substandard. In the same length of time, the second machine produced 600 articles, and 60 were substandard. Construct 95% confidence intervals for

(a) π_1, the true proportion of substandard articles from the first machine.

(b) π_2, the true proportion of substandard articles from the second machine.

(c) The difference between the two proportions $(\pi_1 - \pi_2)$.

8-25 To determine the effectiveness of a certain vitamin supplement, the following data were obtained:

Table of weight increases (in grams) for 2 groups of 3 mice.

Control Group	Treated Group
12	18
19	16
14	23

Assume that $\sigma_1 = \sigma_2$, and that the mice are not paired [i.e., the first row of data (12 and 18) does not come from mice that are related by kinship, or anything else]. Construct a 95% confidence interval for the "vitamin effect," $\mu_2 - \mu_1$.

8-26 Suppose a psychologist runs 6 people through a certain experiment. In order to find the effect on heart rate, he collects the following data:

Heart Rate—Beats per Minute

Person	Before Experiment	After Experiment
Smith	71	84
Jones	67	72
Gunther	71	70
Wilson	78	85
Pestritto	64	71
Seaforth	70	81

Suppose that it is known that people as a whole have an average heart rate approximately normally distributed, with mean 73.

Calculate a 95% confidence interval for the effect of the experiment on heart rate.

8-27 A certain scientist concluded his study in fertility control as follows: "So far one result has emerged from the before-and-after survey, and it is a key measure of the outcome: at the end of 1962, 14.2% of the women in the sample were pregnant, and at the end of 1963 (after the birth-control campaign) 11.4% of the women (in a second independent sample) were pregnant, a decline of about one fifth."

If the samples (both before and after) included 2500 women, what statistical qualification would you add to the above statement, in order to make its meaning clearer?

chapter 9

Hypothesis Testing

9-1 TESTING A SIMPLE HYPOTHESIS

We begin with a very simple example, in order to keep the philosophical issues clear. Suppose that I am gambling with a die, and lose whenever the die shows ace. After 100 throws I notice that I have suffered an inordinate number of losses—27 aces. This makes me suspect that my opponent is using a loaded die; specifically, I begin to wonder whether this is one of the crooked dice recently advertised as giving aces one-quarter of the time.

Is my suspicion well-founded, so that I should make an accusation, and terminate the game? My decision should depend on several factors.

1. How much did I trust my opponent even before I began the game (prior to collecting the evidence)? For example, if I am playing with a sharp-looking character I have just met on a Mississippi steamboat I will be more inclined to terminate the game than if I am playing with an old and trusted friend.

2. What are my potential *losses* involved in making a wrong decision? I may be playing with very attractive odds in my favor; if the die is, after all, a true one, then I will have a good deal to lose if I erroneously conclude that it is crooked and terminate the game.

3. Does the evidence itself (27 aces in 100 tosses) indicate that I am being cheated?

If questions (1) and (2) can be answered, even roughly, then it is useful to put this whole problem into the larger framework of decision theory (Chapter 15). However, in many practical problems, the first two questions cannot easily be answered; using a medical example, what is the cost of making the wrong decision and certifying a drug which has serious side effects? In many instances it is only question (3) than can be answered by the scientist, and it is this *limited but extremely important* question which we address in this chapter.

First, state the two conflicting hypotheses as precisely (mathematically) as possible. The hypothesis that the die is fair is really a statement that the Bernoulli population of all possible throws has a proportion of aces equal to 1/6. This is the hypothesis of "no cheating" or "nothing out of the ordinary." Customarily, it is called the *null hypothesis*,

$$H_0 : \pi = 1/6 = .167 \tag{9-1}$$

The other hypothesis is that the probability of an ace is 1/4; this is customarily called the *alternate hypothesis*,

$$H_1 : \pi = 1/4 = .25 \tag{9-2}$$

Suppose that even before evidence was collected—before we started throwing the die—the following plausible decision rule was suggested. After 100 throws, we should:

$$R \begin{cases} \text{Accept } H_0 \text{ if no more than 20 aces occur} \\ \quad \text{(i.e., if observed } P \leq .20) \\ \text{Reject } H_0 \text{ (i.e., accept } H_1) \text{ if more than} \\ \quad \text{20 aces occur } (P > .20) \end{cases} \tag{9-3}$$

This decision rule is shown in Figure 9-1*a*. The value .20 which separates the two regions is often called the *critical point*, while $P \geq .20$ is referred to as the *critical range*, denoting observed values of P that will lead us to reject H_0. When H_0 is rejected, we call the results *statistically significant*.

Of course, this rule will not always lead to the right decision, because of chance fluctuation (bad luck). We can hope, however, that the probability of error is small. To find out how small we apply probability analysis, as in Figure 9-1*b*.

First, how well will rule R work if H_0 is true? The distribution of P is then concentrated around its mean value .167, with only a small probability that an observed P will be greater than .20, causing R to give the wrong answer ("reject H_0"). We now ask, "How small is the probability of this error?" Recall from Chapter 6 that a sample proportion P has an approximate normal distribution, with

$$\mu_P = \pi = 1/6 = .167$$

and

$$\sigma_P = \sqrt{\frac{\pi(1-\pi)}{n}} = .037 \tag{9-4}$$

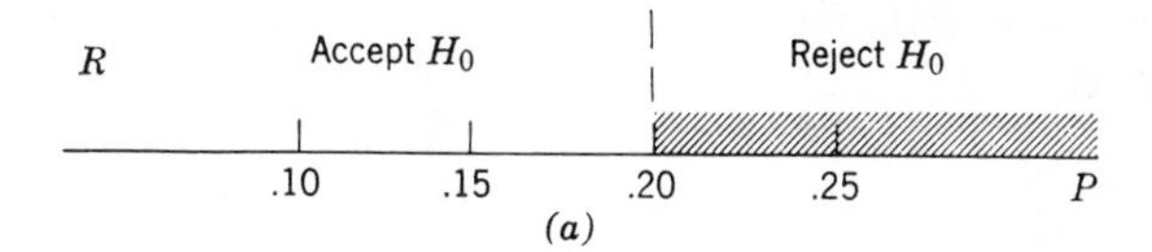

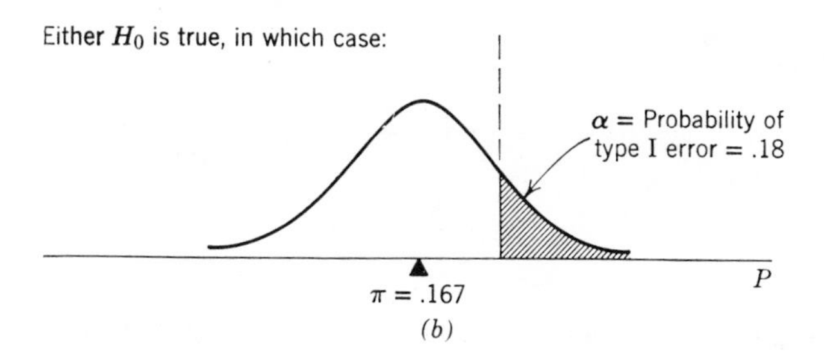

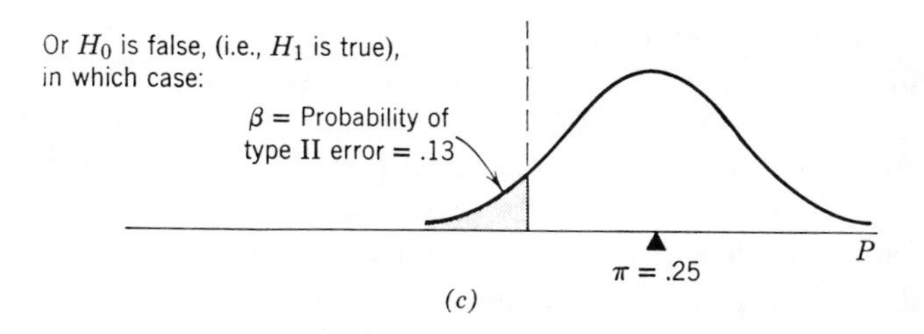

FIG. 9-1 Hypothesis testing. (*a*) Known decision rule, *R*. (*b*) and (*c*) Unknown world, with two possibilities.

This distribution of P is shown in Figure 9-1*b*. The probability of error is easily calculated by evaluating the probability to the right of .20, that is[1]

$$\begin{aligned}\Pr(P > .20/H_0) &= \Pr\left(\frac{P-\mu_P}{\sigma_P} > \frac{.20 - .167}{.037}\right) \\ &= \Pr(Z > .9) \\ &= .18 = \alpha, \quad \text{let us say.}\end{aligned} \tag{9-5}$$

This error of rejecting H_0 when it is true is called a type I error, with its probability denoted α.

On the other hand, suppose that H_0 is false; what is the probability of error then? In this case H_1 is true, and $\pi = .25$; the distribution of P is again approximately normal. According to the rule R, we will make an error in accepting the false H_0 if we observe $P < .20$. The probability of this

[1] In this chapter, we write Pr ($/H_0$) to mean "probability, assuming H_0 is true."

error is therefore calculated by evaluating the normal distribution in Figure 9-1*c* lying to the *left* of .20. Since the mean and standard deviation of this distribution are

$$\mu_P = \pi = .25 \tag{9-6}$$

$$\sigma_P = \sqrt{\frac{\pi(1-\pi)}{n}} = .043 \tag{9-7}$$

it follows that:

$$\Pr(P < .20/H_1) = \Pr\left(\frac{P-\mu_P}{\sigma_P} < \frac{.20-.25}{.043}\right) \tag{9-8}$$

$$= \Pr(Z < -1.15)$$

$$= .13 = \beta, \quad \text{let us say} \tag{9-9}$$

This error of accepting H_0 when it is false is called a type II error, with its probability denoted β.

The terminology of testing is reviewed in Table 9-1. Note that the probabilities in each row must sum to 1; this must follow, so long as we use a rule (like R) which involves the decision either[2] to accept or reject H_0.

We now recall that our decision rule R in (9-1) was determined arbitrarily. We now ask: "Is there a better decision rule, i.e., a better critical point for our test than $P = .20$?" Of course, we should like to make the probabilities of error (α and β) as small as possible, but these two objectives conflict. This is illustrated in Figure 9-2, which is a condensed version of

TABLE 9-1 Possible Errors in Hypothesis Testing

State of the world	Decision: Accept H_0	Decision: Reject H_0
If H_0 is true	Correct decision. Probability $= 1 - \alpha$; corresponds to "confidence level"	Type I error. Probability $= \alpha$; also called "significance level"
If H_0 is false (H_1 true)	Type II error. Probability $= \beta$	Correct decision. Probability $= 1 - \beta$; also called "power"

[2] Of course other more complicated decision rules may be used. For example, the statistician may decide to suspend judgement if the observed P is in the region around .20 (say $.18 < P < .22$). If he observes an ambiguous P in this range, he would then undertake a second stage of sampling—which might yield a clear-cut decision, or might lead to further stages of sequential sampling.

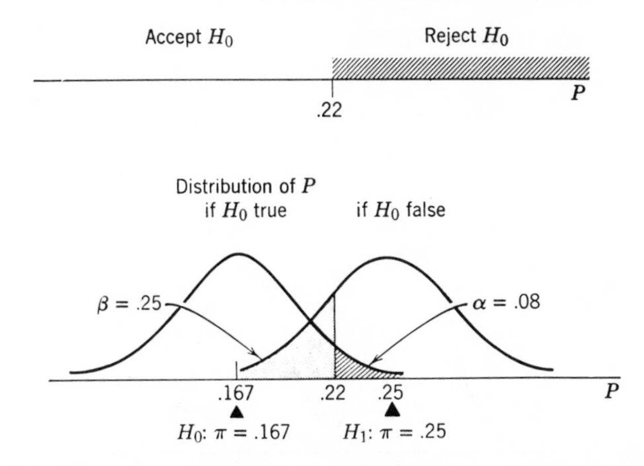

FIG. 9-2 Illustration of how reducing α increases β (compare with Fig. 9-1).

Figure 9-1, except that the critical point has been moved up from .20 to .22. As we hope, this does reduce α; but it also increases β. Moreover, we note that the only way to eliminate α is to move the critical point far enough to the right, but as we do so β approaches 1; i.e., our test becomes "powerless" since we can no longer reject even the most dishonest die. Similarly, it is easy to confirm that any attempt to reduce β (by lowering the critical point below .20) will increase α. In statistics, as in economics, the problem is trading off conflicting objectives.

The only way to reduce both α and β is to increase the sample size. From equation (9-4) it is clear that an increase in n will reduce the spread of P, concentrating its distribution more closely around its central value. Thus if n is increased from 100 to 200, we obtain the result shown in Figure 9-3. The only difference in this test and the one shown in Figure 9-1 is the increase in sample size; note how it reduces both α and β.

These principles are illustrated with an interesting legal analogy. In a murder trial, the jury is being asked to decide between H_0, the hypothesis that the accused is innocent, and the alternate H_1, that he is guilty. A type I error is committed if an innocent man is condemned (innocence is rejected), while a type II error occurs if a guilty man is set free (innocence is accepted). The judge's admonition to the jury that "guilt must be proved (i.e., innocence rejected) beyond a reasonable doubt" means that α should be kept very small. There have been many legal reforms (for example, in limiting the evidence that can be used against an accused man) which have been designed to reduce α, the probability that an innocent man will be condemned. But these same reforms have increased β, the probability that a guilty man will evade punishment. There is no way of pushing α down to zero, and insuring absolutely against convicting an innocent man without letting every defendant

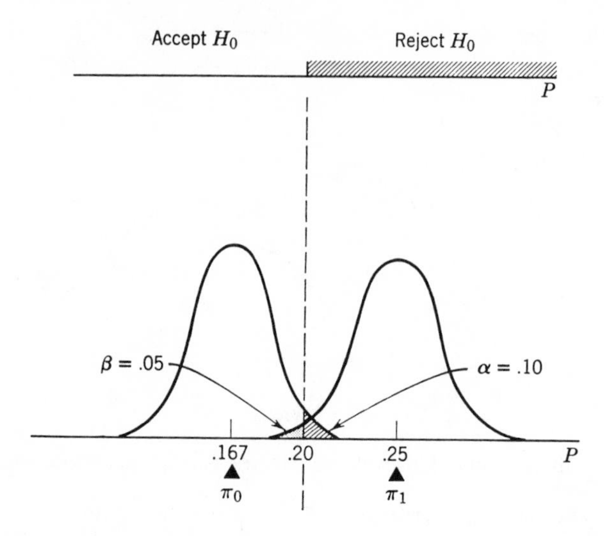

FIG. 9-3 How α and β are both reduced by increasing sample size (compare with Fig. 9-1).

go free, thus raising β to 1 and making the trial meaningless (powerless). It should also be noted that historically α and β have been *both* reduced by improved crime detection—i.e., by increased available evidence brought to bear on H_0.

Returning to the statistical problem, we conclude that (short of raising more funds for increasing sample size) we are left with the problem of how best to balance, or trade off, α and β. Whenever possible the answer should take into consideration the factors mentioned at the outset of the chapter.

1. The relative prior likelihood of the two competing hypotheses. To use an earlier example: if your opponent is a trusted friend, rather than a complete stranger, your greater prior confidence in H_0 will make you more reluctant to reject it; thus you will keep α small.

2. The relative cost of making each type of error. To use the same example: suppose the cost of making a type I error (and accusing an old friend of being a cheat) is high, while the cost of making a type II error is relatively low (you continue to bet against a crooked die—but it is only for peanuts); in these circumstances, your greater concern about making a type I error will lead you to reduce α to a relatively small value, even though β is increased as a consequence. Or, drawing on our legal analogy, we may interpret legal reforms designed to protect the innocent (i.e., reduce α) as a reflection of the judgement that the cost of type I errors (condemning innocent men) exceeds the cost of type II errors (allowing the guilty to go free).

The difficulty is that, in a great deal of scientific inquiry, these questions cannot be answered with any precision; but because type I errors are usually quite serious, α is set at a small value—usually 5% or 1%. Then the test rule is constructed on this basis.

We illustrate with our die-tossing example how hypotheses are typically tested. Three steps are involved.

1. The null hypothesis H_0 and the alternative H_1 are formally stated, as in (9-1) and (9-2). At the same time, the sample size (e.g., 100) and the significance of the test (e.g., $\alpha = 5\%$) are set.

2. We now assume that the null hypothesis H_0 is true. And we ask: "What can we expect of a sample drawn from this kind of world?" This question is answered in our die-tossing example thus: if H_0 is true, then there is a probability of only 5% that we will observe a sample P greater than .228. This critical value (.228) is determined as follows. We note from Appendix Table IV that a Z value of 1.64 cuts a 5% tail off the standard normal distribution. This critical Z value is translated into a P value:

$$\frac{P - \pi}{\sqrt{\dfrac{\pi(1 - \pi)}{n}}} = Z = 1.64 \tag{9-10}$$

and for the H_0 value of $\pi = 1/6$:

$$\frac{P - 1/6}{\sqrt{\dfrac{(1/6)(5/6)}{100}}} = 1.64 \tag{9-11}$$

which yields the critical value of $P = .228$. The resulting test R_* is shown in Figure 9-4. This shows us what to expect of a sample P, if H_0 is true. (At this stage, the probability of a type II error (β) and the power of this test $(1 - \beta)$ may also be calculated, but this is not always done. As an exercise, the reader should confirm that $\beta = .31$ and the power of the test is .69.) With the rule R_* now established, there remains only the last automatic step.

3. The sample is taken, and P observed. We now ask: "Is this P consistent with H_0?" If it is not (i.e., $P > .228$) we reject H_0.

As an example, recall that in our 100 tosses, we rolled 27 aces. This observed $P = .27$ is in such conflict with H_0 that it cannot be "reasonably" attributed to chance and H_0 is rejected.

Summary. Whereas in our first test (R) we arbitrarily specified the critical value (.20) and solved for α, in this more typical hypothesis test (R_*) we specify α (.05) and solve for the critical value. Note that the "95%

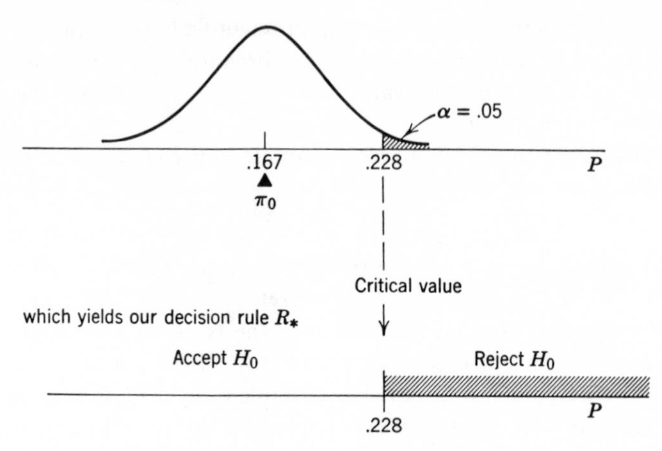

FIG. 9-4 Construction of a test of H_0, the hypothesis that $\pi = .167$, at a 5% significance level ($\alpha = .05$) with a sample $n = 100$.

confidence level" of this test is similar to the concept of 95% confidence used in interval estimation. We set up the test in Figure 9-4 on the assumption that H_0 is true. If this is so, there is a 95% probability that P will be observed below .228, and we will (correctly) accept H_0. Thus we are using a method in which there is a 95% probability that we will be right when H_0 is true.

There is another way of looking at this testing procedure. If we get an observed P exceeding .228, there are two explanations.

1. H_0 is true, but we have been exceedingly unlucky and got a very improbable sample P. (We're born to be losers; even when we bet with odds of 19 to 1 in our favor, we still lose); or
2. H_0 is *not* true after all; the die is crooked, and it is no surprise that we rolled so many aces.

Being reasonable, we opt for the second explanation. Although the first explanation is conceivable, it is not as plausible as the second. But we are left in some doubt; it is just possible that the first explanation is the correct one. For this reason we qualify our conclusion "to be at the 5% significance level (type I error level)."

PROBLEMS

9-1 Fill in the blanks.

Consider the problem facing a radar operator whose job is to detect enemy aircraft. When something irregular appears on the screen, he must decide between

H_0: all is well; only a bit of interference on the screen.
H_1: an attack is coming.

In this case, the type _____ error is a "false alarm," and the type _____ error is a "missed alarm." To reduce both α and β, the electronic equipment is made as reliable and sensitive as possible.

9-2 (a) To test whether a die has a fair number of aces, using $n = 100$ construct a test, at the 1% significance level, of

$$H_0: \pi = .167$$
$$\text{versus } H_1: \pi = .25$$

(b) What is β?

9-3 (a) Construct the appropriate test of
H_0: coin unbiased, versus the alternative H_1: Pr (heads) = .60 using a 25% level of significance. Assume a sample of 100 tosses.
(b) Do the sample results you observed in Problem 3-2 lead you to reject H_0? About how many students in your class will mistakenly reject H_0?
(c) What is β for this test? Interpret.

⇒ 9-4 (a) To test whether a die has a fair number of aces, using $n = 100$ construct an appropriate test, at the 5% significance level, of

$$H_0: \pi = .167$$
$$\text{versus } H_1: \pi = .300$$

(b) What are α and β for this test?
(c) Compared with the test developed in the text in Figure 9-4, is
(1) The critical value different?
(2) α different?
(3) β different?

9-2 COMPOSITE HYPOTHESES

(a) Introduction

In our die-tossing example we have assumed that there is only *one* way in which a die can be crooked (i.e., $\pi = 1/4$). Thus the alternative hypothesis H_1 was a simple one. But usually there is no way of knowing how heavily the die may be biased against us. Thus our alternative hypothesis H_1 (crooked die) would be a *composite* hypothesis, embracing a whole set of possibilities.

$$\begin{aligned} H_1\colon \pi &= .17 \\ \pi &= .18 \\ \pi &= .19 \\ &\vdots \end{aligned} \tag{9-12}$$

including our previous, simple alternative:

$$\begin{aligned} \pi &= .25 \\ &\vdots \end{aligned}$$

To summarize, we wish to test

$$H_0\colon \pi = .167 \tag{9-13}$$

against the composite alternative

$$H_1\colon \pi > .167 \tag{9-14}$$

Since there are many alternatives included in H_1, we can no longer evaluate β as simply as in the previous section. But note that H_0 is still a simple hypothesis; thus the evaluation of α has not been complicated. There is now, therefore, an even stronger case for concentrating on α, which we set at .05; we shall return to an evaluation of the more complicated β values later.

With this significance level of .05 given, the reader should now develop as an exercise the appropriate decision rule for accepting or rejecting H_0. Note that this is identical to rule R_* developed in Figure 9-4. Since the rule is based on the level of significance selected, ($\alpha = .05$ in both cases), it is entirely independent of any considerations of β. But while the formal test may remain the same, there are two major changes in its interpretation.

(b) The Power Function

With a simple alternative H_1, β was a single probability value. With a composite H_1 there are now many possible values of π, each giving a different β. We show three such calculations in Figure 9-5; each involves evaluating the area under a curve, lying to the left of the critical value ($P = .228$). Thus the middle curve shows how the sample P will be distributed if the true π is .25 and yields $\beta = 31\%$. To interpret: if π is in fact .25, then there is a probability of 31% that an observed P will be less than our critical

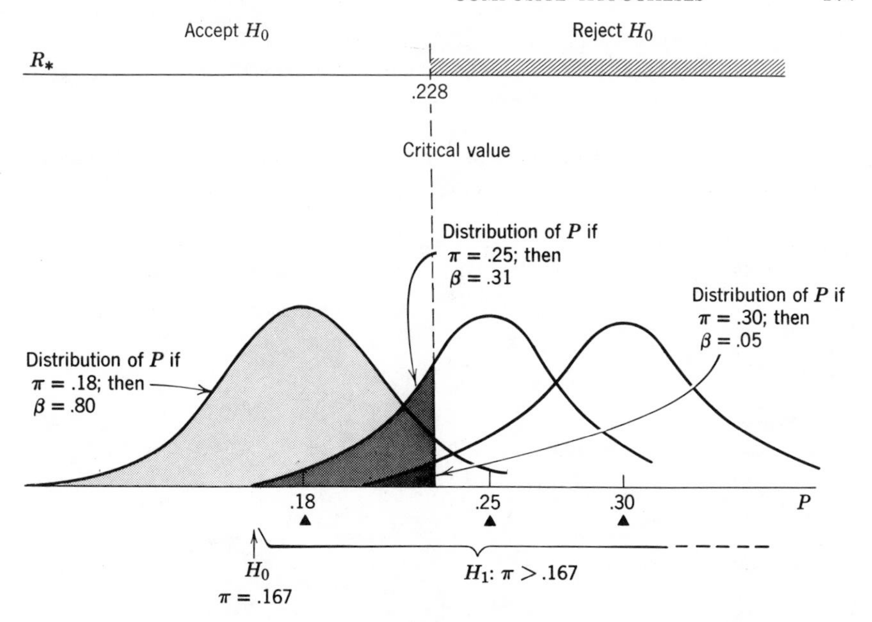

FIG. 9-5 Calculation of β, the probability of type II error, ($n = 100$).

value of .228—and we will erroneously conclude that the die is fair (accept H_0).

Table 9-2 has been constructed using a whole set of possible values of π; the corresponding β values are shown in column 2, and the power of the test $(1 - \beta)$ is shown in column 3; this is the probability that we will correctly reject a crooked die. If a dishonest gambler uses a die which always turns up an ace, he knows he will be quickly found out, and the game abandoned; the more crooked the die he uses, the greater your "power" to uncover him as a cheat. The less crooked the die, (i.e., as we move down this column), the more difficult it becomes to reject it. The dishonest gambler will recognize this, and will prefer to get you to play against a slightly crooked die. The power of this test is thus seen to be our ability to uncover a crooked die; and if it is only slightly biased, our test has little power, and it becomes almost impossible to distinguish between the two conflicting hypotheses. This is confirmed from the last, limiting line in Table 9-2. Here the value of π is H_0, and to reject H_0 would be wrong. The probability of this was $\alpha = 5\%$ by definition.

The "power function" is graphed in Figure 9-6. Clearly we should like a power function that begins very close to the baseline, since its initial height is α, the level of significance, which we wish to keep low. At the same time, we wish the power function to be very steep; the more rapidly it rises, the greater is our power to distinguish between competing hypotheses.

TABLE 9-2 β and Power Function for the Test R_*
(Test of Fair Die, at 5% Level of Significance)

(1) Possible Values of π	(2) Probability of (Erroneously) Accepting H_0 β	(3) Probability of Correctly Rejecting H_0 Power $= 1 - \beta$
.32	.02	.98
.30	.05	.95
.28	.12	.88
.26	.23	.77
.24	.39	.61
.22	.58	.42
.20	.76	.26
.18	.89	.11
.17	.94	.06
.	.	.
.	.	.
.	.	.
Limit (.167)	(.95)	(.05) $= \alpha$

(c) A Warning About Accepting H_0

This introduces a second reason for interpreting our test in this section (with a composite H_1) differently from our test in the previous section (with a simple H_1). It is now possible that this die is only slightly biased ($\pi = .18$). If this is in fact the case, it is very likely ($\beta = .89$) that we will observe P in the range below .228. R_* tells us to accept H_0 (true die); but this is a mistake. At the same time the evidence is not strong enough to reject H_0. What to do?

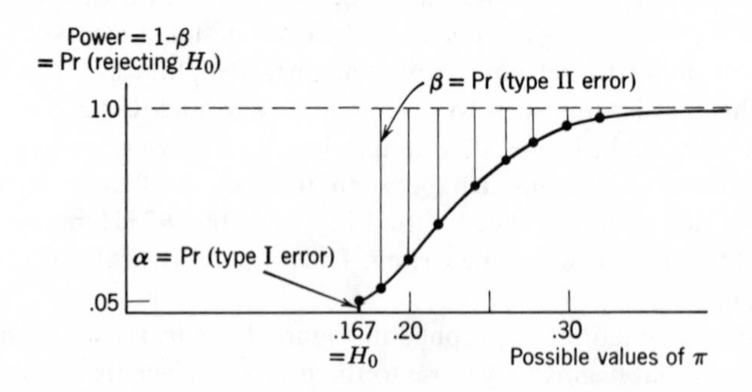

FIG. 9-6 Graph of the power function of Table 9-2, for the test R_* of the fair die at the 5% significance level.

The only other apparent option is to suspend judgement; or formally "not reject H_0." It is only in this way that we can avoid the great risk of incurring a type II error. Earlier, with a simple H_1 (substantially different from H_0), we could live with the risk of type II error; but with our composite alternative this risk becomes prohibitive (β can run up to .95). Thus we prefer to "not reject H_0," "suspend judgement," or conclude that "our sample P is not statistically significant, i.e., P is not significantly greater than 1/6." But we do not accept H_0 outright.

We confirm this from another point of view. Suppose we are using test rule R_* on a biased die ($\pi = .21$); we toss the die and observe $P = .21$. This would *not* be an unlucky result; in fact it is the best luck that we could have hoped for, since our estimate P is exactly on the true value π. If we started out suspecting bias, it has now been confirmed by our sample. There are no grounds whatsoever for concluding that this is a fair die. We cannot accept H_0. Since we also cannot reject H_0, we suspend judgement.[3]

There is another alternative which is generally even more attractive and to this we shall now turn.

(d) Prob-value

The prob-value[4] is defined as

$$\text{Prob-value} \triangleq \Pr\begin{pmatrix}\text{the sample value would be as extreme} \\ \text{as the value we actually observed}/H_0\end{pmatrix} \tag{9-15}$$

[3] This section illustrates a problem involved in accepting H_0 if the sample size is small; note how an increase in sample size, by reducing our standard error, would eventually allow us to reject H_0—presuming, of course, that we continued to observe $P = .21$.

On the other hand, if the sample size is extremely large we can fall into a trap in *rejecting* H_0. To see why, we consider more carefully the question of whether any die is absolutely true, with $\pi = 1/6$ *exactly*. The answer to this must be no; $H_0: \pi = 1/6$ like any other simple hypothesis, must be (slightly) false. And all we have to do to reject it is to take a large enough sample thus reducing our standard error of estimate to the point where even an observed P just slightly different from 1/6 will call for rejection of H_0, i.e., be "statistically significant." But concluding that this die is dishonest misses the point: it is not dishonest enough to be of any practical consequence. We therefore must distinguish between statistical significance, and practical importance.

In conclusion, sample size is obviously an important consideration in hypothesis testing. If the sample is very large, rejecting H_0 may be very dangerous; on the other hand, if the sample is very small, accepting H_0 is dangerous (and this is the more critical problem for economists and other social scientists, with their limited available information).

[4] Short for "probability-value." It is sometimes further contracted to "*p*-value"; we do not use this term, to avoid confusion.

For the gambling example, in tossing a die 100 times and observing the proportion of aces to be $P = .27$, we have

$$\text{prob-value} \stackrel{\Delta}{=} \Pr(P \geq .27/H_0) \tag{9-16}$$

$$= \Pr\left(\frac{P - \mu_P}{\sigma_P} \geq \frac{.27 - .167}{\sqrt{5/3600}}\right)$$

$$= \Pr(Z \geq 2.77)$$

$$= .0028 \tag{9-17}$$

This calculation is very similar to the calculation of α, and is shown in Figure 9-7*a*. We further note that if the observed value of P is extreme, the prob-value is very small. Thus the *prob-value measures the credibility of H_0*. It is an excellent way for the scientist to summarize what the data says about the null hypothesis.

The relation of prob-value to testing H_0 may be seen in Figure 9-7*b*.

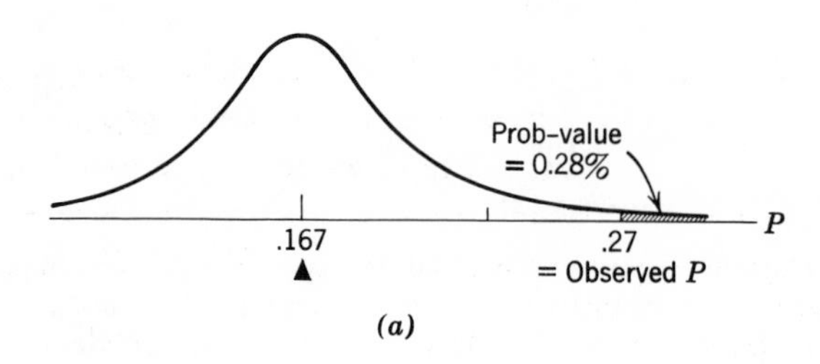

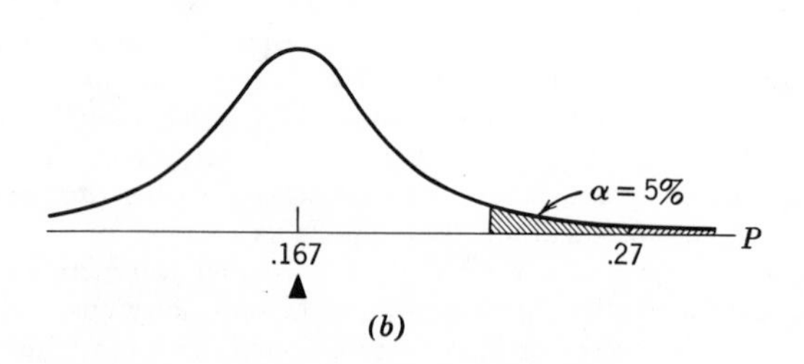

FIG. 9-7 Prob-value for the gambling example; H_0 is $\pi = 1/6$ and sample size is $n = 100$. (*a*) Calculation of prob-value when observed $P = .27$. (*b*) Fig. 9-4 repeated to show relation of prob-value to α: Reject H_0 iff prob-value $< \alpha$.

Since the prob-value is smaller than α, the observed value of P is in the rejection region of the test, i.e.,

$$\boxed{\text{Reject } H_0 \text{ iff prob-value} < \alpha} \tag{9-18}$$

To restate this, we recall that the prob-value is a measure of the credibility of H_0; if this credibility sinks below α, then H_0 must be rejected.

Figure 9-7 shows yet another interpretation: the prob-value is the smallest possible value of α at which H_0 may be rejected.

To conclude, a major criticism of the traditional hypothesis testing of Section 9-1 is that α is set rather arbitrarily, and the simple decision to reject or not reject H_0 does not allow the sample to "tell us" all that it might. Prob-value is therefore the preferred way of stating the result of a hypothesis test. Then each reader can set his own level of significance α at whatever value he deems appropriate, and make his own decision to reject H_0 if the prob-value $<\alpha$. [If the prob-value $> \alpha$, he should suspend judgement for the reasons cited in Section 9-2(c) above.]

Another Example. Suppose that an auto firm has been using brake linings with a stopping distance of 90 feet. The firm is considering a switch to another type of lining, which is similar in all other respects, but alleged to have a shorter stopping distance. In a test run the new linings are installed on 64 cars; the average stopping distance is 87 feet, with a standard deviation of 16 feet. In your job of quality control, you are asked to evaluate whether or not the new lining is better.

Let

μ = average stopping distance for the population of new linings

and test

$$H_0 : \mu = 90$$

against the alternative

$$H_1 : \mu < 90$$

Noting that the observed $\bar{X}$ is 87, you calculate the prob-value, using a method similar to (9-16):

$$\text{prob-value} = \Pr(\bar{X} \leq 87/H_0) \tag{9-19}$$

In other words, this is the probability that $\bar{X}$ will be as extreme as the value you observed, i.e., 3 feet or more below the hypothetical value of 90 feet. Translating (9-19) into Z values, we have

$$\begin{aligned} \text{prob-value} &= \Pr\left(\frac{\bar{X} - \mu_{\bar{X}}}{\sigma_{\bar{X}}} \leq \frac{87 - 90}{16/\sqrt{64}}\right) \\ &= \Pr(Z \leq -1.5) \\ &= .067 \end{aligned} \tag{9-20}$$

You report therefore that there is evidence that the new linings are better, since there is only a 6.7% probability that you would get such extreme test results from an equivalent product. Thus you leave the decision to the vice-president. If he uses a 10% significance level, he will switch to the new linings. But if he uses a 5% significance level, he will not switch.

(e) How to Select H_0

So far we have tested a simple H_0 against both a simple and composite H_1. Cases occasionally occur when both H_0 and H_1 are composite. As an example, suppose we are asking whether American men are more likely to vote Democratic than American women. The null hypothesis (that voting preference is the same) contains many simple hypotheses.

$$\begin{aligned} H_0: \pi_M &= \pi_W = .50 \\ \pi_M &= \pi_W = .51 \\ &\vdots \\ \pi_M &= \pi_W = x, \qquad 0 \leq x \leq 1 \end{aligned}$$

where π_M and π_W represent the proportion of men and women voting Democratic.

Moreover, the alternate hypothesis is even more composite.[5]

$$\begin{aligned} H_1: \pi_M &= .51 \quad \text{and} \quad \pi_W = .50 \\ \pi_M &= .52 \quad \text{and} \quad \pi_W = .50 \\ \pi_M &= .52 \quad \text{and} \quad \pi_W = .49 \\ &\vdots \\ \pi_M &= x \quad \text{and} \quad \pi_W = y \qquad 0 \leq x \leq 1 \\ & \qquad\qquad\qquad\qquad\qquad 0 \leq y \leq 1 \\ & \qquad\qquad\qquad\qquad\qquad x > y \end{aligned}$$

Additional complications are now involved, above and beyond those introduced when only H_1 was composite. Indeed it is difficult to know where to start. The key is to define a new population parameter δ—the difference between voting preferences. Specifically

$$\delta = \pi_M - \pi_W$$

[5] We refer to H_0 as one-dimensional, and H_1 as two-dimensional.

The null hypothesis becomes

$$H_0 : \delta = 0 \tag{9-21}$$

against the composite alternative

$$H_1 : \delta > 0 \tag{9-22}$$

For large samples, a test can now be constructed on the basis of the difference in the sample proportions, $P_M - P_W$.

As this illustration makes clear, the null hypothesis may sometimes be uninteresting, and one that we neither believe nor wish to establish. It is selected because of its simplicity. It is the alternative H_1 that we are trying to establish, and we prove H_1 by rejecting H_0. We can see now why statistics is sometimes called "the science of disproof." H_0 cannot be proven;[6] and H_1 can be proven only by disproving (rejecting) H_0. It follows that if we wish to prove some proposition, we call it H_1 and set up the contrary hypothesis H_0 as the "straw man" we hope to destroy.

Another Example. Suppose that the research engineers in an electronics company claim that they have developed a new television tube superior to the old, which had an average lifetime of 12,400 hours. They ask you to prove its superiority. You wish to establish,

$$H_1 : \mu > 12{,}400$$

where μ is the average lifetime of all new tubes. The "straw man" you hope to destroy is that this tube is no better, i.e.,

$$H_0 : \mu = 12{,}400$$

The new tube is then tested in the hope that the observed sample mean will be significantly greater than 12,400. If it is, then H_0 is rejected and H_1 is established.

This example emphasizes our earlier warning against accepting H_0. Suppose the sample mean $\bar{X}$ is slightly above μ_0, yielding a prob-value of 20%. If the vice president specifies the significance level α at 5%, the evidence is not strong enough to allow us to reject H_0. But we cannot accept H_0 either, for two reasons: (1) we did not believe it in the first place; it was set up simply as a "straw man" we hoped to knock over in order to establish H_1; (2) the tests *suggest* H_0 is wrong, (although not as strongly as we would have liked). We therefore opt to withhold judgement, simply quoting the prob-value, and wait for further evidence.

---

[6] See Section 9-2(c) above.

PROBLEMS

9-5 A certain type of seed has always grown to a mean height of 8.5 inches. A sample of 49 seeds grown under new conditions has a mean height of 8.8 inches and a standard deviation of 1 inch.
(a) At the 5% significance level, test the hypothesis that the new conditions grow no better plants.
(b) Graph the power function of this test.

9-6 Whereas the power function of our die test involved graphing all the $(1 - \beta)$ values in column 3 of Table 9-2, an "operating characteristics curve" (OCC) is defined by graphing all β values (column 2). Draw the operating characteristics curve for this test, and compare it with the power function in Figure 9-6. What is the most desirable shape for an OCC?

9-7 A man makes the implausible claim that the average yearly salary of men in a certain profession is only $6600. A random sample of 150 men in that profession shows a mean salary of $6730 with a standard deviation of $900.
(a) Calculate the prob-value, and interpret.
(b) At a 5% level of significance would you reject the man's claim?
(c) At a 1% level of significance, would you reject his claim? Would you therefore accept it? Explain your answer.

9-8 A coffee shop sells on the average 320 cups of coffee per day, with a standard deviation of 40. After advertising, they find that on 7 days they sell an average of 350 cups.
(a) Has advertising left their business unchanged? Calculate the prob-value.
(b) If the owner of the coffee shop specifies that the type I error of the test (significance level) is to be 5%, do you reject the hypothesis that business is unchanged?
(c) What assumptions have you made implicitly in parts (a) and (b)? Under what conditions are they questionable?
*(d) If coffee sales can be observed for 25 days, what would the average sales have to be in order to justify a statement that business had improved, at the 5% significance level?

9-9 In order to compare the yearly incomes in two professions a survey was made among 100 men in each. In one sample the mean income is $6000 with a standard deviation of $700; in the second sample the mean is $6200 with a standard deviation of $400. To weigh the claim that the mean salary in the second profession is no larger than in the first profession, calculate the prob-value. (*Hint.* Use the theory on differences in means developed in Chapter 8.)

(9-10) Records show that in a random sample of 100 hours a machine produced an hourly average of 678 articles with a standard deviation of 25. After a safety device was installed, in a random sample of 500 hours the machine produced an hourly average of 674 articles with a standard deviation of 5. Pointing to the drop of 4 articles per hour in the sample mean, management claimed the safety device was reducing production. The union countered that the drop of 4 articles was "merely statistical fluctuation."

(a) To objectively summarize the evidence on whether production is left unchanged, calculate the prob-value.

(b) If the arbitration board decides that $\alpha = 1\%$ is a fair level of significance (type I error), do they rule in favor of management or union?

*9-11 At a 5% significance level, test the hypothesis that the following sample is drawn from a population of random digits.

1 2 1 2 0 8 6 1 2 4 5 1 2 4 8 4 4 3 0 2

Suppose that the alternate hypothesis is that there is a bias towards small digits.

*9-12 The output of all machines in a factory is substandard 4% of the time. A machine suspected of being inferior produces X substandard articles in 400. How small would X have to be in order to reject the machine as inferior at the 5% significance level?

9-3 TWO-SIDED TESTS

In the previous section we asked whether men voters were more heavily Democratic than women. Suppose instead we ask whether men voters are more *or less* Democratic than women. In either case we use the same simple null hypothesis

$$H_0 : \delta = 0 \tag{9-23}$$

But where in (9-22) we used a one-sided alternative

$$H_1 : \delta > 0$$

we now must use the two-sided alternative

$$H_1 : \delta \neq 0, \text{ which is equivalent to } \begin{cases} \delta > 0, \text{ or} \\ \delta < 0 \end{cases} \tag{9-24}$$

We reject H_0 if our sample estimate of δ is significantly greater than, *or* less than 0. We test H_0 "from both sides."

As a second illustration, suppose we are again testing the trueness of a die. But instead of testing a suspect die we are betting against, suppose we work in quality control in a die-making factory. We are now just as concerned about a die that shows too few aces, as one which shows too many. Our appropriate test involves:

$$H_0: \pi = .167 \tag{9-25}$$

(9-13) repeated

against the two-sided alternative

$$H_1: \pi \neq .167, \quad \text{i.e.,} \quad \begin{cases} \pi > .167, \text{ or} \\ \pi < .167 \end{cases} \tag{9-26}$$

[compare with (9-14)]. The critical region (for rejecting H_0) must now also be two-sided. For a level of significance $\alpha = 5\%$ this is shown in Figure 9-8*b*;

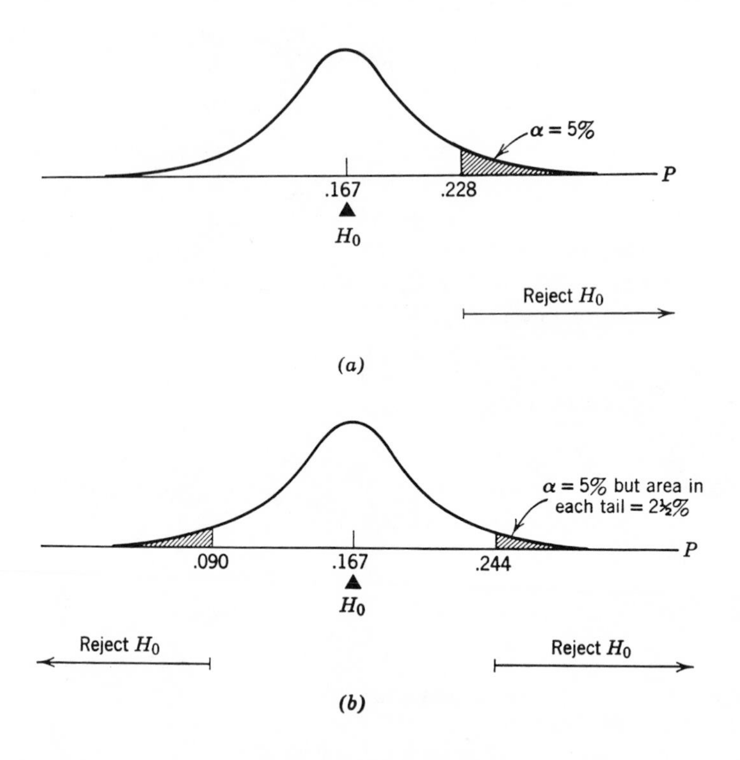

FIG. 9-8 A one-sided and two-sided test of a die compared. (*a*) A one-sided test of $H_0: \pi = .167$ against the alternative $H_1: \pi > .167$ (Fig. 9-4 repeated). (*b*) A two-sided test of $H_0: \pi = .167$ against the alternative $H_1: \pi \neq .167$.

an equal area ($2\frac{1}{2}\%$) is cut off each tail in order to keep the critical region for rejecting H_0 as large as possible. Thus:

$$\text{Reject } H_0 \text{ if } |Z| = \left| \frac{P - \pi_0}{\sqrt{\dfrac{\pi_0(1 - \pi_0)}{n}}} \right| > 1.96 \qquad \text{(9-27)}$$

where π_0 is .167, the null hypothesis value of π. Equation (9-27) simply asks whether P differs from π_0 by a critical amount—on either the high side or the low side. The final question is "How do we recognize when to use a two-tailed test or a one-tailed test?" The one-tailed test is recognized by an asymmetrical phrase like "more than, less than, at least, no more than, better, worse, . . . " and so on. Thus our first test of whether the probability of an ace on the gambler's die was *more than* one sixth, required a one-sided test.

PROBLEMS

9-13 Test $H_0{:}\pi = 1/2$ versus $H_1{:}\pi \neq 1/2$ where π is the probability of tossing a thumbtack "point up." Use a 5% level of significance, and use the sample observations of Problem 3-1.
(a) After 10 tosses.
(b) After 100 tosses.

9-14 Referring to Problem 9-7, suppose that the man's claim that $\mu_0 = 6600$ is no longer implausibly low, i.e., suppose the alternate hypothesis is two-sided: $H_1{:}\mu > 6600$ *or* $\mu < 6600$.
Using now a two-sided test of H_0, and also a two-sided prob-value, answer the same questions as in Problem 9-7.

9-4 THE RELATION OF HYPOTHESIS TESTS TO CONFIDENCE INTERVALS

(a) Two-sided Hypothesis Tests

In this section we shall reach a very important conclusion: a confidence interval can be used to test *any* hypothesis; in fact, the two procedures are equivalent. We illustrate with an example.

Suppose a firm has been producing a light bulb with an average life of 800 hours. It wishes to test a new bulb. A sample of 25 new bulbs has an average life of 810 hours ($\bar{X}$), with a standard deviation of 30 hours (s). Noting that because of our small sample we should use the t, rather than the normal distribution, we can either

1. Test the hypothesis

$$H_0 : \mu_0 = 800 \tag{9-30}$$

against the alternative

$$H_1 : \mu \neq 800 \tag{9-31}$$

H_0 may be accepted[7] at the 5% level of significance if

$$|\text{observed } t| = \left| \frac{\quad - \mu_0}{s/\sqrt{n}} \right| \leq t_{.025} \tag{9-32}$$

i.e., if $\mu_0 - 2.06\, s/\sqrt{n} \leq \bar{X} \leq \mu_0 + 2.06\, s/\sqrt{n}$.

Given our sample s, along with our hypothesis μ_0, this condition becomes

$$788 \leq \bar{X} \leq 812 \tag{9-33}$$

Since our observed $\bar{X}$ (810) does fall within this interval, μ_0 is acceptable. This is shown in Figure 9-9*a*.

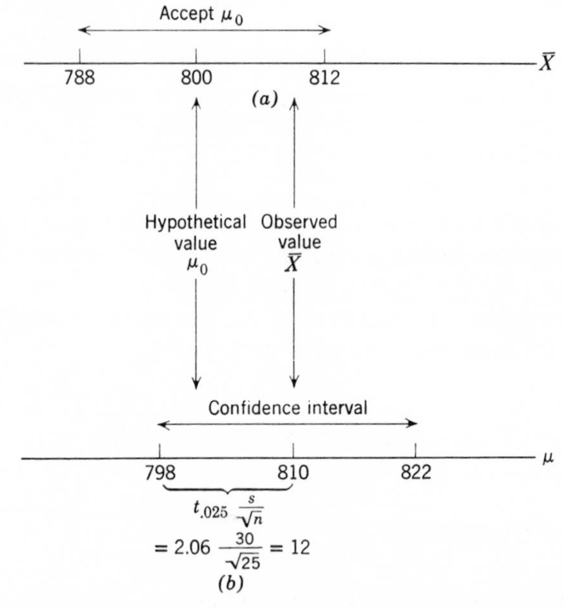

FIG. 9-9 Comparison of two-sided hypothesis test with confidence interval (using a sample with $\bar{X} = 810$ and $s = 30$). (*a*) Test of $H_0 : \mu_0 = 800$ versus $H_1 : \mu \neq 800$. (*b*) Confidence interval for μ.

[7] More specifically "H_0 should not be rejected." To simplify the exposition in this section and avoid double negatives, we shall use "accept H_0," rather than "do not reject H_0"—although as we have pointed out earlier, the latter (weaker) conclusion may be the only one justified.

2. Alternatively, the sample result could be used to construct a confidence interval for μ. Using the same 95% level of confidence, this confidence interval is defined in (8-15) as:

$$\bar{X} \pm t_{.025} \frac{s}{\sqrt{n}} \tag{9-34}$$

$$810 \pm 2.06 \frac{30}{\sqrt{25}}$$

or

$$798 < \mu < 822 \tag{9-35}$$

This is shown in Figure 9-9*b*.

The observed $\bar{X}$ of 810 falls in the acceptable region defined in the hypothesis test in Figure 9-9*a*; hence μ_0 is acceptable. At the same time, in Figure 9-9*b* we note that μ_0 falls within the confidence interval.

This is the key point: if and only if μ_0 falls within this confidence interval, will it be an acceptable hypothesis. This is clear from the diagram, since the interval we use is the same length in both cases: it is constructed by adding and subtracting precisely the same error allowance ($t_{.025}\ s/\sqrt{n} = 12$). Provided the sample mean $\bar{X}$ and μ_0 differ by less than this, μ_0 will fall in the confidence interval, and will also be an acceptable hypothesis. This holds for any μ_0. (To confirm, note that $\mu_0 = 797.6$ would be just barely contained in the confidence interval at the bottom; at the same time this hypothetical value would shift our acceptable region to the left in the top diagram to the point where our sample $\bar{X} = 810$ would just barely remain in that region. But any smaller hypothetical value of μ will fall outside our confidence region and be rejected.)

It can be proven, in general, that[8]

H_0 is accepted if and only if the relevant confidence interval contains H_0	(9-36)

[8] For a general algebraic proof (rather than geometric interpretation) for (9-36), consider the basis of both the confidence interval and hypothesis test. (We illustrate with the normal test of $\bar{X}$, but our remarks are equally valid for most tests.) With 95% probability,

$$\left| \frac{\bar{X} - \mu}{\sigma/\sqrt{n}} \right| < 1.96 \tag{9-37}$$

In deciding whether to accept the null hypothesis μ_0, we first fix μ_0, and then see whether the observed $\bar{X}$ satisfies this inequality.

In constructing a confidence interval, we first observe $\bar{X}$; then the values of μ which satisfy (9-37) form our confidence interval. μ_0 will be in the confidence interval if and only if the hypothesis μ_0 is accepted, for in both cases we have

$$\left| \frac{\bar{X} - \mu_0}{\sigma/\sqrt{n}} \right| < 1.96$$

noting, of course, that the level of confidence (e.g., 95%) must match the level of type I error (level of significance, 5%).

*(b) One-sided Hypothesis Tests

Equation (9-36) remains true for a one-sided test of a hypothesis, provided, of course, that we use a one-sided confidence interval, as shown in Figure 9-10. Using the same sample result, we see that the observed $\bar{X}$ of 810 falls in the acceptable region defined in the hypothesis test in Figure 9-10*a*; hence μ_0 is acceptable. At the same time, in Figure 9-10*b* we note that μ_0 falls within the confidence interval. This illustrates once more that H_0 is accepted if and only if the confidence interval contains H_0.

The reasons for one-sided hypothesis tests have been established at length in this chapter. These same reasons justify the use of one-sided confidence intervals too. Suppose, for example that the federal government is considering construction of a multipurpose dam in a river basin. Suppose

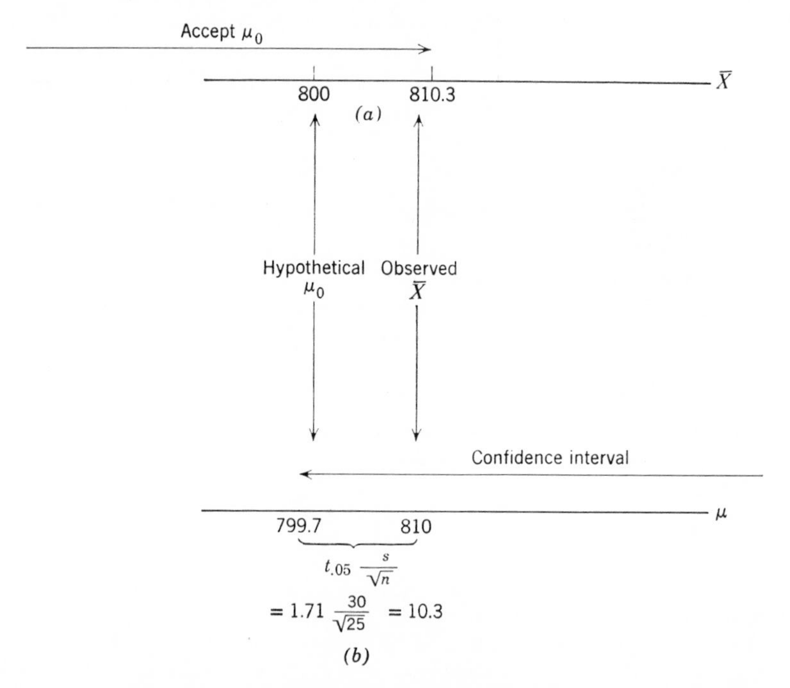

FIG. 9-10 Comparison of one-sided hypothesis test and confidence interval (using same sample result as Fig. 9-9). (*a*) Test of $H_0: \mu_0 = 800$ versus $H_1: \mu > 800$. (*b*) Confidence interval for μ.

further that the cost of this installation is $100 million. The problem is: would the benefits from the project exceed this cost?

To get an idea of irrigation benefits, suppose we run a careful calculation of the operation of a random sample of 25 farmers in the river basin, and estimate that the net profit (per 100 acres) will increase on the average by $810 (with a standard deviation of $30). To simplify the exposition, we have used the same numbers as in Figures 9-9 and 9-10, except that $\bar{X}$ and μ now refer to the average increase in profit.

The best point estimate of μ (average profit increase) is 810. But if we use this in our benefit calculations, we will take no account of its reliability; i.e., it may be way too high, or way too low. Now consider the alternative estimate of $799.7, the critical point in our one-sided confidence interval in Figure 9-10. We can be 95% confident that this figure understates. We don't know by how much, but this doesn't matter; the point is that we are almost certain that this underestimates benefits. Suppose we use similar underestimates of other benefits (flood control, recreation, etc.) and that these[9] sum to $110 million. We can now be very confident that benefits exceed costs, since at each stage we have consciously underestimated benefits. From a policy point of view this is a much stronger conclusion than that the "best estimate" of benefits is $120 million, since the reliability of this estimate remains a mystery. (This strategy clearly has a major drawback. An understatement of benefits may reduce the estimated benefits below cost—in which case we would have to start all over again.)

Thus, by "cooking the case" against our conclusion, it is strengthened. Economists often apply this general philosophy in another way by selecting adverse assumptions in order to strengthen a policy conclusion; they may use one-sided confidence intervals in the future for the same reason.

(c) The Confidence Interval as a General Technique

The reader may ask: "Doesn't (9-36) reduce hypothesis testing to a very simple adjunct of interval estimation?" In a sense this is true. Whenever a confidence interval has been constructed, it can immediately be used to test any null hypothesis: the hypothesis is accepted if and only if it is in the confidence interval. To emphasize this point, we can restate (9-36) in an equivalent form:

A confidence interval may be regarded as just the set of acceptable hypotheses.	(9-38)

[9] I.e., the present value of these accumulated benefits. Issues such as the appropriate rate of discount, and/or the extent to which benefits must exceed costs to justify the project are also important considerations; but we concentrate here on the statistical issues.

The next question is whether, in view of this, our study of hypothesis testing in this chapter has been a waste of time. Why not simply construct the (single) appropriate confidence interval, and use this to test any null hypothesis that anyone may suggest? There is a good deal of validity to this conclusion; nevertheless, our brief study of hypothesis testing has been necessary for the following reasons:

1. Historically, hypothesis testing has been frequently used in physical and social science research. This technique must be understood to be evaluated; specifically the nature of type I and type II error and the warnings about accepting H_0 must be understood.
2. Certain hypotheses have no corresponding simple confidence interval, and are consequently tested on their own.
3. The calculation of a prob-value provides additional information not available if the hypothesis is tested from a confidence interval.
4. Hypothesis testing plays an important role in statistical decision theory, developed in Chapter 15.

PROBLEMS

9-15 Three different sources claim that the average income in a certain profession is \$7200, \$6000, and \$6400 respectively. You find from a sample of 16 persons in the profession that their mean salary is \$6030 and the standard deviation is \$570.
(a) At the 5% significance level, test each of the three hypotheses, one at a time.
(b) Construct a 95% confidence interval for μ. Then test each of the 3 hypotheses by simply noting whether it is included in the confidence interval.

(9-16) A sample of 8 students made the following marks: 3, 9, 6, 6, 8, 7, 8, 9. Assume the population of marks is normal. At a 5% level of significance, which of the following hypotheses about the mean mark (μ) would you reject?
(a) $\mu_0 = 8$.
(b) $\mu_0 = 6.3$.
(c) $\mu_0 = 4$.
(d) $\mu_0 = 9$.

*9-17 As in the second example of Section 9-2(e), suppose a standard process of manufacturing television tubes has a mean of 12,400 hours. The engineers have found a new process which they hope is better than the old standard. To establish this, a sample of 100 tubes from a new process has a mean of 12,760 hours, and a standard deviation of 4000 hours.

(a) Construct a one-sided confidence interval for the new μ.

(b) Calculate the prob-value associated with the null hypothesis of no improvement.

(c) At the 5% level of significance, do you reject the null hypothesis?

9-5 CONCLUSIONS

Hypothesis testing is a technique that must be used with great care, for several reasons. First, the construction of a confidence interval is usually preferred to an hypothesis test; the interval gives a clearer picture of the observed sample result, whereas a test merely indicates whether or not the sample is statistically significant.

Second, there are real problems—especially with a small sample—in accepting an implausible H_0; instead, the prob-value of the test should be calculated. This provides a clear and immediate picture of how well the statistical results match H_0, leaving the rejection decision to the reader.

Finally, rejection of H_0 does not answer the question "Is there any practical economic (as opposed to statistically significant) difference between our sample result and H_0?" This is the broader question of decision theory, developed in Chapter 15.

Review Problems

9-18 Four coins are tossed together 144 times. The average number of heads is 2.2. To answer a gambler who fears the coins are biased towards heads, calculate the prob-value associated with the null hypothesis of fair coins.

9-19 A sample of 784 men and 820 women in 1962 showed that 30 percent of the men and 22 percent of the women stated they were against the John Birch Society. The majority had no opinion.

(a) Letting π_M and π_W be the population proportion of men and women respectively who are against the Society, construct a 95% confidence interval for the difference $(\pi_M - \pi_W)$.

(b) What is the prob-value for the null hypothesis that $(\pi_M - \pi_W) = 0$?

(c) At the 5% significance level, is the difference between men and women statistically significant? (i.e., do you reject the null hypothesis)?

(d) Would you judge this difference to be of *sociological* significance?

(9-20) Of 400 randomly selected townspeople in a certain city, 184 favored a certain presidential candidate. Of 100 randomly selected students in the same city, 40 favored the candidate.

(a) To judge whether the student population and town population have the same proportion favoring the candidate, calculate the prob-value.

(b) Is the difference in the students and townspeople statistically significant, at the 5% level?

9-21 To complete a certain task a sample of 100 workers in one plant took an average of 12 minutes, and a standard deviation of 2.5 minutes. A sample of 100 workers in a second plant took an average of 11 minutes, and a standard deviation of 2.1 minutes.

(a) Construct a 95% confidence interval for the difference in the two population means.

(b) Calculate the prob-value for the null hypothesis that the two population means are the same.

(c) Is the difference in the two sample means statistically significant at the 5% level?

9-22 By talking to a random sample of 50 students, suppose you find that 27 percent support a certain candidate for student government. To what extent does this invalidate the claim that only 20% of all the students support the candidate?

chapter 10

Analysis of Variance

10-1 INTRODUCTION

In the last three chapters we have made inferences about one population mean; moreover, in Section 8-1 we extended this to the difference in two population means. Now we compare r means, using techniques commonly called analysis of variance.[1] Since the development of this technique becomes complicated and mathematical, we shall give a plausible, intuitive description of what is involved, rather than rigorous proofs.

10-2 ONE-FACTOR ANALYSIS OF VARIANCE

As an example, suppose that three machines (A, B, and C) are being compared. Because these machines are operated by men, and for other inexplicable reasons, output per hour is subject to chance fluctuation. In the hope of "averaging out" and thus reducing the effect of chance fluctuation, a random sample of 5 hours is obtained from each machine and set out in Table 10-1, along with the mean of each sample.

Of the many questions which might be asked, the simplest are set out in Table 10-2.

[1] To keep the argument simple, we assume (among other things) that there is an equal size sample (n) drawn from each of the r populations. While such balanced samples are typical in the experimental sciences (such as biology and psychology), they are often impossible in the nonexperimental sciences (e.g., economics and sociology). While analysis of variance can be extended to take account of these circumstances, regression analysis (dealt with in Chapters 11 to 14) is an equally good—and often preferred—technique. But regardless of its limitations, analysis of variance is an enlightening way of introducing regression.

TABLE 10-1 Sample Output of Three Machines

Machine, or Sample Number	Sample from Machine i					$\bar{X}_i$
$i = 1$	48.4	49.7	48.7	48.5	47.7	48.6
$= 2$	56.1	56.3	56.9	57.5	55.1	56.4
$= 3$	52.1	51.1	51.6	52.1	51.1	51.6

Average $\bar{X} = \bar{\bar{X}} = 52.2$

TABLE 10-2

Question	How It Is Answered
(a) Are the machines different?	Analysis of Variance Table (test of hypothesis)
(b) *How much* are the machines different?	Multiple comparisons (simultaneous confidence intervals)

(a) Hypothesis Test

The first question is "Are the machines really different?" That is, are the sample means $\bar{X}_i$ in Table 10-1 different because of differences in the underlying population means μ_i (where μ_i represents the lifetime performance of machine i). Or may these differences in $\bar{X}_i$ be reasonably attributed to chance fluctuations alone? To illustrate, suppose we collect three samples from one machine, as shown in Table 10-3. As expected, sample statistical fluctuations cause small differences in sample means even though the μ's are

TABLE 10-3 Three Samples of the Output of One Machine

Sample Number	Sample Values					$\bar{X}_i$
$i = 1$	51.7	53.0	52.0	51.8	51.0	51.9
$= 2$	52.1	52.3	52.9	53.6	51.1	52.4
$= 3$	52.8	51.8	52.3	52.8	51.8	52.3

$\bar{\bar{X}} = 52.2$

identical. So the question may be rephrased, "Are the differences in $\bar{X}$ of Table 10-1 of the same order as those of Table 10-3 (and thus attributable to chance fluctuation), or are they large enough to indicate a difference in the underlying μ's?" The latter explanation seems more plausible; but how do we develop a formal test?

As before, the hypothesis of "no difference" in the population means becomes the null hypothesis,

$$H_0: \mu_1 = \mu_2 = \mu_3 \tag{10-1}$$

The alternate hypothesis is that some (but *not* necessarily all) of the μ's are different,

$$H_1: \mu_i \neq \mu_j \qquad \text{for some } i \text{ and } j \tag{10-2}$$

To develop a plausible test of this hypothesis we first require a numerical measure of the degree to which the sample means differ. We therefore take the three sample means in the last column of Table 10-1 and calculate their variance. Using formula (2-6) (and being very careful to note that we are calculating the variance of the sample means and *not* the variance of all values in the table), we have

$$\begin{aligned} s_{\bar{X}}^2 &= \frac{1}{(r-1)} \sum_{i=1}^{r} (\bar{X}_i - \bar{\bar{X}})^2 \\ &= \tfrac{1}{2} \sum [(48.6 - 52.2)^2 + (56.4 - 52.2)^2 + (51.6 - 52.2)^2] \\ &= 15.5 \end{aligned} \tag{10-3}$$

where r = number of rows (i.e., the number of sample means), and

$$\bar{\bar{X}} = \text{average } \bar{X} = \frac{1}{r} \sum_{i=1}^{r} \bar{X}_i = 52.2 \tag{10-4}$$

Yet $s_{\bar{X}}^2$ does not tell the whole story; for example, consider the data of Table 10-4, which has the same $s_{\bar{X}}^2$ as Table 10-3, yet more erratic machines that produce large chance fluctuations within each row. The implications of

TABLE 10-4 Samples of the Production of Three Different Machines

Machine	Sample Output from Machine i					$\bar{X}_i$
$i = 1$	54.6	45.7	56.7	37.7	48.3	48.6
$= 2$	53.4	57.5	54.3	52.3	64.5	56.4
$= 3$	56.7	44.7	50.6	56.5	49.5	51.6
						$\bar{\bar{X}} = 52.2$

this are shown in Figure 10-1. In Figure 10-1*a*, the machines are so erratic that all sample outputs could be drawn from the same population—i.e., the differences in sample means may be explained by chance. On the other hand, the (same) differences in sample means can hardly be explained by chance in Figure 10-1*b*, because the machines in this case are *not* erratic.

We now have our standard of comparison. In Figure 10-1(b) we conclude the μ's are different—and reject H_0—because the variance in sample means ($s_{\bar{X}}^2$) is large *relative to* the chance fluctuation.

How can we measure this chance fluctuation? Intuitively, we seem to be interpreting it as the spread (or variance) of observed values *within* each sample. Thus we compute the variance within the first sample in Table 10-1,

$$s_1^2 = \frac{1}{(n-1)} \sum_{j=1}^{n} (X_{1j} - \bar{X}_1)^2 = \frac{(48.4 - 48.6)^2 + \cdots}{4}$$

$$= .52 \tag{10-5}$$

where X_{1j} is the jth observed value in the first sample.

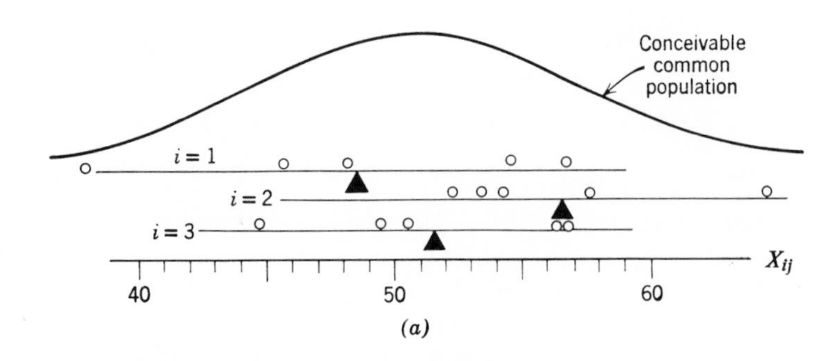

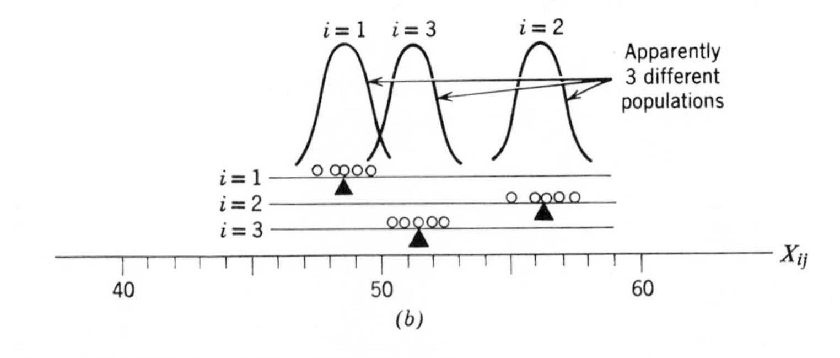

FIG. 10-1 (*a*) Graph of Table 10-4. (*b*) Graph of Table 10-1. The populations appear to be different.

Similarly we compute the variance or chance fluctuation within the second (s_2^2) and third samples (s_3^2). The simple average of these

$$s_p^2 = \frac{1}{r} \sum_{i=1}^{r} s_i^2 = \frac{.52 + .87 + .25}{3} = .547 \tag{10-6}$$

becomes the measure of chance fluctuation—and is referred to as "pooled variance." From each of the r samples, we have a sample variance with $(n - 1)$ degrees of freedom, so that the pooled variance s_p^2 has $r(n - 1)$ degrees of freedom. The key question can now be stated. Is $s_{\bar{X}}^2$ large relative to s_p^2?

In practice, we examine the ratio

$$\boxed{F = \frac{n s_{\bar{X}}^2}{s_p^2}} \tag{10-7}$$

called the "variance ratio." n is introduced into the numerator so that, whenever H_0 is true, this ratio will have, on the average, a value near 1; however, because of statistical fluctuation, it will sometimes be above one, and sometimes below.

If H_0 is not true (and the μ's are not the same) then $n s_{\bar{X}}^2$ will be relatively large compared to s_p^2, and the F value in (10-7) will be greater than 1. Formally, H_0 is rejected if the computed value of F is significantly greater than 1.

Before developing this test further, we interpret (10-7) from another point of view. Suppose that our samples are drawn from three normal populations with the same variance; (in fact, these assumptions are necessary for the formal test below). If in addition, H_0 is true, and the three population means are the same, then the division of our data into three samples is meaningless. All observations could be viewed as one large sample drawn from a single population. Now consider three alternative ways of estimating σ^2, the variance of that population.

1. The most obvious way is to estimate it by computing the variance of the one large sample.

2. The second way is to estimate it by averaging the variances within each of the 3 samples as in (10-5) and (10-6). This is the s_p^2 in the denominator of (10-7).

3. Infer σ^2 from $s_{\bar{X}}^2$, the observed variance of sample means. Recall from Chapter 6 how the variance of sample means is related to the variance of the population:

$$\sigma_{\bar{X}}^2 = \frac{\sigma^2}{n} \tag{10-8}$$

(6-12) repeated

or

$$\sigma^2 = n\sigma_{\bar{X}}^2 \tag{10-9}$$

This suggests estimating σ^2 as $ns_{\bar{X}}^2$, which is recognized as the numerator of (10-7). We note that we are estimating population variance by "blowing up" the observed variance of the sample means.

To recapitulate: if H_0 is true, we can estimate σ^2 by three valid methods. Considering only the last two, we note that one appears in the numerator of (10-7), the other in the denominator; they should be about equal, and their ratio close to 1. [This establishes why n was introduced into the numerator of (10-7).] But if H_0 is not true, the denominator will still reflect only chance fluctuation, but the numerator will be a blow-up of the differences between means; and this ratio will consequently be large.

The formal test of H_0, like any other test, requires knowledge of the distribution of the observed statistic—in this case F—if H_0 is true. This is shown in Figure 10-2. The critical $F_{.05}$ value, cutting off 5% of the upper tail of the distribution is also shown. Thus, if H_0 is true there is only a 5% probability that we would observe an F value exceeding 3.89, and consequently reject H_0. It is conceivable, of course, that H_0 is true and we were very unlucky; but we choose the more plausible explanation that H_0 is false.

To illustrate this procedure, let us reconsider the three sets of sample results shown in Tables 10-1, 10-3, and 10-4, and in each case ask whether the machines exhibit differences that are statistically significant. In other words, in each case we test $H_0: \mu_1 = \mu_2 = \mu_3$ against the alternative that they are not equal. For the data in Table 10-3, an evaluation of (10-7) yields:

$$F = \frac{ns_{\bar{X}}^2}{s_p^2} = \frac{.35}{.547} = .64 \tag{10-10}$$

Since this is below the critical $F_{.05}$ value of 3.89, we conclude that the observed differences in means can reasonably be explained by chance fluctuations. (This is no surprise; recall that we generated these three samples in Table 10-3 from the same machine.)

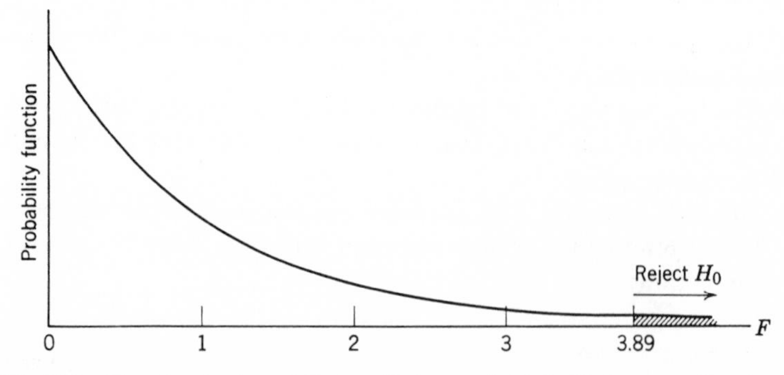

FIG. 10-2 The distribution of F when H_0 is true (with 2, 12 degrees of freedom).

For the data in Table 10-4, the F ratio is

$$F = \frac{77.4}{35.7} = 2.17 \tag{10-11}$$

In this case, the difference between sample means (and consequently the numerator) is much greater. But so is the chance fluctuation (reflected in a large denominator). Again, the F value is less than the critical value 3.89.

However, for the data in Table 10-1, the F ratio is

$$F = \frac{77.4}{.547} = 141 \tag{10-12}$$

In this case, the difference in sample means is very large relative to the chance fluctuation, making the F ratio far exceed the critical value 3.89, so that H_0 is rejected.

These three formal tests confirm our earlier intuitive conclusions. Table 10-1 provides the only case in which we conclude that the underlying populations have different means.

(b) The F Distribution

This distribution is so important for later applications, it is worth considering in some detail. The F distribution shown in Figure 10-2 is only one of many; there is a different distribution depending on degrees of freedom $(r - 1)$ in the numerator, and degrees of freedom $[r(n - 1)]$ in the denominator. Intuitively, we can see why this is so. The more degrees of freedom in calculating both numerator and denominator, the closer these two estimates of variance will likely be to their target σ^2; thus the more closely their ratio will concentrate around 1. This is illustrated in Figure 10-3.

We could present a whole set of F tables, each corresponding to a different combination of degrees of freedom. For purposes of practical testing, however, only the critical 5% or 1% points are required, and are set out in Table VII in the Appendix. From this table, we confirm the critical point of 3.89 used in Figure 10-2.

(c) The ANOVA Table

This section is devoted to a summary shorthand of how these calculations are usually done. The model is summarized in Table 10-5. We confirm in column 2 that all samples are assumed drawn from normal populations with the same variance σ^2—but, of course, means that may, or may not, differ. (Indeed it is the possible differences in means that are being tested).

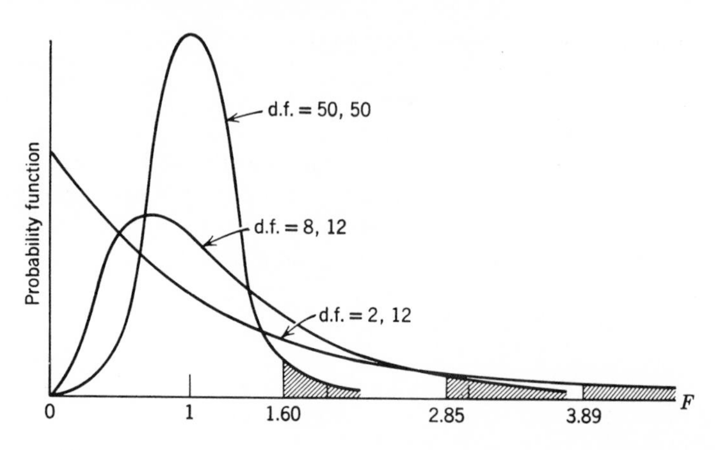

FIG. 10-3 The F distribution, with various degrees of freedom in numerator and denominator. Note how the critical point (for rejecting H_0) moves toward 1 as degrees of freedom increase.

The resulting calculations are conveniently laid out in Table 10-6, called an ANOVA table—an obvious shorthand for ANalysis Of VAriance. This is mostly a bookkeeping arrangement, with the first row showing calculations of the numerator of the F ratio, and the second row the denominator; in (b) part of this table we evaluate the specific example of the three machines in Table 10-1.

In addition, this table provides two handy intermediate checks on our calculations. One is on degrees of freedom in column 3. The other is on sums

TABLE 10-5 Summary of Assumptions

(1) Population	(2) Assumed Distribution	(3) Observed Sample Values	
1	$N(\mu_1, \sigma^2)$	X_{1j}	$(j = 1 \cdots n)$
2	$N(\mu_2, \sigma^2)$	X_{2j}	$(j = 1 \cdots n)$
3	$N(\mu_3, \sigma^2)$	X_{3j}	$(j = 1 \cdots n)$
.			
.			
.			
In general:			
i	$N(\mu_i, \sigma^2)$	X_{ij}	$(j = 1 \cdots n)$

H_0: $\mu_1 = \mu_2 = \mu_i$, for any i
H_1: these means are not all equal

TABLE 10-6

(a) ANOVA Table, General

(1) Source of Variation	(2) Variation; Sum of Squares (SS)	(3) d.f.	(4) Variance; Mean Sum of Squares (MSS)	(5) F ratio
Between rows; "EXPLAINED" by differences in $\bar{X}_i$	$n\sum_{i=1}^{r}(\bar{X}_i - \bar{\bar{X}})^2 = SS_r$	$(r-1)$	$MSS_r = SS_r/(r-1)$ $= ns_{\bar{X}}^2$	$\frac{\text{explained variance}}{\text{unexplained variance}} = F$
Within rows; residual variation, resulting from chance fluctuation, "UNEXPLAINED"	$\sum_{i=1}^{r}\sum_{j=1}^{n}(X_{ij} - \bar{X}_i)^2 = SS_u$	$r(n-1)$	$MSS_u = SS_u/r(n-1)$ $= s_p^2$	
Total	$\sum_i\sum_j(X_{ij} - \bar{\bar{X}})^2$	$(nr-1)$		

(b) ANOVA Table, for Sample Values Shown in Table 10-1

(1) Source of Variation	(2) Variation	(3) d.f.	(4) Variance	(5) F ratio
Between machines; "EXPLAINED"	154.8	2	77.4	$\frac{77.4}{.547} = 141$
Within machines; "UNEXPLAINED"	6.56	12	.547	
Total	161 √	14 √		

of squares in column 2; the sum of squares *between* rows plus the sum of squares *within* rows adds up to the total sum of squares.[2] When any variation is divided by the appropriate degrees of freedom, the result is variance.

The variance *between* rows is "explained" by the fact that the rows may come from different parent populations (e.g., machines perform differently). The variance *within* rows is "unexplained" because it is the random or chance variation that cannot be systematically explained (by differences in machines). Thus F is sometimes referred to as the variance ratio.

$$F = \frac{\text{explained variance}}{\text{unexplained variance}} \tag{10-17}$$

---

[2] Proved as follows. The difference, or deviation of any observed value (X_{ij}) from the mean of all observed values ($\overline{\overline{X}}$), can be broken down into two parts.

Total deviation = explained deviation + unexplained deviation

$$(X_{ij} - \overline{\overline{X}}) = (\overline{X}_i - \overline{\overline{X}}) + (X_{ij} - \overline{X}_i) \tag{10-13}$$

Thus, using Table 10-1 as an example, the third observation in the second sample (56.9) is 4.7 greater than $\overline{\overline{X}} = 52.2$. This total deviation can be broken down into

$$(56.9 - 52.2) = (56.4 - 52.2) + (56.9 - 56.4)$$

$$4.7 = 4.2 + .5$$

Thus most of this total deviation is explained by the machine (4.2), while very little (.5) is unexplained, due to random fluctuations. Clearly (10-13) must always be true, since the two occurrences of $\overline{X}_i$ cancel.

Square both sides of (10-13) and sum over all i and j:

$$\sum_i \sum_j (X_{ij} - \overline{\overline{X}})^2 = \sum_i \sum_j (\overline{X}_i - \overline{\overline{X}})^2 + 2 \sum_i \sum_j (\overline{X}_i - \overline{\overline{X}})(X_{ij} - \overline{X}_i) + \sum_i \sum_j (X_{ij} - \overline{X}_i)^2 \tag{10-14}$$

On the right side, the middle (cross product) term is

$$2 \sum_{i=1}^{r} \left[(\overline{X}_i - \overline{\overline{X}}) \underbrace{\sum_{j=1}^{n} (X_{ij} - \overline{X}_i)} \right], \text{ which must be zero since}$$

$\Updownarrow$

the algebraic sum of deviations about the mean is always zero.

Furthermore, the first term on the right side of (10-14) is:

$$\sum_{j=1}^{n} \underbrace{\left[\sum_{i=1}^{r} (\overline{X}_i - \overline{\overline{X}})^2 \right]}_{\text{independent of } j} = n \sum_{i=1}^{r} (\overline{X}_i - \overline{\overline{X}})^2 \tag{10-15}$$

Substituting these two conclusions back into (10-14), we have:

$$\sum_i \sum_j (X_{ij} - \overline{\overline{X}})^2 = n \sum_i (\overline{X}_i - \overline{\overline{X}})^2 + \sum_i \sum_j (X_{ij} - \overline{X}_i)^2 \tag{10-16}$$

Total variation = explained variation + unexplained variation.

This suggests a possible means of strengthening this F test. Suppose that these three machines are sensitive to differences in temperature. Why not introduce temperature explicitly into the analysis? If some of the previously unexplained variation can now be explained by temperature, the denominator of (10-17) will be reduced. With the larger F value that results we will have a more powerful test of the machines (i.e., we will be in a stronger position to reject H_0). Thus the introduction of other explanations of variance will assist us in detecting whether one specific influence (machine) is important. This brings us to two-way ANOVA in Section 10-3.

*(d) Confidence Intervals

The difficulties with hypothesis tests cited in Chapter 9 hold true in the ANOVA case as well. It may not be too enlightening to ask whether population means differ; by increasing sample size enough, such a difference can nearly always be established—even though it is too small to be of any practical or economic importance. Again, it may be more important to find out "by *how much* do population means differ?"

If we wanted to compare only two machines in Table 10-1, this would be an easy question to answer: just construct a confidence interval for $(\mu_1 - \mu_2)$ using $(\bar{X}_1 - \bar{X}_2)$ and the t distribution:

$$(\mu_1 - \mu_2) = (\bar{X}_1 - \bar{X}_2) \pm t_{.025} s_p \sqrt{\frac{1}{n} + \frac{1}{n}} \qquad \text{(10-18)}$$

(8-17) repeated

In (8-17), s_p^2 was the variance pooled from the two samples. However, it is more reasonable to use all the information available, and pool the variance from all three samples as in (10-6), obtaining $s_p^2 = .547$ with $4 + 4 + 4 = 12$ degrees of freedom. Thus the 95% confidence interval is

$$(\mu_1 - \mu_2) = (48.6 - 56.4) \pm 2.179\sqrt{.547}\sqrt{\tfrac{1}{5} + \tfrac{1}{5}}$$

Similar confidence intervals for $(\mu_1 - \mu_3)$ and for $(\mu_2 - \mu_3)$ may be constructed, for a total of three intervals [or $\binom{r}{2}$ for r populations]; in our example, these intervals are:

$$\begin{aligned} (\mu_1 - \mu_2) &= -7.8 \pm 1.0 \quad \text{(a)} \\ (\mu_1 - \mu_3) &= -3.0 \pm 1.0 \quad \text{(b)} \\ (\mu_2 - \mu_3) &= +4.8 \pm 1.0 \quad \text{(c)} \end{aligned} \qquad \text{(10-19)}$$

The results of this piece-by-piece approach are summarized in Table 10-7.

TABLE 10-7 Differences in Population Means ($\mu_i - \mu_I$) Estimated from Sample Means ($\bar{X}_i - \bar{X}_I$). 95% Level of Confidence in Each Interval Estimate

i \ I	1	2	3
1	0	-7.8 ± 1.0	-3.0 ± 1.0
2		0	4.8 ± 1.0
3			0

*(e) Simultaneous Confidence Intervals: Multiple Comparisons

There is just one difficulty with the above approach. Although we can be 95% confident of each individual statement [e.g., 10-19(a)], we can be far less confident that the whole *system* of statements (10-19) is true; there are three ways in which this could go wrong.

The level of confidence in the system (10-19) would be reduced to $(.95)^3 = .857$, if the three individual statements were independent. But in fact they are not; for example, they all involve the common term s_p. Thus if our observed s_p is high, all three interval estimates in (10-19) will be wide as a consequence. The problem is how to allow for this dependence in order to obtain the correct *simultaneous* confidence coefficient for the whole system. In fact, this problem is usually stated the other way around: how much wider must the *individual* intervals in (10-19) be in order to yield a 95% level of confidence that all are simultaneously true?

Of the many solutions, we quote without proof the simplest, due to Scheffé;[3] with 95% confidence, *all* the following statements[4] are true.

$$(\mu_1 - \mu_2) = (\bar{X}_1 - \bar{X}_2) \pm \sqrt{F_{.05}}\, s_p \sqrt{\frac{(r-1)}{n}2} \qquad \text{(a)}$$

$$(\mu_1 - \mu_3) = (\bar{X}_1 - \bar{X}_3) \pm \sqrt{F_{.05}}\, s_p \sqrt{\frac{(r-1)}{n}2} \qquad \text{(b)} \qquad \text{(10-20)}$$

$$(\mu_2 - \mu_3) = (\bar{X}_2 - \bar{X}_3) \pm \sqrt{F_{.05}}\, s_p \sqrt{\frac{(r-1)}{n}2} \qquad \text{(c)}$$

---

[3] H. Scheffé, *The Analysis of Variance*, p. 66-73, New York: John Wiley, 1959.

[4] And some other statements as well—as we shall see in (10-26). In fact if we were interested *only* in the three comparisons of means in (10-20), our interval estimates could be made slightly narrower.

where

$F_{.05}$ = the critical value of F (with $r - 1$ and $r(n-1)$ d.f.) leaving 5% in the upper tail.

s_p^2 = the pooled sample variance, as calculated in Table 10-6 or equation (10-6)

r = number of rows (means) to be compared.

n = each sample size.

We note the similarity of statements (10-20) and (10-19). For the machines in Table 10-1, the actual simultaneous confidence intervals are

$$\begin{aligned} \mu_1 - \mu_2 &= (48.6 - 56.4) \pm \sqrt{3.89}\,(.74)\sqrt{\tfrac{2}{5}(2)} \\ &= -7.8 \pm 1.3 && \text{(a)} \\ \mu_1 - \mu_3 &= -3.0 \pm 1.3 && \text{(b)} \qquad (10\text{-}21) \\ \mu_2 - \mu_3 &= 4.8 \pm 1.3 && \text{(c)} \end{aligned}$$

These calculations are summarized in Table 10-8. As expected, the width of the confidence interval is greater than in Table 10-7 (compare 1.3 versus 1.0). Indeed, it is this increased width (vagueness) that makes us 95% confident that *all* statements are true.

As a bonus, this theory can be used to make any number of comparisons of means, called "contrasts." A "contrast of means" is defined as a linear combination, or weighted sum, with weights that add to zero:

$$\sum_{i=1}^{r} C_i \mu_i$$

provided

$$\sum_{i=1}^{r} C_i = 0 \qquad (10\text{-}22)$$

TABLE 10-8 Differences in Population Means ($\mu_i - \mu_I$) Estimated from Sample Means ($\bar{X}_i - \bar{X}_I$). 95% Level of Confidence in *All* Interval Estimates. (Compare with Table 10-7.)

i \ I	1	2	3
1	0	−7.8 ± 1.3	−3.0 ± 1.3
2		0	+4.8 ± 1.3
3			0

For example, the simplest contrast is the difference

$$\mu_1 - \mu_2 = (+1)\mu_1 + (-1)\mu_2 + (0)\mu_3 \tag{10-23}$$

It was this contrast that was estimated in (10-21a). Another interesting contrast is the difference between μ_1 and the average of μ_2 and μ_3:

$$\mu_1 - \frac{(\mu_2 + \mu_3)}{2} = (+1)\mu_1 + (-\tfrac{1}{2})\mu_2 + (-\tfrac{1}{2})\mu_3 \tag{10-24}$$

There is no limit to the number of contrasts. It is no surprise that each contrast of the population means will be estimated by the same contrast of the sample means, plus or minus an error allowance. (10-21a) is one example. As another example, the contrast of means given in (10-24) is estimated as

$$\mu_1 - \tfrac{1}{2}\mu_2 - \tfrac{1}{2}\mu_3 = (\bar{X}_1 - \tfrac{1}{2}\bar{X}_2 - \tfrac{1}{2}\bar{X}_3) \pm \sqrt{F_{.05}}\, s_p \sqrt{\frac{(r-1)}{n}\frac{3}{2}} \tag{10-25}$$

The general statement, from which (10-20) and (10-25) were derived, is

With 95% confidence, *all* contrasts are bracketed by the bounds:

$$\sum C_i\mu_i = \sum C_i\bar{X}_i \pm \sqrt{F_{.05}}\, s_p \sqrt{\frac{(r-1)(\sum C_i^2)}{n}} \tag{10-26}$$

provided only that $\sum C_i = 0$ to satisfy the definition of "contrast." As before s_p^2 is pooled variance, and $F_{.05}$ is the critical value of F.

When we examine (10-26) more carefully, we discover that this defines a set of 95% simultaneous confidence intervals which includes not only the three statements in (10-20) but also statements like (10-25), and indeed an infinite number of contrasts that can be constructed. The student may justifiably wonder "How can we be 95% confident of an infinite number of statements?" The answer is: because these statements are dependent. Thus, for example, once we have made the first two statements in (10-21), our intuition tells us that the third is likely to follow. Moreover, once these three statements are made, intervals like (10-25) tend to follow, and can be added with little damage to our level of confidence. As the number of statements or contrasts grows and grows, each new statement tends to become simply a restatement of contrasts already specified, and essentially no damage is done to our level of confidence. Thus, it can be mathematically confirmed that the entire (infinite) set of contrasts in (10-26) are all simultaneously estimated at a 95% level of confidence.

PROBLEMS

10-1 A sample of 4 workers was drawn at random from two different industries, with their average annual income (in $00) recorded, as follows:

Industry A	66	62	65	63
Industry B	58	56	53	61

(a) Using first a t-test (as in Chapter 8) and then an ANOVA F-test, calculate whether or not there is a statistically significant difference in income at the 5% level.

(b) Are the t and F tests exactly equivalent? Can you see why the t^2 distribution is often referred to as the F distribution with 1 degree of freedom in the numerator?

*(c) Using first the t distribution (8-17), and then the F distribution (10-20), construct a 95% confidence interval for the difference in mean incomes in the two industries.

⇒ 10-2 Twelve plots of land are randomly divided into 3 groups. The first is held as a control group while 2 fertilizers A and B are applied to the other 2 groups. Yield is observed to be:

Control, C	60	64	65	55
A	75	70	66	69
B	74	78	72	68

(a) At a 5% significance level, does fertilizer affect yield?

*(b) Construct a table of differences in means, similar to Table 10-8, starring the differences that are statistically significant.

*(c) Can you be 95% confident that the two fertilizers have a different effect?

*(d) What is the difference between a contrast of means, and a weighted average of means?

⇒ 10-3 You have observed the income (Y) of a sample of men and women in a certain occupation to be:

Women	Men
48	60
56	70
50	62
54	48

(a) At a 5% level of significance, can you reject the null hypothesis that mean income is the same for men and women?

*(b) Construct a 95% confidence interval for the difference in the two means.

Since this problem is important later in Chapter 13 we state its solution.

(a) $\overline{Y}_1 = 52 \qquad \overline{Y}_2 = 60 \qquad \overline{\overline{Y}} = 56$

ANOVA Table

Source	Variation	d.f.	variance	
Between sexes	128	1	128	$\frac{128}{48} = 2.67 = F$
Residual	288	6	48	
Total	416	7		

F is less than the critical value of 5.99, thus not statistically significant.

*(b) Evaluate the first equation in (10-20); or, more simply (10-18), noting that $t_{.025} = \sqrt{F_{.05}}$

$$
\begin{aligned}
(\mu_1 - \mu_2) &= (52 - 60) \pm 2.45\sqrt{48}\sqrt{2/4} \\
&= -8 \pm 12
\end{aligned}
$$

This also confirms the answer in (a); since this interval includes zero, this is not statistically significant.

*10-4 Referring to the machine example of Table 10-1 and ANOVA Table 10-6(b), use equation (10-26) to incidentally solve the following problem:

Suppose one factory is to be outfitted entirely with machines of the first type. Suppose a second factory is to be outfitted with machines of the second and third types, in the proportions 30% and 70%. Find a 95% confidence interval for the difference in mean production for the 2 factories.

10-5 From each of three large classes, 50 students were sampled, with the following results:

Class	Average Grade $\overline{X}$	Standard Deviation, s
A	68	11
B	73	12
C	70	8

Test whether the classes are equally good at a 5% significance level.

10-3 TWO-FACTOR ANALYSIS OF VARIANCE

(a) The ANOVA Table

We have already seen that the F test on the differences in machines given in (10-17) would be strengthened if the unexplained variance could be reduced. We suggested, for example, that if some unexplained variance is due to temperature, this might be taken into account; or if some unexplained variance is due to the human factor, we shall see how this might be adjusted for. Suppose that the sample outputs given in Table 10-4 were produced by five different machinists—with each machinist producing one of the sample values on each machine. This data, reorganized according to a two-way classification (by machine *and* operator), is shown in Table 10-9. It is necessary to complicate our notation somewhat. We are now interested in the average of each operator ($\bar{X}_{.j}$, each column average) as well as the average of each machine ($\bar{X}_{i.}$, each row average[5]).

Now the picture is clarified; some operators are efficient (the first and fourth), some are not. The machines are not that erratic after all; there is just a wide difference in the efficiency of the operators. If we can explicitly adjust for this, it will reduce our unexplained (or chance) variation in the denominator of (10-17); since the numerator will remain unchanged, the F ratio will be larger as a consequence, perhaps allowing us to reject H_0. To sum up, it appears that another influence (difference in operators) was responsible for a

TABLE 10-9 Samples of Production (X_{ij}) of Three Different Machines (as given in Table 10-4, but now arranged according to machine operator)

Operator → / Machine ↓	$j = 1$	2	3	4	5	Machine Average $\bar{X}_{i.}$
$i = 1$	56.7	45.7	48.3	54.6	37.7	48.6
2	64.5	53.4	54.3	57.5	52.3	56.4
3	56.7	50.6	49.5	56.5	44.7	51.6
Operator average $\bar{X}_{.j}$	59.3	49.9	50.7	56.2	44.9	$\bar{\bar{X}} = 52.2$

[5] The dot indicates the subscript over which summation occurs. For example, the dot suppresses the subscript j in $\bar{X}_{i.} = \frac{1}{n}\sum_j X_{ij}$.

TABLE 10-10 Two-Way ANOVA-General

(1) Source	(2) Variation; Sum of Squares (SS)	(3) d.f.	(4) Variance; Sum of Squares (MSS)	(5) F
Between rows; EXPLAINED by differences in machines, i.e. differences in $\overline{X}_{i.}$	$SS_r = c\sum_{i=1}^{r}(\overline{X}_{i.} - \overline{\overline{X}})^2$	$r - 1$	$MSS_r = \frac{SS_r}{r-1} = cs^2_{\overline{X}_{i.}}$	$\frac{MSS_r}{MSS_u}$
Between columns; EXPLAINED by differences in operators, i.e., differences in $\overline{X}_{.j}$	$SS_c = r\sum_{j=1}^{c}(\overline{X}_{.j} - \overline{\overline{X}})^2$	$c - 1$	$MSS_c = \frac{SS_c}{c-1} = rs^2_{\overline{X}}$	$\frac{MSS_c}{MSS_u}$
UNEXPLAINED, i.e., residual variation, resulting from chance fluctuation.	$SS_u = \sum_{i=1}^{r}\sum_{j=1}^{c}(X_{ij} - \overline{X}_{i.} - \overline{X}_{.j} + \overline{\overline{X}})^2$	$(r-1)(c-1)$	$MSS_u = \frac{SS_u}{(r-1)(c-1)} = s_p^2$	
Total	$SS = \sum_{i=1}^{r}\sum_{j=1}^{c}(X_{ij} - \overline{\overline{X}})^2$	$rc - 1$		

lot of extraneous noise in our simple one-way analysis in the previous section; by removing this noise, we hope to get a much more powerful test of our machines.

The analysis is an extension of the one-factor ANOVA, and is summarized in Table 10-10. Of course, the small letter c represents the number of columns in Table 10-9, and replaces n in Table 10-4. As before, the component sources of variation shown in column 2 sum to the total variation at the bottom of this column, i.e.,

$$\underset{\text{Total variation}}{\sum_{i=1}^{r}\sum_{j=1}^{c}(X_{ij}-\bar{\bar{X}})^2} = \underset{\text{machine (row) variation}}{c\sum_{i=1}^{r}(\bar{X}_{i.}-\bar{\bar{X}})^2} + \underset{\text{operator (column) variation}}{r\sum_{j=1}^{c}(\bar{X}_{.j}-\bar{\bar{X}})^2}$$

$$+ \underset{\text{random variation}}{\sum_{i=1}^{r}\sum_{j=1}^{c}(X_{ij}-\bar{X}_{i.}-\bar{X}_{.j}+\bar{\bar{X}})^2} \quad (10\text{-}27)$$

We note that operator variation is defined like machine variation; the only difference is that this is defined as the variation exhibited by *column* means. (10-27) is established by a complex set of manipulations, parallel to those used to establish (10-16) in the simpler case. (The last term—the random variation—in (10-27) may seem a bit puzzling; it will be interpreted below.)

(b) Testing Hypotheses

With the total variation broken down into components in (10-27), we can now test whether there is a significant difference in machines, *or* whether there is a significant difference in operators; in either test the extraneous influence of the other factor will be taken into account.

On the one hand, we test for differences in machines by constructing the ratio

$$F = \frac{\text{MSS}_r}{\text{MSS}_u} = \frac{\text{variance explained by machines}}{\text{unexplained variance}} \quad (10\text{-}28)$$

which, if H_0 is true, has an F distribution. Thus, if the observed F calculated in (10-28) exceeds the critical F value we may reject the null hypothesis, concluding that there is a difference in population row means.

Our calculations are shown in full in Table 10-11, whence (10-28) is evaluated as:

$$F = \frac{77.4}{5.9} = 13.1 \quad (10\text{-}29)$$

Since this exceeds the critical[6] F value of 4.46, we reject the null hypothesis

[6] 2 and 8 d.f., and 5% significance.

TABLE 10-11 Two-Way ANOVA, for Observations Given in Table 10-9

(1) Source	(2) Variation; (SS)	(3) d.f.	(4) Variance; (MSS)	(5) F	(6) Critical F
Between machines	154.8	2	77.4	13.1	4.46
Between operators	381.6	4	95.4	16.2	3.84
Residual variation	47.3	8	5.9		
Total	583.7 √	14 √			

that the machines are similar. We now compare this with our F test in (10-11), where we could not reject the null hypothesis. The numerator remains unchanged, but the chance variation in the denominator is much smaller, since the effect of differing operators has been netted out. This has given us greater statistical leverage,[7] allowing rejection of the null hypothesis.

Similarly, we might test the null hypothesis that the operators perform equally well. Once again F is the ratio of an explained to an unexplained variance; but this time, of course, the numerator is the variance estimated from column differences. Thus

$$F = \frac{\text{variance explained by operators}}{\text{unexplained variance}} = \frac{MSS_c}{MSS_u} = \frac{95.4}{5.9} = 16.2 \quad \text{(10-30)}$$

In this case, the "machine" noise has been isolated; as a consequence we get a strong test of how operators compare. Since our observed F value of 16.2 exceeds the critical F value[8] of 3.84, we reject the null hypothesis, concluding that machinists do differ.

There is one issue that we passed over quickly, that still requires clarification. In our one-factor test we calculated unexplained variation by looking at the spread of n observed values within a category, or cell, e.g., within a whole row in Table 10-4. But in the two-way test (Table 10-9) we have split our observations columnwise, as well as rowwise; this has left us with only

[7] Strictly speaking, we have a stronger test because we have gained more by reducing unexplained variance than we have lost because our degrees of freedom in the denominator have been reduced by 4. (The student will observe that if we are short of degrees of freedom—i.e., if we are near the top of F Table VII, loss of degrees of freedom may be serious.)
[8] Different than in the previous test since degrees of freedom are now 4 and 8.

one observation in each cell. Thus, for example, there is only a single observation (57.5) of how much output is produced by operator 4 on machine 2. Variation can no longer be computed within that cell. What should we do?

We ask, "If there were no random error, how would we predict the output of operator 4 on machine 2?" We note, informally, that this is a better-than-average machine ($\bar{X}_{2.} = 56.4$) and a relatively efficient operator ($\bar{X}_{.4} = 56.2$). On both counts we would predict output to be above average. This strategy can easily be formalized to predict $\hat{X}_{2,4}$. We can do this for each cell, with the random element estimated as the difference in our observed value (X_{ij}) and the corresponding predicted value ($\hat{X}_{ij}$). This yields a whole set of random elements, whose sum of squares is precisely the unexplained variation[9] SS_u (the last term in equation (10-27), also appearing in column 2 of Table 10-10); divided by d.f., this becomes the unexplained variance used in the denominator of both tests in this section.

One final warning: in computing predicted output $\hat{X}_{ij}$, we assume that there is no interaction between the two factors as would occur, for example, if certain operators like some machines, and dislike others; such interaction would require a more complex model, and more sample observations. The

[9] Predicted value $\hat{X}_{ij}$ is defined as

$$\hat{X}_{ij} = \bar{\bar{X}} + \text{adjustment reflecting machine performance} + \text{adjustment reflecting operator performance} \tag{10-31}$$

$$= \bar{\bar{X}} + (\bar{X}_{i.} - \bar{\bar{X}}) + (\bar{X}_{.j} - \bar{\bar{X}}) \tag{10-32}$$

Specifically, in our example

$$\hat{X}_{24} = 52.2 + (56.4 - 52.2) + (56.2 - 52.2)$$
$$= 52.2 + 4.2 + 4.0 = 60.4$$

Thus, our prediction of the performance of operator 4 on machine 2 is calculated by adjusting average performance (52.2) by the degree to which this machine is above average (4.2) and the degree to which this operator is above average (4.0).

Cancelling $\bar{\bar{X}}$ values in (10-32):

$$\hat{X}_{ij} = \bar{X}_{i.} + \bar{X}_{.j} - \bar{\bar{X}} \tag{10-33}$$

and the random element, being the difference between the observed and expected, becomes:

$$X_{ij} - \hat{X}_{ij} = X_{ij} - \bar{X}_{i.} - \bar{X}_{.j} + \bar{\bar{X}} \tag{10-34}$$

We emphasize that this random element is output left unexplained after adjustment for both machine i and operator j.

In our example

$$X_{24} - \hat{X}_{24} = 57.5 - 60.4 = -2.9 \tag{10-35}$$

Thus, this observed output is 2.9 units below what we expected, and must be left unexplained—the result of random influences.

Unexplained variation (SS_u) is recognized to be the sum of squares of all random elements as defined in (10-34).

two-way analysis of variance developed in this section is based on the assumption that interaction does not exist.

*(c) Multiple Comparisons

Turning from hypothesis tests to confidence intervals, we may write a statement for two-factor ANOVA which is quite similar to (10-26):

With 95% confidence, all contrasts in row means fall within the bounds:

$$\sum C_i\mu_i = \sum C_i\bar{X}_{i.} \pm \sqrt{F_{.05}}\, s_p \sqrt{\frac{(r-1)}{c}\left(\sum C_i^2\right)} \tag{10-36}$$

where

$F_{.05}$ = the critical value of F, with $(r-1)$ and $(r-1)(c-1)$ d.f.

$s_p = \sqrt{\text{MSS}_u}$, as calculated in Table 10-10, column 4

r = number of rows

c = number of columns

Note that (10-36) differs from (10-26) because unexplained variance s_p^2 is now smaller, making the confidence interval more precise.

As an example, consider the machines of Table 10-9, analyzed in ANOVA Table 10-11. With 95% confidence, all the following statements are true:

$$\mu_1 - \mu_2 = (48.6 - 56.4) \pm \sqrt{4.46}\sqrt{5.9}\sqrt{\tfrac{2}{5}(2)}$$

i.e.,

$$\begin{aligned} \mu_1 - \mu_2 &= -7.8 \pm 4.5^* \\ \mu_1 - \mu_3 &= -3.0 \pm 4.5. \\ \mu_2 - \mu_3 &= 4.8 \pm 4.5^* \end{aligned} \tag{10-37}$$

and all other possible contrasts

[Intervals that do not overlap zero are starred to indicate their statistical significance: thus H_0 (no difference in means) would be rejected in these cases—another illustration of how confidence intervals may be used to test hypotheses.]

Of course, we could contrast the column means equally well, by simply interchanging r and c in equation (10-36). As an example, how do the operators of Table 10-9 compare, when analyzed in ANOVA Table 10-11? With 95% confidence, all the following statements are true:

$$\mu_1 - \mu_2 = (59.3 - 49.9) \pm \sqrt{3.84}\,\sqrt{5.9}\,\sqrt{\tfrac{4}{3}(2)}$$

i.e.,

$$\left.\begin{aligned}
\mu_1 - \mu_2 &= 9.4 \pm 7.8^* \\
\mu_1 - \mu_3 &= 8.6 \pm 7.8^* \\
\mu_1 - \mu_4 &= 3.1 \pm 7.8 \\
\mu_1 - \mu_5 &= 14.4 \pm 7.8^* \\
\mu_2 - \mu_3 &= -0.8 \pm 7.8 \\
&\;\vdots
\end{aligned}\right\} \tag{10-38}$$

and all other possible contrasts, of the form

$$\sum C_j\mu_j = \sum C_j\bar{X}_{.j} \pm 5.5\sqrt{\sum C_j^2}$$

For example,

$$\frac{\mu_1 + \mu_3 + \mu_4}{3} - \frac{\mu_2 + \mu_5}{2} = (55.4 - 47.4) \pm 5.5\sqrt{5/6}$$

$$= 8.0 \pm 5.0^*$$

This last contrast might be of interest if workers 1, 3, and 4 are men, and workers 2 and 5 are women; thus the average difference in men and women has been estimated, as a bonus.

The first part of equation (10-38)—all differences in means—can be presented more concisely, in Table 10-12.

TABLE 10-12 Differences in Operator Means $\mu_j - \mu_J$ [Estimated from the sample means $(\bar{X}_{.j} - \bar{X}_{.J})$. To construct 95% simultaneous confidence intervals, take the listed value ±7.8. Statistically significant differences are starred.]

j \ J	1	2	3	4	5
1		9.4*	8.6*	3.1	14.4*
2			−.8	−6.3	5.0
3				−5.5	5.8
4					11.3*
5					

PROBLEMS

10-6 To refine the experimental design of Problem 10-2, suppose the twelve plots of land are on 4 farms (3 plots on each). Moreover, you suspect that there may be a difference in fertility between farms. You now retabulate the data in Problem 10-2, according to fertilizer *and* farm as follows.

Fertilizer \ Farm	1	2	3	4
Control C	60	64	65	55
A	69	75	70	66
B	72	74	78	68

(a) Reanalyze whether or not the fertilizers differ, at the 5% significance level.

(b) Is there, after all, a difference in the fertility of the four farms? (Use a 5% significance level.)

*(c) Construct a table of differences in fertilizers similar to Table 10-12, starring differences that the statistically significant; also construct a table of differences in farms.

(10-7) Three men work on an identical task of packing boxes. The number of boxes packed by each in 3 given hours is shown in the table below.

Hour \ Man	A	B	C
11–12 A.M.	21	18	21
1–2 P.M.	22	22	25
4–5 P.M.	17	16	18

(a) Test whether each factor is statistically significant at the 5% level.

*(b) For the factors which are statistically significant, construct a table of simultaneous 95% confidence intervals as in Table 10-12.

10-8 Five children were tested for pulse rate before and after a certain television program, with the following results:

	Before	After
A	96	104
B	102	112
C	108	112
D	89	93
E	85	89

(a) Test whether pulse rate changes, at the 5% significance level.

*(b) Construct a 95% confidence interval for the change in pulse rate for the population of all children.

10-9 Rework Problem 10-8 using the following technique (matched t-test) First, tabulate the change in pulse rate:

Before (X)	After (Y)	Difference $D = (Y - X)$	
96	104	+8	A sample to estimate $\Delta = \mu_Y - \mu_X$
102	112	+10	
108	112	+4	
89	93	+4	
85	89	+4	

The sample of D's fluctuates around the true difference Δ. Now apply equation (8-15).

chapter 11

Introduction to Regression

Our first example of statistical inference (in Chapter 7) was estimating the mean of a single population. This was followed (Chapter 8) by a comparison of two population means. Finally (Chapter 10) r population means were compared, using analysis of variance. We now consider the question "Can the analysis be improved upon if the r populations do not fall in unordered categories, but are ranked numerically?"

For example, it is easy to see how the analysis of variance could be used to examine whether wheat yield depended on 7 different *kinds* of fertilizer.[1] Now we wish to consider whether yield depends on 7 different *amounts* of fertilizer; in this case, fertilizer application is defined in a numerical scale. If yield (Y) that follows from various fertilizer applications (X) is plotted, a scatter similar to Figure 1-11 might be observed. From this scatter it is clear

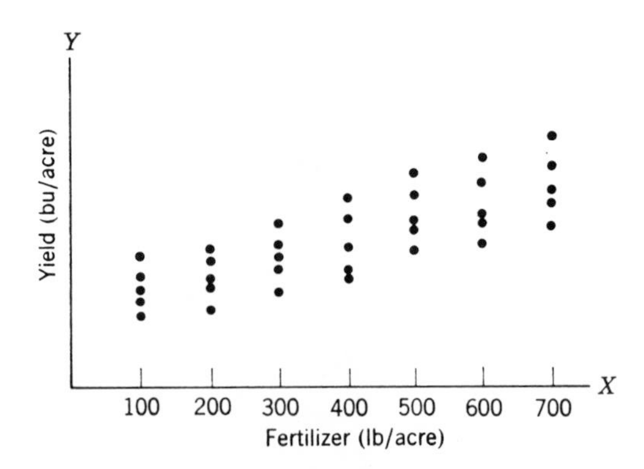

FIG. 11-1 Observed relation of wheat yield to fertilizer application.

[1] By extending Problem 10-2.

that the amount of fertilizer does matter. Moreover, it should be possible to define *how* fertilizer affects yield—i.e., define an equation describing the dependence of Y on X. Estimating an equation is of course equivalent geometrically to fitting a curve through this scatter, the so-called regression of Y on X. This regression will be a simple mathematical model, useful as a brief and precise description, or as a means of predicting the yield Y for a given amount of fertilizer X. Regression is the most useful of all statistical techniques. As another example, in economics it provides a means of defining how the quantity of a good demanded depends on its price, or how consumption depends on income.

This chapter is devoted exclusively to how a straight line may best be fitted. The characteristics of this line (e.g., its slope) may be subjected to statistical tests of significance; but these issues are deferred to Chapter 12. Furthermore, it is possible that Y is related to X in a more complicated nonlinear way; but these issues are not dealt with here. Instead we assume that the appropriate description is a straight line.

11-1 AN EXAMPLE

Since wheat yield depends on fertilizer, it is referred to as the "dependent" variable Y; since fertilizer application is not dependent on yield, but instead is determined by the experimenter, it is referred to as the "independent" variable X. Suppose funds are available for only seven experimental observations, so that the experimenter sets X at seven different values, taking only one observation Y in each case, shown in Figure 11-2 and Table 11-1.

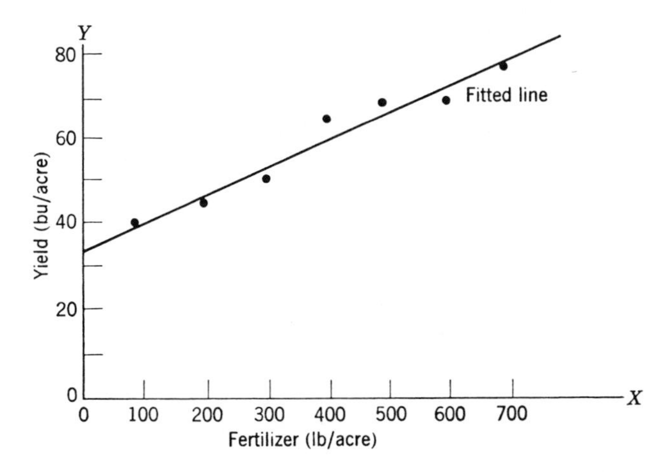

FIG. 11-2 Observed wheat yields at various levels of fertilizer application.

TABLE 11-1 Experimental Data Relating Yield of Wheat to the Amount of Applied Fertilizer, as in Figure 11-2

X Fertilizer (lb/acre)	Y Yield (bu/acre)
100	40
200	45
300	50
400	65
500	70
600	70
700	80

We first of all note that if the points were exactly in a line, as in Figure 11-3*a*, then the fitted line could be drawn in with a ruler "by eye" perfectly accurately. Even if the points were *nearly* in a line, as in Figure 11-3*b*, fitting by eye would be reasonably satisfactory. But in the highly scattered case, as in Figure 11-3*c*, fitting by eye is too subjective and too inaccurate. Furthermore, fitting by eye requires plotting all the points first. If there were 100

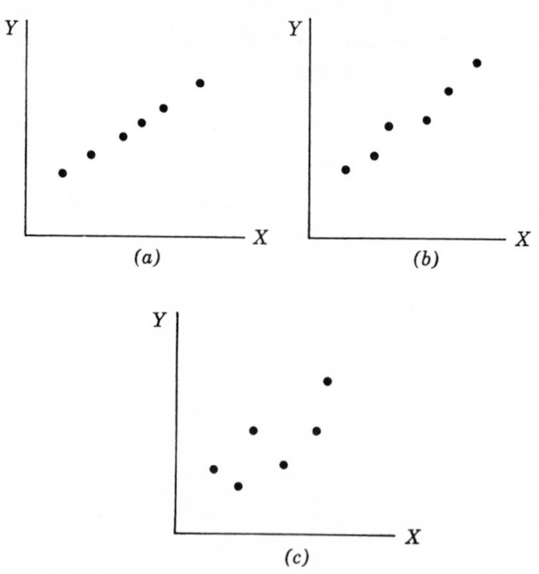

FIG. 11-3 Various degrees of scatter.

experimental observations this would be very tedious, and an algebraic technique which an electronic computer could solve would be preferable.

The following sections set forth various algebraic methods for fitting a line, successively more sophisticated and satisfactory.

11-2 POSSIBLE CRITERIA FOR FITTING A LINE

It is time to ask more precisely "What is a good fit?" The answer surely is, "a fit that makes the total error small." One typical error is shown in Figure 11-4. It is defined as the vertical distance from the observed Y to the fitted line—i.e., $(Y_i - \hat{Y}_i)$, where $\hat{Y}_i$ is the "fitted value of Y" or the ordinate of the line. We note that the error is positive when the observed Y_i is above the line and negative when the observed Y_i is below the line.

1. As our first tentative criterion, consider a fitted line which minimizes the total of all these errors,

$$\sum_{i=1}^{n}(Y_i - \hat{Y}_i) \tag{11-1}$$

But this criterion works badly. Using this criterion, the two lines shown in Figure 11-5 fit the observations equally well, even though the fit in Figure 11-5*a* is intuitively a good one, and the fit in Figure 11-5*b* is a very bad one. The problem is one of sign; in both cases positive errors just offset negative errors, leaving their sum equal to zero. This criterion must be rejected, since it provides no distinction between bad fits and good ones.

2. There are two ways of overcoming the sign problem. The first is to

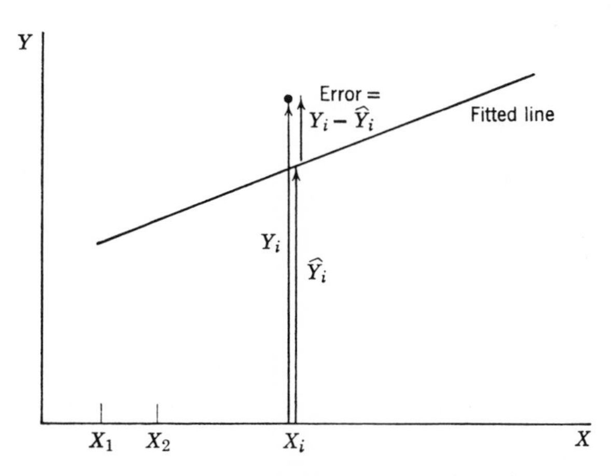

FIG. 11-4 Error in fitting points with a line.

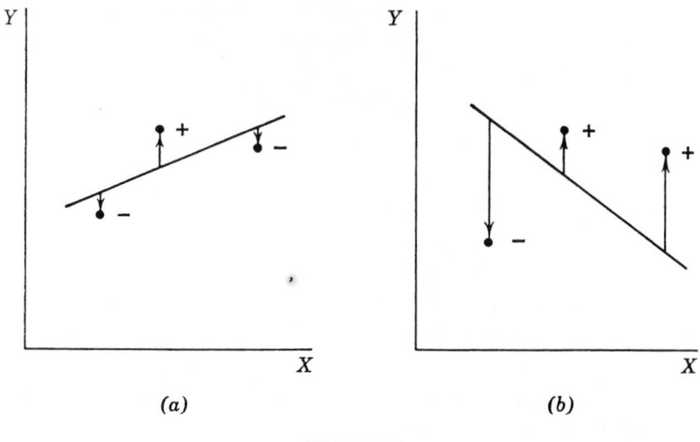

FIG. 11-5

minimize the sum of the *absolute* values of the errors,

$$\sum |Y_i - \hat{Y}_i| \tag{11-2}$$

Since large positive errors are not allowed to offset large negative ones, this criterion would rule out bad fits like Figure 11-5*b*. However, it still has a drawback. It is evident in Figure 11-6, that the fit in part *b* satisfies this criterion better than the fit in part *a*; ($\sum |Y_i - \hat{Y}_i|$ is 3, rather than 4). In fact, the reader can satisfy himself that the line in part *b* joining the two end points satisfies this criterion better than *any* other line. But it is not a good common-sense solution to the problem, because it pays no attention whatever to the middle point. The fit in part *a* is preferable because it takes account of all points.

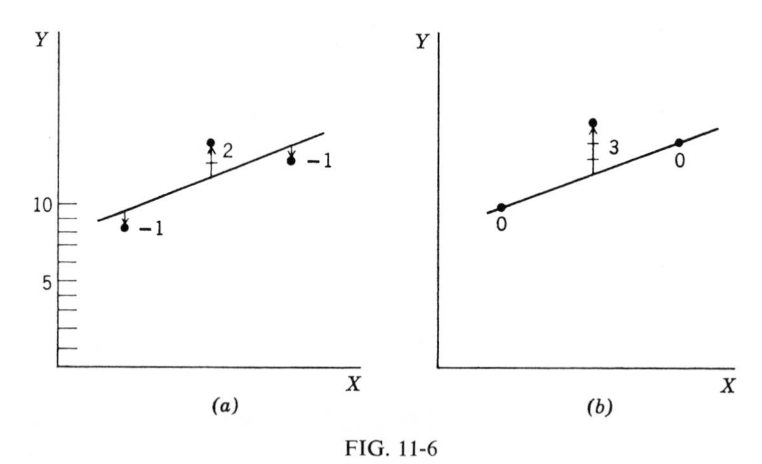

FIG. 11-6

3. As a second way to overcome the sign problem, we finally propose to minimize the sum of the squares of the errors,

$$\sum (Y_i - \hat{Y}_i)^2 \tag{11-3}$$

This is the famous "least squares" criterion; its justifications include:

(a) Squaring overcomes the sign problem by making all errors positive.
(b) Squaring emphasizes the large errors, and in trying to satisfy this criterion large errors are avoided if at all possible. Hence all points are taken into account, and the fit in Figure 11-6*a* is selected by this criterion in preference to Figure 11-6*b*.
(c) The algebra of least squares is very manageable.
(d) There are two important theoretical justifications for least squares, developed in the next chapter.

11-3 THE LEAST SQUARES SOLUTION

Our scatter of observed X and Y values from Table 11-1 is graphed in Figure 11-7*a*. Our objective is to fit a line

$$Y = a_0 + bX \tag{11-4}$$

This involves three steps:

Step 1. Translate X into deviations from its mean; i.e., define a new variable x, so that:

$$x = X - \bar{X} \tag{11-5}$$

In Figure 11-7*b* we show how this involves a geometric translation of axis similar to the procedure developed in Section 5-3, where both axes were translated to study covariance. The new x value becomes positive or negative depending on whether X was above or below $\bar{X}$. There is no change in the Y values. The intercept a differs from the original a_0, but the slope b remains the same.

One of the advantages of measuring X_i as deviations from their central value is that we can more explicitly ask the question "How is Y affected when X is unusually large, or unusually small?" In addition, the mathematics will be simplified because the sum of the new x values equals zero,[2]

$$\sum x_i = 0 \tag{11-6}$$

[2] Proof:

$$\sum x_i = \sum (X_i - \bar{X})$$
$$= \sum X_i - n\bar{X}$$

Noting that the mean $\bar{X}$ is defined as $\frac{\sum X_i}{n}$, it follows that $\sum X_i = n\bar{X}$ and

$$\sum x_i = n\bar{X} - n\bar{X} = 0 \qquad \text{(11-6) proved}$$

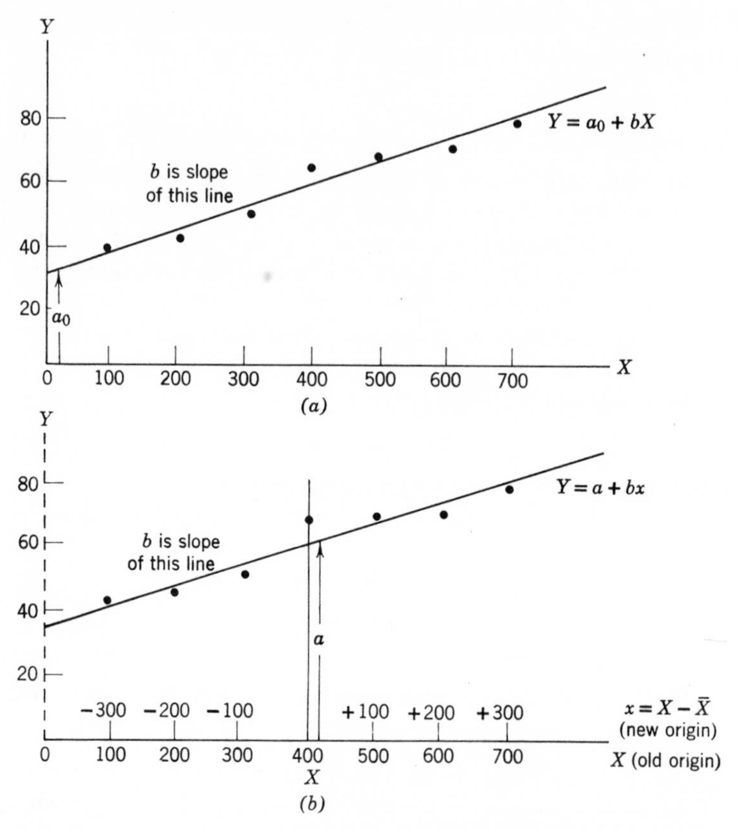

FIG. 11-7 Translation of axis. (*a*) Regression, using original variables. (*b*) Regression, translating X.

Step 2. Fit the line in Figure 11-7*b*; i.e., fit the line

$$Y = a + bx \tag{11-7}$$

to this scatter by selecting the values for a and b that satisfy the least squares criterion, i.e., select those values of a and b that minimize

$$\sum (Y_i - \hat{Y}_i)^2 \tag{11-8}$$

Since the fitted value $\hat{Y}_i$ is on our estimated line (11-7)

$$\hat{Y}_i = a + bx_i \tag{11-9}$$

When this is substituted into (11-8), the problem becomes one of selecting a and b to minimize the sum of squares,

$$S(a, b) = \sum (Y_i - a - bx_i)^2 \tag{11-10}$$

The notation $S(a, b)$ is used to emphasize that this expression depends on a and b. As a and b vary (i.e., as various lines are tried), $S(a, b)$ will vary too, and we ask at what value of a and b it will be a minimum. This will give us our optimum (least squares) line.

The simplest minimization technique is calculus, and it will be used in the next paragraph. [Readers without calculus can minimize (11-10) with the more cumbersome algebra of Appendix 11-1, and rejoin us where the resulting theorem is stated below.]

Minimizing $S(a, b)$ requires setting its partial derivatives with respect to a and b equal to zero. In the first instance, setting the partial derivative with respect to a equal to zero:

$$\frac{\partial}{\partial a}\sum (Y_i - a - bx_i)^2 = \sum 2(-1)(Y_i - a - bx_i)^1 = 0 \qquad (11\text{-}11)$$

Dividing through by -2 and rearranging:

$$\sum Y_i - na - b\sum x_i = 0 \qquad (11\text{-}12)$$

Noting that $\sum x_i = 0$ by (11-6), we can solve for a.

$$a = \frac{\sum Y_i}{n}, \qquad \text{or} \qquad a = \bar{Y} \qquad (11\text{-}13)$$

Thus our least squares estimate of a is simply the average value of Y; referring to Figure 11-7, we see that this ensures that our fitted regression line must pass through the point $(\bar{X}, \bar{Y})$, which may be interpreted as the center of gravity of the sample of n points.

It is also necessary to set the partial derivative of (11-10) with respect to b equal to zero,

$$\frac{\partial}{\partial b}\sum (Y_i - a - bx_i)^2 = \sum 2(-x_i)(Y_i - a - bx_i)^1 = 0 \qquad (11\text{-}14)$$

$$\sum x_i(Y_i - a - bx_i) = 0 \qquad (11\text{-}15)$$

Rearranging,

$$\sum Y_i x_i - a\sum x_i - b\sum x_i^2 = 0$$

Noting that $\sum x_i = 0$, we can solve for b.

$$b = \frac{\sum Y_i x_i}{\sum x_i^2} \qquad (11\text{-}16)$$

TABLE 11-2 Least Squares Calculations

(1)	(2)	(3)	(4)	(5)	(6)	(7)	(8)
X_i	Y_i	$x_i = X_i - \overline{X}$ $= X_i - 400$	$Y_i x_i$	x_i^2	$\hat{Y}_i = a + bx_i$ $= 60 + .068x_i$	$Y_i - \hat{Y}_i$	$(Y_i - \hat{Y}_i)^2$
100	40	−300	−12,000	90,000	39.60	.40	.16
200	45	−200	−9,000	40,000	46.40	−1.40	1.96
300	50	−100	−5,000	10,000	53.20	−3.20	10.24
400	65	0	0	0	60.00	5.00	25.00
500	70	100	7,000	10,000	66.80	3.20	10.24
600	70	200	14,000	40,000	73.60	−3.60	12.96
700	80	300	24,000	90,000	80.40	−.40	.16
$\Sigma X_i = 2{,}800$	$\Sigma Y_i = 420$	$\Sigma x_i = 0 \surd$	$\Sigma Y_i x_i = 19{,}000$	$\Sigma x_i^2 = 280{,}000$			$\Sigma (Y_i - \hat{Y}_i)^2 = 60.72$
$\overline{X} = \frac{\Sigma X_i}{n}$	$\overline{Y} = \frac{\Sigma Y_i}{n}$			$b = \frac{\Sigma Y_i x_i}{\Sigma x_i^2}$			$s^2 = \frac{1}{n-2} \Sigma (Y_i - \hat{Y}_i)^2$
$= \frac{2{,}800}{7}$	$= \frac{420}{7}$			$= \frac{19{,}000}{280{,}000}$			$= 12.144$
$= 400$	$= 60$						and $s = 3.48$
	$\therefore a = 60$			$\therefore b = .068$			

Our results[3] in (11-13) and (11-16) are important enough to restate as:

Theorem

With x values measured as deviations from their mean, the least squares values of a and b are

$$a = \bar{Y} \tag{11-13}$$

$$b = \frac{\sum Y_i x_i}{\sum x_i^2} \tag{11-16}$$

For the example problem in Table 11-1, a and b are calculated in the first five columns in Table 11-2; (the last three columns may be ignored until the next chapter). It follows that the least-squares equation is:

$$Y = 60 + .068x \tag{11-17}$$

This fitted line is graphed in Figure 11-7*b*.

Step 3. If desired, this regression can now be retranslated back into our original frame of reference in Figure 11-7*a*. Express (11-17) in terms of the original X values:

$$\begin{aligned} Y &= 60 + .068(X - \bar{X}) \\ &= 60 + .068(X - 400) \\ &= 60 + .068X - 27.2 \\ Y &= 32.8 + .068X \end{aligned} \tag{11-18}$$

This fitted line is graphed in Figure 11-7*a*.

A comparison of (11-17) and (11-18) confirms that the slope of our fitted regression ($b = .068$) remains the same; the only difference is in the intercept. Moreover, we note how easily the original intercept ($a_0 = 32.8$) may be recovered.

An estimate of yield for any given fertilizer application is now easily derived from our least squares equation (11-18). For example, if 350 lb of fertilizer is to be applied, our best estimate of yield is

$$Y = 32.8 + .068\ (350) = 56.6 \text{ bushels/acre}$$

The alternative least squares equation (11-17) yields exactly the same result. When $X = 350$, then $x = -50$, and

$$Y = 60 + .068\ (-50) = 56.6$$

[3] To be perfectly rigorous, we could have shown that when the partial derivatives are set equal to zero, we actually do have a minimum sum of squares—rather than a maximum, saddle point or local minimum.

PROBLEMS

(Save your work in the next three chapters, for future reference.)

11-1 Suppose a random sample of 5 families had the following income and savings:

Family	Income Y	Savings S
A	$8,000	$600
B	11,000	1200
C	9,000	1000
D	6,000	700
E	6,000	300

(a) Estimate and graph the regression line of savings S on income Y.
(b) Interpret the intercepts a and a_0.

11-2 Use the data of Problem 11-1 to regress consumption C on income Y. (Economists define $C = Y - S$.)

11-3 To interpret the regression slope b, use equation (11-18) to answer the following questions.
(a) About how much is the yield increased for every pound of fertilizer applied?
(b) If wheat were worth $2 per bushel and fertilizer cost $.25 per pound, would it be economical to apply fertilizer?
(c) To what price approximately would fertilizer have to drop to make it economical?
[The answer to (a) is simply the slope b. Economists refer to b as the "marginal" effect of fertilizer x on yield Y.]

⇒ 11-4 If we translated both X and Y into deviations x and y (just as X was translated in Figure 11-7*b*), then
(a) What would the new y-intercept be? Would the slope remain the same? Does not this imply that the fitted regression equation is simply

$$y = bx$$

(b) Prove that $\sum x_i y_i = \sum x_i Y_i$, hence we may alternatively write b in terms of deviations as

$$b = \frac{\sum x_i y_i}{\sum x_i^2}$$

*11-5 (Requires calculus.) Suppose X is left in its original form, rather than being translated into x (deviations from the mean).

(a) Write out the sum of squared deviations as in (11-10), in terms of a_0 and b.
(b) Set equal to zero the partial derivatives with respect to a_0 and b, thus obtaining two so-called "normal" equations.
(c) Evaluate these two normal equations using the data in Problem 11-1 and solve for a_0 and b. Do you get the same answer?
(d) Compare the two alternative methods of solution.

*11-6 Suppose four firms had the following profits and research expenditures.

Firm	Profit, P (thousands of dollars)	Research Expenditure, R (thousands of dollars)
1	50	40
2	60	40
3	40	30
4	50	50

(a) Fit a regression line of P on R.
(b) Does this regression line "show how research generates profits?" Criticize.

APPENDIX 11-1 AN ALTERNATIVE DERIVATION OF LEAST SQUARES ESTIMATES OF a AND b, WITHOUT CALCULUS

Before estimating a and b, it is necessary to solve the theoretical problem of minimizing an ordinary quadratic function of one variable b, of the form

$$f(b) = k_2 b^2 + k_1 b + k_0 \tag{11-19}$$

where k_2, k_1, k_0 are constants, with $k_2 > 0$.

With a little algebraic manipulation, (11-19) may be written as

$$f(b) = k_2\left(b + \frac{k_1}{2k_2}\right)^2 + \left(k_0 - \frac{k_1^2}{4k_2}\right) \tag{11-20}$$

Note that b appears in the first term, but not in the second. Therefore our hope of minimizing the expression lies in selecting a value of b to minimize the first term. Being a square and hence never negative, the first term will be minimized when it is zero, that is, when

$$b + \frac{k_1}{2k_2} = 0 \tag{11-21}$$

then

$$b = \frac{-k_1}{2k_2} \tag{11-22}$$

This result is shown graphically in Figure 11-8. To restate, *a quadratic function of the form* (11-19) *is minimized by setting*

$$b = -\frac{\text{(coefficient of first power)}}{2\text{ (coefficient of second power)}} \tag{11-23}$$

With this theorem in hand, let us return to the problem of selecting values for a and b to minimize

$$S(a, b) = \sum [(Y_i - a) - bx_i]^2 \tag{11-24}$$

It will be useful to manipulate this, as follows:

$$S(a, b) = \sum [(Y_i - a)^2 - 2b(Y_i - a)x_i + b^2x_i^2] \tag{11-25}$$

$$= \sum (Y_i - a)^2 - 2b \sum (Y_i - a)x_i + b^2 \sum x_i^2 \tag{11-26}$$

In the middle term, consider

$$\sum (Y_i - a)x_i = \sum Y_ix_i - a \sum x_i$$
$$= \sum Y_ix_i + 0$$

Using this to rewrite the middle term of (11-26) we have

$$S(a, b) = \sum (Y_i - a)^2 - 2b \sum Y_ix_i + b^2 \sum x_i^2 \tag{11-27}$$

This is a useful recasting of (11-24), because the first term contains a alone, while the last 2 terms contain b alone. To find the value of a which minimizes (11-27) only the first term is relevant. This may be written

$$\sum (Y_i - a)^2 = \sum Y_i^2 - 2a \sum Y_i + na^2$$

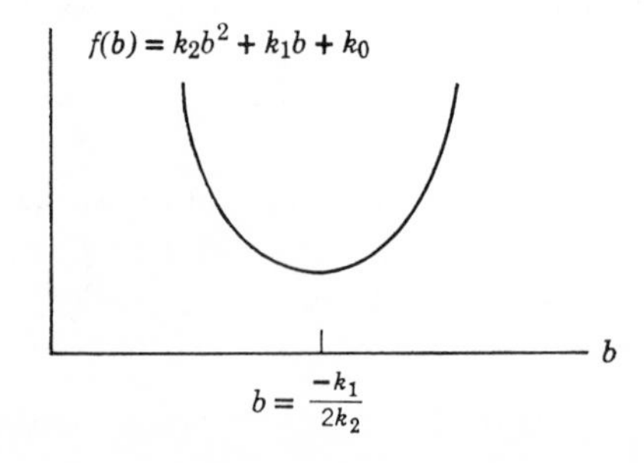

FIG. 11-8 The minimization of a quadratic function.

According to (11-23), this is minimized when

$$a = \frac{-(-2\sum Y_i)}{2n} = \frac{\sum Y_i}{n} = \bar{Y} \qquad \text{(11-13) proved}$$

To find the value of b which minimizes (11-27), only the last two terms are relevant. According to (11-23), this is minimized when

$$b = \frac{-(-2\sum Y_i x_i)}{2\sum x_i^2} = \frac{\sum Y_i x_i}{\sum x_i^2} \qquad \text{(11-16) proved}$$

chapter 12

Regression Theory

12-1 THE MATHEMATICAL MODEL

So far we have only mechanically fitted a line. This yielded a and b, which are descriptive statistics of the sample, (like $\bar{X}$ in Chapter 2); now we wish to make inferences about the parent population (like our inferences about μ in Chapter 7). Specifically we must consider the mathematical model which allows us to run tests of significance on a and b.

Turning back to the example in Section 11-1, suppose that the experiment could be repeated many times at a fixed value of x. Even though fertilizer application is fixed from experiment to experiment, we would not observe exactly the same yield each time. Instead, there would be some statistical fluctuation of the Y's, clustered about a central value. We can think of the many possible values of Y forming a population; the probability function of Y for a given x we shall call[1] $p(Y/x)$. Moreover, there will be a similar probability function for Y at any other experimental level of x. One possible sequence of Y populations is shown in Figure 12-1a. There would obviously be mathematical problems involved in analyzing such a population. To keep the problem manageable, we make a reasonable set of assumptions about the regularity of these populations, as shown in Figure 12-1b. We assume the probability functions $p(Y_i/x_i)$ have

1. The same variance σ^2 for all x_i; and
2. Means $E(Y_i)$ lying on a straight line, known as the true regression line:

$$E(Y_i) = \alpha + \beta x_i \tag{12-1}$$

[1] Remember that our notation conventions are different from Chapters 4 to 7. Now a capital letter denotes an original observation and a small letter denotes its deviation from the mean.

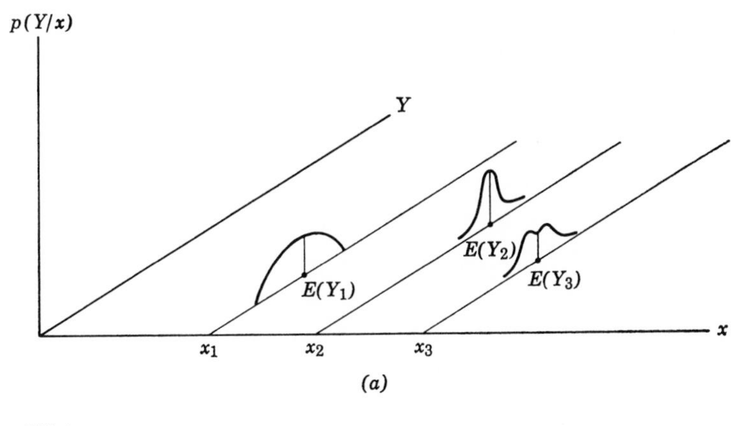

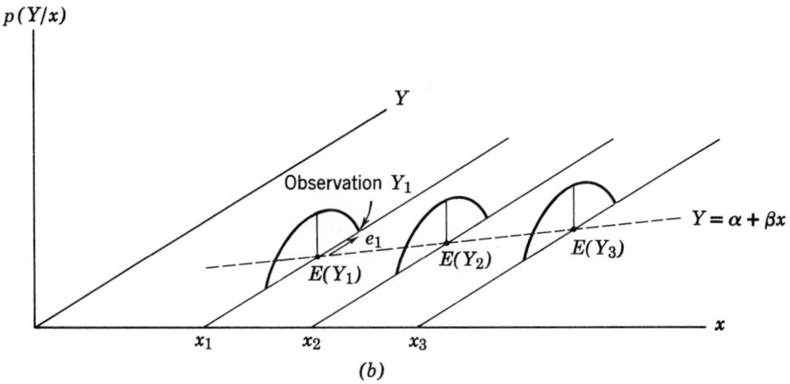

FIG. 12-1 (*a*) General populations of Y, given x. (*b*) The special form of the populations of Y assumed in simple linear regression.

The population parameters α and β specify the line; they are to be estimated from sample information. We also assume that

3. The random variables Y_i are statistically independent. For example, a large value of Y_1 does not tend to make Y_2 large; i.e., Y_2 is "unaffected" by Y_1.

These assumptions may be written more concisely as:

> The random variables Y_i are statistically independent, with
>
> $$\text{mean} = \alpha + \beta x_i$$
>
> $$\text{and variance} = \sigma^2 \tag{12-2}$$

On occasion, it is useful to describe the deviation of Y_i from its expected value as the error or disturbance term e_i, so that the model may alternatively be written

$$Y_i = \alpha + \beta x_i + e_i \tag{12-3}$$

where the e_i are independent random variables, with

$$\left.\begin{array}{c} \text{mean} = 0 \\ \text{and variance} = \sigma^2 \end{array}\right\} \tag{12-4}$$

We note that the distributions of Y and e are identical, except that their means differ. In fact, the distribution of e is just the distribution of Y translated onto a zero mean.

No assumption is made yet about the *shape* of the distribution of e (normal, or otherwise). We therefore refer to assumptions (12-4) as the "weak set"; we shall derive as many results as possible from these, before adding a more restrictive normality assumption later.

12-2 THE NATURE OF THE ERROR TERM

Now let us consider in more detail the "purely random" part of Y_i, the error or disturbance term e_i. Why does it exist? Or, why doesn't a precise and exact value of Y_i follow, once the value of x_i is given?

The error may be regarded as the sum of two components:

(a) Measurement Error

There are various reasons why Y may be measured incorrectly. In measuring wheat yield, there may be an error due to sloppy harvesting or inaccurate weighing. If the example is a study of the consumption of families at various income levels, the measurement error in consumption might consist of budget and reporting inaccuracies.

(b) Stochastic Error

This occurs because of the inherent irreproducibility of biological and social phenomena. Even if there were no measurement error, continuous repetition of our wheat experiment using exactly the same amount of fertilizer would result in different yields; these differences are unpredictable and are called stochastic differences. They may be reduced by tighter experimental control—for example, by holding constant soil conditions, amount of water, etc. But *complete* control is impossible—seeds, for example, cannot be duplicated. Stochastic error may be regarded as the influence on Y of many omitted variables, each with an individually small effect.

In the social sciences, controlled experiments are usually not possible. For example, an economist cannot hold U.S. national income constant for several years while he examines the effect of interest rate on investment. Since he cannot neutralize extraneous influences by holding them constant, his best alternative is to take them explicitly into account, by regressing Y on x *and* the extraneous factors. This is a useful technique for reducing stochastic error; it is called "multiple regression" and is discussed fully in the next chapter.

12-3 ESTIMATING α AND β

Suppose that our true regression $Y = \alpha + \beta x$ is the dotted line shown in Figure 12-2. This will remain unknown to the statistician, whose job it is to estimate it as best he can by observing x and Y. Suppose at the first level x_1, the stochastic error e_1 takes on a negative value, as shown in the diagram; he will observe the Y and x combination at P_1. Similarly, suppose his only other two observations are P_2 and P_3, resulting from positive values of e.

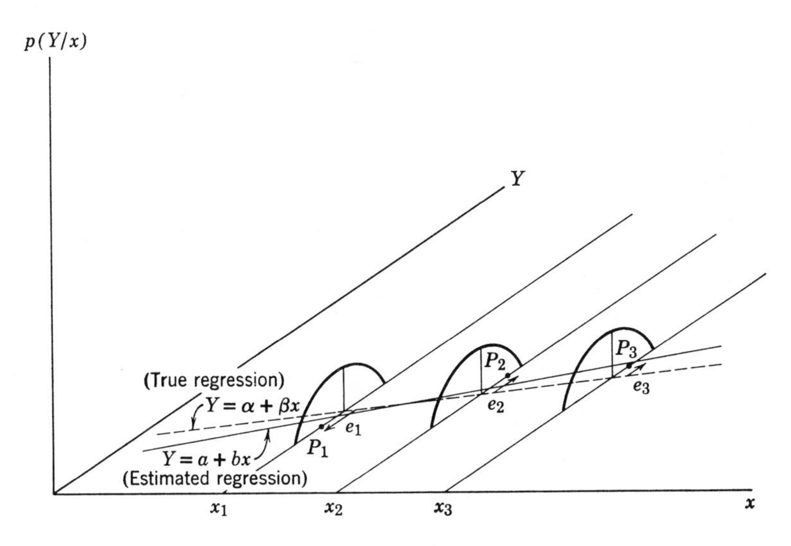

FIG. 12-2 True (population) regression and estimated (sample) regression.

Further, suppose the statistician estimates the true line by fitting a least squares line $Y = a + bx$, applying the method of Chapter 11 to the only information he has—points P_1, P_2, and P_3. He would then come up with the solid estimating line in this figure. This is a critical diagram; before proceeding, the reader should be sure he can clearly distinguish between the true regression and its surrounding e distribution on the one hand, and the estimated regression line on the other.

Unless the statistician is very lucky indeed, it is obvious that his estimated line will not be exactly on the true population line. The best he can hope for is that the least squares method of estimation will be close to the target. Specifically, we now ask: "How is the estimator a distributed around its target α, and b around its target β?"

12-4 THE MEAN AND VARIANCE OF a AND b

We shall show that the random estimators a and b have the following moments:

$$E(a) = \alpha \tag{12-5}$$

$$\text{var}\,(a) = \frac{\sigma^2}{n} \tag{12-6}$$

$$E(b) = \beta \tag{12-7}$$

$$\text{var}\,(b) = \frac{\sigma^2}{\sum x_i^2} \tag{12-8}$$

where σ^2 is the variance of the error (the variance of Y). We note from (12-5) and (12-7) that both a and b are unbiased estimators of α and β. Because of its greater importance we shall concentrate on the slope estimator b, rather than a, for the rest of the chapter.

Proof of (12-7) and (12-8). The formula for b in (11-16) may be rewritten as

$$b = \sum \left\{\frac{x_i}{k}\right\} Y_i \tag{12-9}$$

where

$$k = \sum x_i^2 \tag{12-10}$$

Thus

$$b = \sum w_i Y_i = w_1 Y_1 + w_2 Y_2 \cdots w_n Y_n \tag{12-11}$$

where

$$w_i = \frac{x_i}{k} \tag{12-12}$$

Since each x_i is a fixed constant, so is each w_i. Thus from (12-11) we establish the important conclusion,

b is a weighted sum (i.e., a linear combination) of the random variables Y_i	(12-13)

Hence by (5-31) we may write

$$E(b) = w_1E(Y_1) + w_2E(Y_2) \cdots + w_nE(Y_n) = \sum w_iE(Y_i) \quad (12\text{-}14)$$

Moreover, noting that the variables Y_i are assumed independent, by (5-34) we may write

$$\text{var}\,(b) = w_1^2 \text{ var } Y_1 + \cdots + w_n^2 \text{ var } Y_n = \sum w_i^2 \text{ var } Y_i \quad (12\text{-}15)$$

For the mean, from (12-14) and (12-1)

$$E(b) = \sum w_i[\alpha + \beta x_i] \quad (12\text{-}16)$$

$$= \alpha \sum w_i + \beta \sum w_i x_i \quad (12\text{-}17)$$

and noting (12-12)

$$E(b) = \frac{\alpha}{k} \sum x_i + \frac{\beta}{k} \sum (x_i) x_i \quad (12\text{-}18)$$

but $\sum x_i$ is zero, according to (11-6). Thus

$$E(b) = 0 + \frac{\beta}{k} \sum x_i^2$$

From (12-10)

$$E(b) = \beta \quad \text{(12-7) proved}$$

For the variance, from (12-15) and (12-2)

$$\text{var}\,(b) = \sum w_i^2 \sigma^2 \quad (12\text{-}19)$$

$$= \sum \frac{x_i^2}{k^2} \sigma^2 \quad (12\text{-}20)$$

$$= \frac{\sigma^2}{k^2} \sum x_i^2 \quad (12\text{-}21)$$

Again noting (12-10),

$$\text{var}\,(b) = \frac{\sigma^2}{\sum x_i^2} \quad \text{(12-8) proved}$$

A similar derivation of the mean and variance of a is left as an exercise, completing the proof. We observe from (12-12) that in calculating b, the weight w_i attached to the Y_i observation is proportional to the deviation x_i.

Hence the outlying observations will exert a relatively heavy influence in the calculation of b.

*12-5 THE GAUSS-MARKOV THEOREM

This is the major justification of using the least squares method in the linear regression model.

> *Gauss-Markov Theorem.* Within the class of linear unbiased estimators of β (or α), the least squares estimator has minimum variance. (12-22)

This theorem is important because it follows even from the weak set of assumptions (12-4), and hence requires no assumption of the shape of the distribution of the error term. A proof may be found in most mathematical statistics texts.

To interpret this important theorem, consider b, the least squares estimator of β. We have already seen in (12-13) that it is a linear estimator, and we restrict ourselves to linear estimators because they are easy to analyze and understand. We restrict ourselves even further, as shown in Figure 12-3; within this set of linear estimators we consider only the limited class that are unbiased. The least squares estimator not only is in this class, according to (12-7), but of all the estimators in this class it has the minimum variance. It is often, therefore, referred to as the "best linear unbiased estimator."

The Gauss-Markov theorem has an interesting corollary. As a special case of regression, we might ask what happens if we are explaining Y, but

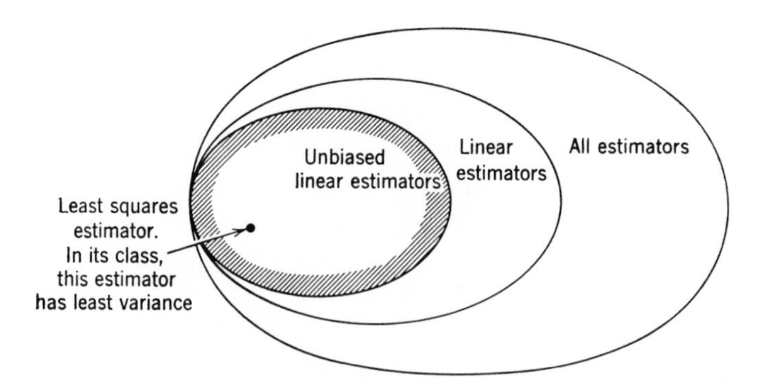

FIG. 12-3 Diagram of the restricted class of estimators considered in the Gauss-Markov theorem.

$\beta = 0$ in (12-2), so that no independent variable x comes into play. From (12-2), α is the mean of the Y population (μ). Moreover, from (11-13) its least squares estimator is $\bar{Y}$. Thus, the least squares estimator of a population mean (μ) is the sample mean ($\bar{Y}$), and the Gauss-Markov theorem fully applies: the sample mean is the best linear unbiased estimator of a population mean.

It must be emphasized that the Gauss-Markov theorem is restricted, applying only to estimators that are both linear and unbiased. It follows that there may be a biased or nonlinear estimator that is better (i.e., has smaller variance) than the least squares estimator. For example, to estimate a population mean, the sample median is a nonlinear estimator. It is better than the sample mean for certain kinds of nonnormal populations. The sample median is just one example of a whole collection of nonlinear statistical methods known as "distribution-free" or "nonparametric" statistics. These are expressly designed for inference when the population cannot be assumed to be normally distributed.

12-6 THE DISTRIBUTION OF b

With the mean and variance of b established in (2-7) and (2-8), we now ask: "What is the shape of the distribution of b?" If we add (for the first time) the strong assumption that the Y_i are normal, and recall that b is a linear combination of the Y_i, it follows that b will also be normal from (6-13). But even without assuming the Y_i are normal, as sample size increases the distribution of b will usually approach normality; this can be justified by a generalized form[2] of the central limit theorem (6-15).

We are now in a position to graph the distribution of b in Figure 12-4, in order to develop a clear intuitive idea of how this estimator varies from sample to sample. First, of course, we note that (12-7) established that b is an unbiased estimator, so that the distribution of b is centered on its target β.

The interpretation of the variance (12-8) is more difficult. Suppose that the experiment had been badly designed with the X_i's close together. This makes the deviations x_i small; hence $\sum x_i^2$ small. Therefore $\sigma^2/\sum x_i^2$, the variance of b from (12-8) is large and b is a comparatively unreliable estimator. To check the intuitive validity of this, consider the scatter diagram in Figure 12-5a. The bunching of the X's means that the small part of the line being

[2] The central limit theorem (6-15) concerned the normality of the sample mean $\bar{X}$. In Problem 6-8 it was seen to apply equally well to the sample sum S. It applies also to a *weighted* sum of random variables such as b in (12-13), under most conditions. See for example, D. A. S. Fraser, *Nonparametric Statistics*, New York: John Wiley, 1957. Similarly the normality of a is justified.

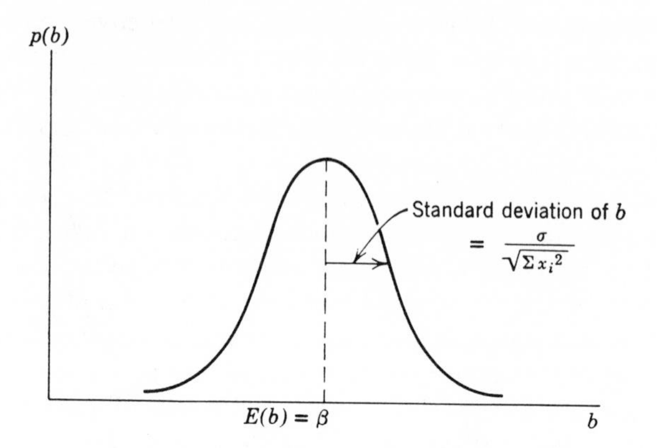

FIG. 12-4 The probability distribution of the estimator b.

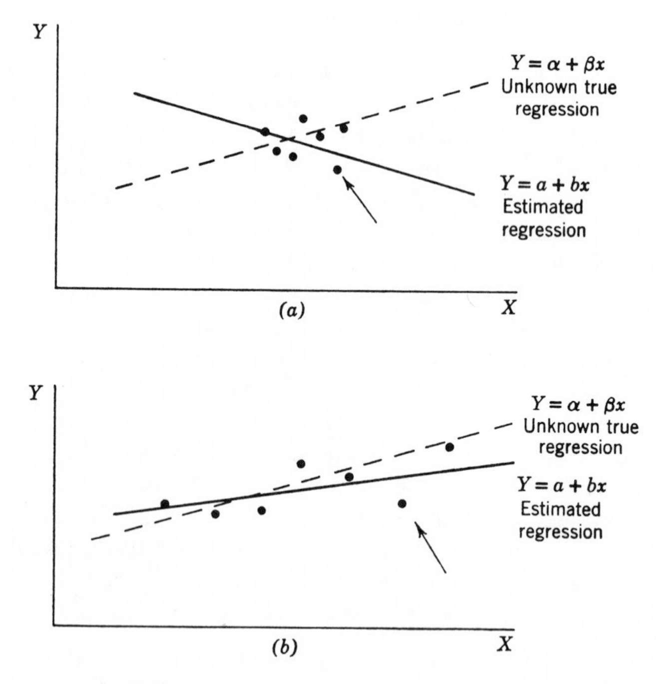

FIG. 12-5 (*a*) Unreliable estimate when X_i are very close. (*b*) More reliable fit because X_i are spread out.

investigated is obscured by the error e, making the slope estimate b very unreliable. In this specific instance, our estimate has been pulled badly out of line by the errors—in particular, the one indicated by the arrow.

By contrast, in Figure 12-5*b* we show the case where the X's are reasonably spread out. Even though the error e remains the same, the estimate b is much more reliable, because errors no longer exert the same leverage.

As a concrete example, suppose we wish to examine how sensitive Canadian imports (Y) are to the international value of the Canadian dollar (x). A much more reliable estimate should be possible using the period 1948 to 1962 when the Canadian dollar was flexible (and took on a range of values) than in the period before or since when this dollar was fixed (and only allowed to fluctuate within a very narrow range).

12-7 CONFIDENCE INTERVALS AND TESTING HYPOTHESES ABOUT β

With the mean, variance and normality of the estimator b established, statistical inferences about β are now in order. Our argument will be similar to the inference about μ in Section 8-2. First standardize the estimator b, obtaining

$$Z = \frac{b - \beta}{\sqrt{\dfrac{\sigma^2}{\sum x_i^2}}} \tag{12-23}$$

where $Z \sim N(0, 1)$.

Since σ^2, the variance of Y is generally unknown, it is estimated with

$$s^2 = \frac{1}{n-2} \sum [Y_i - \hat{Y}_i]^2 \tag{12-24}$$

where $\hat{Y}_i$ is the fitted value of Y on the *estimated* regression line: i.e.

$$\hat{Y}_i = a + bx_i \tag{12-25}$$

s^2 is often referred to as "residual variance," a term similarly used in ANOVA. The divisor $(n - 2)$ is used in (12-24) rather than n in order to make s^2 an unbiased estimator[3] of σ^2. When this substitution of s^2 for σ^2 is made, the estimator b is no longer normal, but instead has the slightly more spread-out

[3] As argued in the footnote to equation (8-11). But in the present calculation of s^2, *two* estimators a and b are required; thus there remain two fewer degrees of freedom for s^2. Hence $(n - 2)$ is the divisor in s^2, and also the degrees of freedom of the subsequent t distribution in (12-26).

t distribution:

$$t = \frac{b - \beta}{\sqrt{\dfrac{s^2}{\sum x_i^2}}} \tag{12-26}$$

For the t distribution to be strictly valid, we require the strong assumption that the distribution of Y_i is normal. From (12-26) we may now proceed to construct a confidence interval or test an hypothesis.

(a) Confidence Intervals

Again letting $t_{.025}$ denote the t value which leaves $2\frac{1}{2}\%$ of the distribution in the upper tail,

$$\Pr(-t_{.025} < t < t_{.025}) = .95 \tag{12-27}$$

Substituting for t from (12-26)

$$\Pr\left[-t_{.025} < \frac{b - \beta}{\sqrt{\dfrac{s^2}{\sum x_i^2}}} < t_{.025}\right] = .95 \tag{12-28}$$

The inequalities within the bracket may be reexpressed

$$\Pr\left[b - t_{.025}\frac{s}{\sqrt{\sum x_i^2}} < \beta < b + t_{.025}\frac{s}{\sqrt{\sum x_i^2}}\right] = .95 \tag{12-29}$$

which yields

The 95% confidence interval[4] for β:

$$\beta = b \pm t_{.025}\frac{s}{\sqrt{\sum x_i^2}} \tag{12-30}$$

where $t_{.025}$ has $(n - 2)$ degrees of freedom.

For our example of wheat yield in the previous chapter, the confidence interval for β (the effect of fertilizer on yield) is computed as follows. s is evaluated in the last three columns of Table 11-2. Also noting the values

[4] Using a similar argument, and noting (12-6), the 95% confidence interval for α is:

$$\alpha = a \pm t_{.025}\frac{s}{\sqrt{n}} \tag{12-31}$$

We note that this is very similar to the confidence interval for μ given by equation (8-15).

for b and $\sum x_i^2$ calculated in that table, our 95% confidence interval (12-30) becomes

$$\beta = .068 \pm 2.571 \frac{3.48}{\sqrt{280{,}000}}$$

$$= .068 \pm .017$$

$$.051 < \beta < .085 \qquad (12\text{-}32)$$

Testing hypotheses. A two-sided test of any hypothesis may be carried out simply by noting whether or not the confidence interval (12-30) contains that hypothesis. For example, the hypothesis typically tested is the null hypothesis.

$$H_0 : \beta = 0 \qquad (12\text{-}33)$$

i.e., using our example, that fertilizer has no effect on yield. In testing this against the two-sided alternative

$$H_1 : \beta \neq 0 \qquad (12\text{-}34)$$

H_0 must be rejected at a 5% significance level, since the null value of zero is not contained in (12-32).

Since fertilizer is expected to favorably affect yield, it seems more appropriate to test (12-33) against the one-sided alternative:

$$H_1 : \beta > 0 \qquad (12\text{-}35)$$

The first step is to calculate the t statistic under the assumption that H_0 is true. From (12-26) and (12-33) this reduces to

$$t = \frac{b}{\sqrt{\dfrac{s^2}{\sum x_i^2}}} \qquad (12\text{-}36)$$

and in our example:

$$= \frac{.068}{3.48/\sqrt{280{,}000}} = 10.3 \qquad (12\text{-}37)$$

Since this observed value exceeds the critical $t_{.05}$ value of 2.015, H_0 is rejected in favor of the conclusion that fertilizer favorably affects yield.

12-8 PREDICTION INTERVAL FOR Y_0

If we plan one new application of 550 pounds of fertilizer ($x_0 = 150$) how do we predict the resulting yield?

The best point estimate will be the corresponding fitted Y value on our

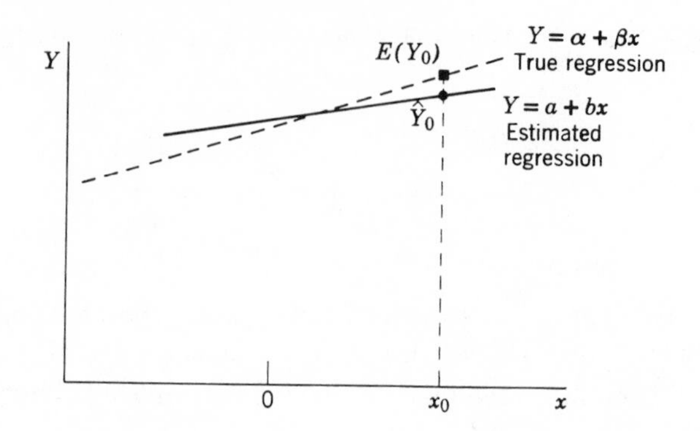

FIG. 12-6 How the estimator $\hat{Y}_0$ is related to the target $E(Y_0)$.

estimated regression line, i.e.:

$$\hat{Y}_0 = a + bx_0 \tag{12-38}$$

$$= 60 + .068(150) = 70.2 \text{ bu/acre} \tag{12-39}$$

But as a point estimate, this will almost certainly involve some error because, for example, of errors made in calculating a and b. Figure 12-6 illustrates the effect of these errors; the true regression is shown, along with an estimated regression. Note how the fitted $\hat{Y}_0$ in this case underestimates. In Figure 12-7 the true regression is again shown along with several estimated regressions fitted from several possible sets of sample data. The fitted value is sometimes too low, sometimes too high, but on the average, just right.

The important observation in Figure 12-7 is that if x_0 were further to

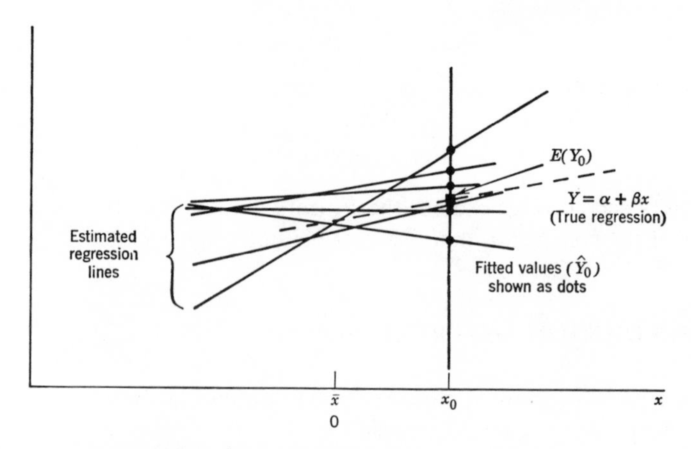

FIG. 12-7 $\hat{Y}_0$ as an unbiased estimator of $E(Y_0)$.

the right, then our estimates would be spread out over an even wider range. On the other hand if x_0 were further to the left and closer to its central value of zero then our estimates would be less spread out. Moreover, it is the error in b that causes this; thus such an error in the slope b will do little harm in a prediction, given an average amount of fertilizer; but any prediction of the effect of an extreme amount of fertilizer will be thrown badly into error. Formally, it may be shown that[5] the 95% prediction interval for an individual Y observation is

$$\boxed{Y_0 = \hat{Y}_0 \pm t_{.025}\, s\sqrt{\frac{1}{n} + \frac{x_0^2}{\sum x_i^2} + 1}} \tag{12-42}$$

where $t_{.025}$ has $(n - 2)$ d.f.

For example, we can now construct a prediction interval for yield if 550 lb/acre of fertilizer were applied. With a 95% chance of being right we predict:

$$Y_0 = 70.2 \pm 2.571(3.48)\sqrt{\frac{1}{7} + \frac{150^2}{280{,}000} + 1}$$

$$60.3 \leq Y_0 \leq 80.1 \tag{12-43}$$

This prediction interval is shown in Figure 12-8. Moreover, the same calculation for all possible x_0 values yields the dotted band of prediction intervals, expanding as x_0 moves farther away from its central value of zero.

It should be emphasized that x_0 may be *any* value of x. If x_0 lies *among* the observed values $x_1 \cdots x_n$, the process is called "interpolation." If x_0 *is* one of the values $x_1 \cdots x_n$, the process might be called, "using also the other values of x to sharpen our knowledge of this one population at x_0." If x_0 is beyond x_1 or x_n, then the process is called "extrapolation." The techniques

[5] Without going into the proof of (12-42), we sketch its plausibility. The variance involved in a prediction is roughly

$$\text{var} = \text{var}\,(a) + \text{var}\,(b)x_0^2 + \text{var}\,(Y) \tag{12-40}$$

that is, the variance of a, plus the variance of b weighted with x_0^2, plus the inherent variance of any Y observation. This last source of error must be included of course; even if α and β were known exactly, the prediction of Y_0 would still be subject to error.

Into (12-40) we substitute (12-6), (12-8), and (12-2)

$$\text{var} = \frac{\sigma^2}{n} + \frac{\sigma^2}{\sum x_i^2}(x_0^2) + \sigma^2$$

$$= \sigma^2\left(\frac{1}{n} + \frac{x_0^2}{\sum x_i^2} + 1\right) \tag{12-41}$$

When s is substituted for σ, (12-42) follows.

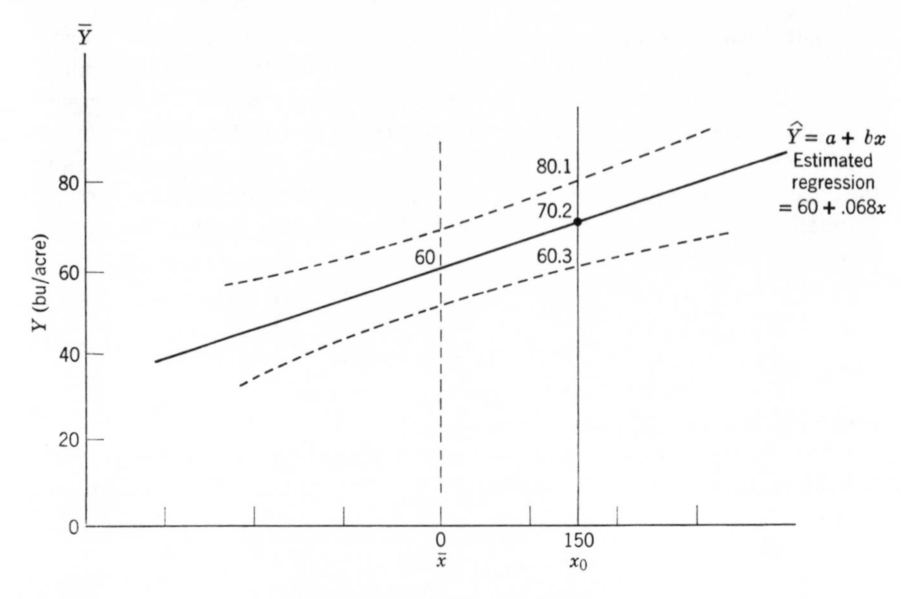

FIG. 12-8 Prediction interval for Y_0.

developed in this section may be used for extrapolation, but only with considerable caution—as we shall see in the next section.

PROBLEMS

12-1 Construct a 95% confidence interval for the regression coefficient β in
  (a) Problem 11-1.
  (b) Problem 11-2.

12-2 Which of the following hypotheses does the data of Problem 11-1 prove to be unacceptable at the 5% level of significance?
  (a) $\beta = 0$
  (b) $\beta = 1/2$
  (c) $\beta = .1$
  (d) $\beta = -.1$

12-3 At the 1% level of significance, use the data of Problem 11-1 to test the hypothesis that savings does not depend on income, against the alternative hypothesis that savings increases with income.

12-4 Using the data of Problem 11-1, what is your 95% prediction interval for the savings of a family with an income of
  (a) \$6,000

(b) \$8,000
(c) \$10,000
(d) \$12,000
(e) Which of these four intervals is least precise? Most precise?
(f) How is the answer to (b) related to the confidence interval found from (12-31)?

12-5 Suppose you are trying to explain how the interest rate (i) affects investment (I) in the U.S. Would you prefer to take observations of i and I over a period in which the authorities were trying to hold interest constant, or a period in which it is allowed to vary widely?

12-9 DANGERS OF EXTRAPOLATION

There are two dangers in extrapolation, which we might call "mathematical" and "practical." In both cases, there is no sharp division between safe interpolation and dangerous extrapolation. Rather, there is *continually* increasing danger of misinterpretation as x_0 gets further and further from its central value.

(a) Mathematical Danger

It was emphasized in the previous section that prediction intervals get larger as x_0 moves away from zero. This is true, even if all the assumptions underlying our mathematical model hold exactly.

(b) Practical Danger

In practice it must be recognized that a mathematical model is never absolutely correct. Rather, it is a useful approximation. In particular, one cannot take seriously the hypothesis that the population means are strung out in an *exactly* straight line. If we consider the fertilizer example, it is likely that the true relation increases initially, but then bends down eventually as a "burning point" is approached, and the crop is overdosed. This is illustrated in Figure 12-9 which is an extension of Figure 11-2 with the scale appropriately reduced. In the region of interest, from 0 to 700 pounds, the relation is *practically* a straight line, and no great harm is done in assuming the linear model. However, if the linear model is extrapolated far beyond this region of experimentation, the result becomes meaningless.

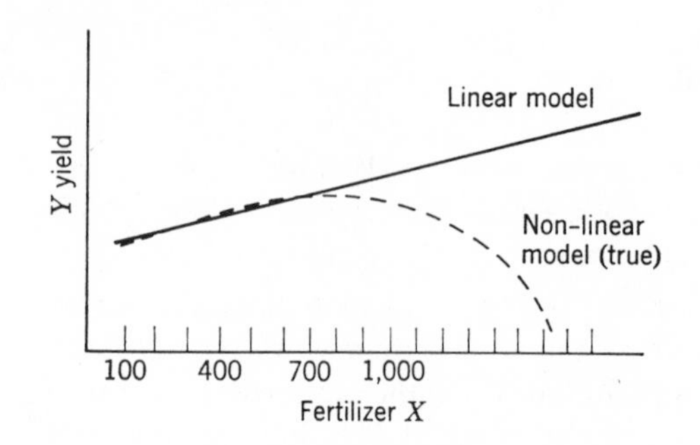

FIG. 12-9 Comparison of linear and nonlinear models.

There are "nonlinear" models available, if they seem more appropriate. Moreover statistical tests are available to help determine whether or not they are appropriate. These topics are covered in more advanced texts.

*12-10 MAXIMUM LIKELIHOOD ESTIMATION

Sections 12-1 to 12-5 including the Gauss-Markov justification of least squares required no assumption of the normality of the error term (i.e., normality of Y). In Sections 12-6 to 12-9, the normality assumption was required only for small sample estimation—and this because of a quite general principle that small sample estimation requires a normally distributed parent population to validate the t distribution. In these last two sections we make the strong assumption of a normally distributed error throughout. On this premise, we derive the maximum likelihood estimates of α and β, i.e., those hypothetical population values of α and β more likely than any others to generate the sample values we observed. These MLE of α and β turn out to be the least squares estimates; thus maximum likelihood provides a second justification for using least squares.

Before addressing the algebraic derivation, it is best to clarify what is going on with a bit of geometry. Specifically, why should the maximum likelihood line fit the data well? To simplify, assume a sample of only three observations (P_1, P_2, P_3).

First, let us try out the line shown in Figure 12-10*a*. (Before examining it carefully, we note that it seems to be a pretty bad fit for our three observed points.) Temporarily, suppose this were the true regression line; then the distribution of errors would be centered around it as shown. The likelihood that such a population would give rise to the samples we observed is the probability density that we would get the particular set of three e values shown in this diagram. The probability density of the three values is shown as the ordinates above the points P_1, P_2, and P_3. Because our three observations are by assumption statistically independent, the likelihood of all three (i.e., the probability density of getting the sample we observed), is the product of these three ordinates. This likelihood seems relatively small, mostly because the very small ordinate of P_1 reduces the product value. Our intuition that this is a bad estimate is confirmed; such a hypothetical population is not very likely to generate our sample values. We should be able to do better.

In Figure 12-10*b* it is evident that we can do much better. This hypothetical population is more likely to give rise to the sample we observed.

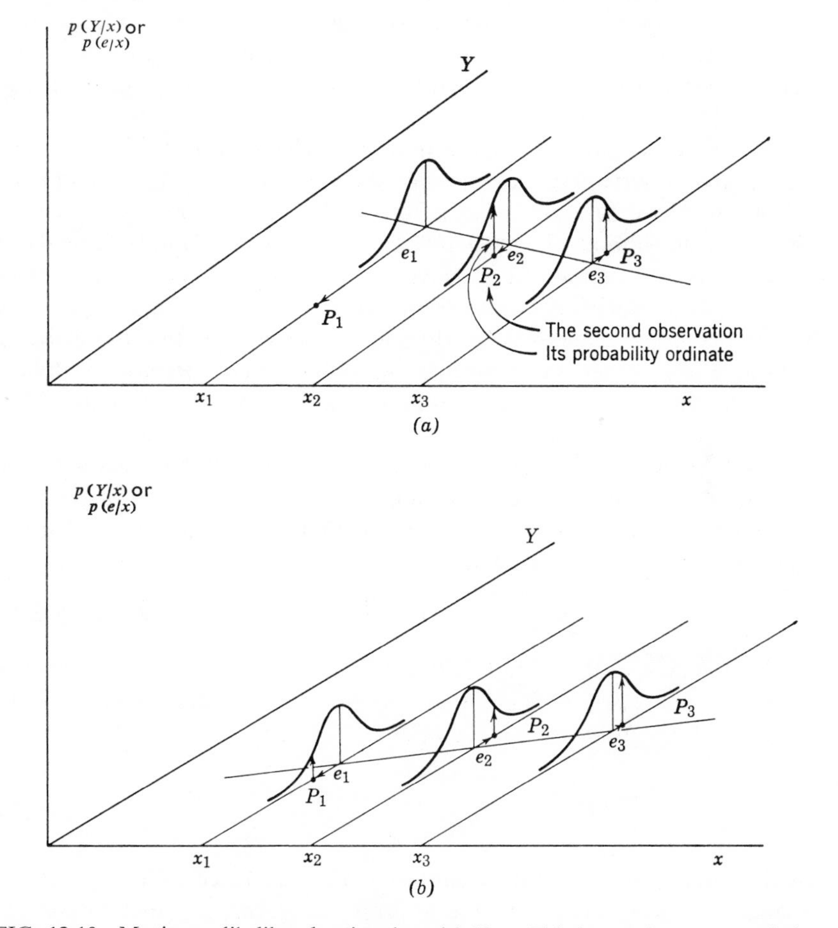

FIG. 12-10 Maximum likelihood estimation. (*a*) *Note*. This is *not* the true population; it is only a hypothetical population that the statistician is considering. But it is not very likely to generate the observed P_1, P_2, P_3. (*b*) Another hypothetical population; this is more likely to generate P_1, P_2, P_3.

The disturbance terms are collectively smaller, with their probability density being greater as a consequence.

The MLE technique is seen to involve speculating on various possible populations. How likely is each to give rise to the sample we observed? Geometrically, our problem would be to try them all out, by moving the population through all its possible values—i.e., *by moving the regression line and its surrounding e distribution through all possible positions in space.* Each position involves a different set of trial values for α and β. In each case the likelihood of observing P_1, P_2, P_3 would be evaluated. For our MLE we choose that hypothetical population which maximizes this likelihood. It is evident that little further adjustment is required in Figure 12-10*b* to arrive at the MLE. This procedure seems to result, intuitively, in a good fit; moreover, since it seems similar to the least squares fit, it is no surprise that we shall be able to show that the two coincide.

There are two other points worth noting. The MLE is derived from our three sample observations; another set of sample observations would almost certainly give rise to another MLE of α and β. The second point is more subtle. The likelihood of any population yielding our sample depends on not only the size of the *e* terms involved, but also on the shape of the *e* distribution—in particular σ^2, the variance of *e*. However, it can be shown that the maximum likelihood *line* does not depend on σ^2. In other words, if we assume σ^2 is larger, the geometry will look different, because *e* will have a flatter distribution; but the end result will be the same maximum likelihood line.

While geometry has clarified the method, it hasn't provided a precise means of arriving at the specific maximum likelihood estimate. This must be done algebraically. For generality, suppose that we have a sample of size *n*, rather than just 3. We wish to know

$$p(Y_1, Y_2 \cdots Y_n) \tag{12-44}$$

the likelihood or probability density of the sample we observed—expressed as a function of the possible population values of α, β, and σ^2. First, consider the probability density of the first value of Y, which is

$$p(Y_1) = \frac{1}{\sqrt{2\pi\sigma^2}} e^{-(1/2\sigma^2)[Y_1 - (\alpha + \beta x_1)]^2} \tag{12-45}$$

This is simply the normal distribution of Y_1, with its mean $(\alpha + \beta x_1)$ and variance (σ^2) substituted into the appropriate positions. [In terms of the geometry of Figure 12-10, $p(Y_1)$ is the ordinate above P_1.] The probability density of the second Y value is similar to (12-45), except that the subscript 2 replaces 1 throughout, and so on, for all the other observed Y values.

The independence of the Y values justifies multiplying all these probabilities together to find (12-44). Thus

$$p(Y_1, Y_2, \ldots, Y_n)$$
$$= \left[\frac{1}{\sqrt{2\pi\sigma^2}} e^{-(1/2\sigma^2)[Y_1-(\alpha+\beta x_1)]^2}\right]\left[\frac{1}{\sqrt{2\pi\sigma^2}} e^{-(1/2\sigma^2)[Y_2-(\alpha+\beta x_2)]^2}\right]\cdots$$
$$= \prod_{i=1}^{n}\left[\frac{1}{\sqrt{2\pi\sigma^2}} e^{-(1/2\sigma^2)[Y_i-(\alpha+\beta x_i)]^2}\right] \qquad (12\text{-}46)$$

where $\prod_{i=1}^{n}$ represents the product of n factors. Using the familiar rule for exponentials,[6] the product in (12-46) can be reexpressed by summing exponents

$$P(Y_1, Y_2, \ldots, Y_n) = \left(\frac{1}{\sqrt{2\pi\sigma^2}}\right)^n e^{\Sigma(-1/2\sigma^2)[Y_i-(\alpha+\beta x_i)]^2} \qquad (12\text{-}47)$$

Recall that the observed Y's are given. We are speculating on various values of α, β, and σ^2. To emphasize this, we rename (12-47) the likelihood function

$$L(\alpha, \beta, \sigma^2) = \frac{1}{(2\pi\sigma^2)^{n/2}} e^{-(1/2\sigma^2)\Sigma[Y_i-\alpha-\beta x_i]^2} \qquad (12\text{-}48)$$

We now ask, which values of α and β make L largest? The only place α and β appear is in the exponent; moreover, maximizing a function with a negative exponent involves minimizing the exponent. Hence our problem is to choose α and β in order to

$$\text{minimize} \sum [Y_i - \alpha - \beta x_i]^2 \qquad (12\text{-}49)$$

Moreover, this provides the maximum likelihood solution for α and β, regardless of the value of σ. This is the proposition suggested in the geometrical analysis in Figure 12-10; no matter what is assumed about the spread of the distribution, the maximum likelihood line is not affected by it.

But an even more important conclusion follows from comparing equation (12-49) with equation (11-10). *Maximum likelihood estimates are identical to least squares estimates.* The selection of least squares estimates a and b to minimize (11-10) is identical to the selection of maximum likelihood estimates of α and β to minimize (12-49). The only difference is that we've called our estimates different names. This establishes our other important theoretical justification of the least squares method: it is the estimate that follows from applying maximum likelihood techniques to a model with normally distributed error.

[6] $e^a \cdot e^b = e^{a+b}$ for any a and b.

*12-11 THE CHARACTERISTICS OF THE INDEPENDENT VARIABLE

So far it has been assumed that the independent variable x takes on a given set of fixed values (for example, fertilizer application was set at certain specified levels). But in many cases x cannot be controlled in this way. Thus if we are examining the effect of rainfall on yield, it must be recognized that x (rainfall) is a random variable, completely outside our control. The surprising thing is that the same MLE follows whether x is fixed *or* a random variable, if we assume [as well as (12-4)], that

1. The distribution of x does not depend on α, β, or σ^2. (12-50)
2. The distribution of e is independent of x, being $N(0, \sigma^2)$ for every x_i. (12-51)

The likelihood of our sample now involves the probability of observing both x and Y. If the x_i are independent, the likelihood function is

$$L = [p(x_1)p(Y_1/x_1)][p(x_2)p(Y_2/x_2)] \cdots \tag{12-52}$$

Because of the normality assumption (12-51),

$$L = p(x_1)\frac{1}{\sqrt{2\pi\sigma^2}}e^{-(1/2\sigma^2)(Y_1-\alpha-\beta x_i)^2}p(x_2)\frac{1}{\sqrt{2\pi\sigma^2}}e^{-(1/2\sigma^2)(Y_2-\alpha-\beta x_2)^2}\cdots \tag{12-53}$$

Collecting the exponents

$$L = p(x_1)p(x_2)\cdots\frac{1}{(2\pi\sigma^2)^{n/2}}e^{-(1/2\sigma^2)\Sigma(Y_i-\alpha-\beta x_i)^2} \tag{12-54}$$

Since $p(x)$ does not depend on the parameters α, β, and σ^2 according to (12-50), the problem of maximizing this likelihood function with respect to these parameters reduces to the minimization of the same exponent as before. This holds true, in fact, even if the x_i are not independent, and are determined by a joint probability distribution; then (12-54) becomes:

$$L(\alpha, \beta, \sigma^2) = p(x_1, x_2, \ldots, x_n)\frac{1}{(2\pi\sigma^2)^{n/2}}e^{-(1/2\sigma^2)\Sigma(Y_i-\alpha-\beta x_i)^2} \tag{12-55}$$

again requiring the same (least squares) minimization of the exponent.

We conclude that MLE and least squares coincide regardless of whether the independent variable x is fixed, or a random variable—if x is independent of the error and parameters in the equation being estimated. This greatly generalizes the application of the regression model.

chapter 13

Multiple Regression

13-1 INTRODUCTORY EXAMPLE

Suppose that the fertilizer and wheat yield observations in Chapter 11 were taken at several different agricultural experiment stations across the country. Even if soil conditions and temperatures were essentially the same in all these areas, we still might ask, "Can't part of the fluctuation in Y (i.e., the disturbance term e) be explained by varying levels of rainfall in different areas?" A better prediction of wheat yield may be possible if *both* fertilizer and rainfall are examined. Notice how this argument is similar to the one used in two factor ANOVA: if the error e can be reduced by taking rainfall R into account, we will get a better explanation of how the other variables are related. The observed levels of rainfall are shown in Table 13-1, along with the original observations of wheat yield and fertilizer from Table 11-1.

TABLE 13-1 Observed Wheat Yield, Fertilizer Application, and Rainfall

Y Wheat Yield (bu/acre)	X Fertilizer (lb/acre)	Z Rainfall (inches)
40	100	36
45	200	33
50	300	37
65	400	37
70	500	34
70	600	32
80	700	36

13-2 THE MATHEMATICAL MODEL

The multiple regression technique used to describe how a dependent variable is related to two or more independent variables is in fact only an extension of the simple regression analysis of the previous two chapters. Yield Y is now to be regressed on the two independent variables, or "regressors," fertilizer X and rainfall Z. Let us suppose it is reasonable to argue that the model is of the form

$$E(Y_i) = \alpha + \beta x_i + \gamma z_i \tag{13-1}$$

with both regressors x and z measured as deviations from their means. Geometrically this equation is a plane[1] in the three-dimensional space shown in Figure 13-1. For any given combination of rainfall and fertilizer (x_i, z_i), the expected yield $E(Y_i)$ is the point on this plane directly above, shown as a hollow dot. Of course, the observed value of Y, shown as a solid dot, is very unlikely to fall precisely on this plane. For example, our particular observed Y_i at this fertilizer/rainfall combination is somewhat greater than its expected value, and is shown as the solid dot lying directly above this plane. The difference between any observed and expected value of Y_i is the

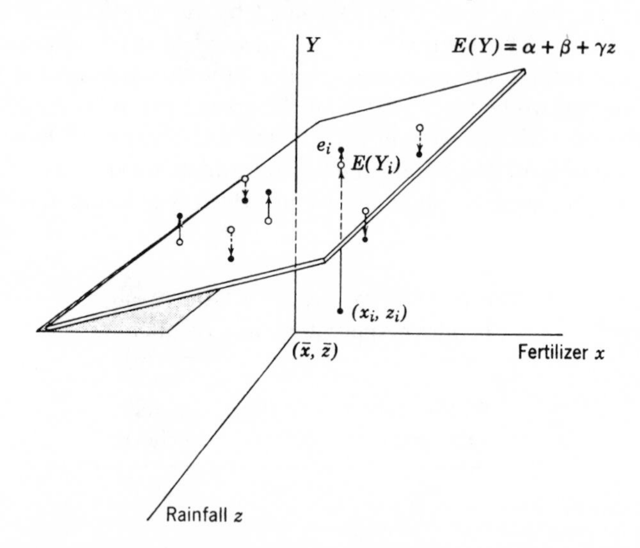

FIG. 13-1 Scatter of observed points about the true regression plane.

[1] It is a plane because it is linear in x and z. Looked at from another point of view we could say that (13-1) is linear in α, β, and γ. In fact, this latter linearity assumption is the more important of the two, since we are involved in estimating α, β, and γ; it is this assumption that keeps our estimating equations (13-4) linear.

stochastic or error term e_i. Thus any observed value Y_i may be expressed as its expected value plus this disturbance term

$$Y_i = \alpha + \beta x_i + \gamma z_i + e_i \tag{13-2}$$

with our assumptions about e the same as in Chapter 12.

β is geometrically interpreted as the slope of the plane as we move in a direction parallel to the (x, Y) plane, i.e., keep z constant; thus β is the marginal effect of fertilizer x on yield Y. Similarly γ is geometrically interpreted as the slope of the plane as we move in a direction parallel to the (z, Y) plane, i.e., keep x constant; hence γ is the marginal effect of z on Y.

13-3 LEAST SQUARES ESTIMATION

Least squares estimates are derived by selecting the estimates of α, β, and γ that minimize the sum of the squared deviations between the observed Y's and the fitted Y's; i.e., minimize[2]

$$\sum (Y_i - a - bx_i - cz_i)^2 \tag{13-3}$$

where a, b, and c are our estimators of α, β, and γ. This is done with calculus by setting the partial derivatives of this function with respect to a, b, and c

---

[2] Maximum likelihood estimates of α, β, and γ are derived in the same way as in the simple regression case; again this coincides with least squares. Geometrically, this involves trying out all possible hypothetical regression planes in Figure 13-1, and selecting that one that is most likely to generate the solid-dot sample values we actually observed.

But first, note that Figure 13-1 involves 3 parameters (α, β, and γ), and 3 variables (Y, x, and z). However, there is one additional variable in our system—$p(Y/x, z)$—which has not yet been plotted. It may appear that there is no way of forcing 4 variables into a three-dimensional space, but this is not so. For example, economists often plot 3 variables (labor, capital, and output) in a two-dimensional labor-capital space by introducing the third output variable as a system of isoquants. Those for whom this is a familiar exercise should have little trouble in graphing four variables [Y, x, z, and $p(Y/x, z)$] in a three-dimensional (Y, x, and z) space by introducing the fourth variable [$p(Y/x, z)$] as a system of isoplanes. Each of these isoplanes represents (Y, x, z) combinations that are equiprobable (i.e., for which the probability density of Y is constant). Thus the complete geometric model is the regression plane shown in Figure 13-1, with isoprobability planes stacked above and below it. Our assumptions about the error term (12-4) gaurantee that the isoprobability planes will be parallel to the true regression plane.

For MLE, we introduce the additional assumption that the error configuration is normal. Then we shift around a hypothetical regression plane along with its associated set of parallel isoprobability planes. In each position the probability density of the observed sample of points is evaluated by examining the isoprobability plane on which each point lies, and multiplying these together. That hypothetical regression which maximizes this likelihood is chosen. The algebra resembles the simple case in Section 12-10; it is easy to show that this results in minimizing the sum of squares (13-3).

TABLE 13-2 Least Squares Estimates for Multiple Regression of Y on X and Z

Y_i	X_i	Z_i	$x_i = X_i - \bar{X}$	$z_i = Z_i - \bar{Z}$	$Y_i x_i$	$Y_i z_i$	x_i^2	z_i^2	$x_i z_i$	$\hat{Y}_i = 60 + .06893x + .6028z$
40	100	36	−300	1	−12,000	40	90,000	1	−300	39.9
45	200	33	−200	−2	−9,000	−90	40,000	4	400	45.0
50	300	37	−100	2	−5,000	100	10,000	4	−200	54.3
65	400	37	0	2	0	130	0	4	0	61.2
70	500	34	100	−1	7,000	−70	10,000	1	−100	66.3
70	600	32	200	−3	14,000	−210	40,000	9	−600	72.0
80	700	36	300	1	24,000	80	90,000	1	300	81.3
$\sum Y_i = 420$	$\sum X_i = 2800$	$\sum Z_i = 245$	$\sum x_i = 0$	$\sum z_i = 0$	$\sum Y_i x_i = 19,000$	$\sum Y_i z_i = -20$	$\sum x_i^2 = 280,000$	$\sum z_i^2 = 24$	$\sum x_i z_i = -500$	
$a = \bar{Y} = 60$	$\bar{X} = 400$	$\bar{Z} = 35$								

Estimating equations (13-4)
$$\begin{cases} 19,000 = 280,000b - 500c \\ -20 = -500b + 24c \end{cases}$$

Solution
$$\begin{cases} b = .06893 \\ c = .6028 \end{cases}$$

Thus our regression is
$$Y = a + bx + cz$$
$$= 60 + .06893x + .6028z$$

Or, in terms of the original X and Z
$$Y = 60 + .06893(X - \bar{X}) + .6028(Z - \bar{Z})$$
$$= 60 + .06893(X - 400) + .6028(Z - 35)$$

$$\boxed{Y = 11.3307 + .06893X + .6028Z}$$

equal to zero, (or algebraically by a technique similar to that used in Appendix 11-1). The result is the following three estimating equations:

$$
\begin{aligned}
a &= \bar{Y} \\
\sum Y_i x_i &= b \sum x_i^2 + c \sum x_i z_i \\
\sum Y_i z_i &= b \sum x_i z_i + c \sum z_i^2
\end{aligned}
\tag{13-4}
$$

Again, note that the intercept estimate a is the mean of Y. The second and third equations may be solved for b and c. These calculations are shown in Table 13-2, and yield the fitted multiple regression equation.

PROBLEMS

13-1 Suppose a random sample of 5 families yielded the following data (an extension of Problem 11-1)

Family	Savings S	Income Y	Assets W
A	\$ 600	\$ 8,000	\$12,000
B	1,200	11,000	6,000
C	1,000	9,000	6,000
D	700	6,000	3,000
E	300	6,000	18,000

(a) Estimate the multiple regression equation of S on Y and W.
(b) Does the coefficient of Y differ from the answer to Problem 11-1(a)? Which coefficient better illustrates the relation of S to Y?
(c) For a family with assets of \$5000 and income of \$8000, what would you predict savings to be?
(d) Calculate the residual sum of squares, and residual variance s^2.
(e) Are you satisfied with the degrees of freedom you have for s^2 in this problem? Explain.

13-2 Suppose a random sample of 5 families yielded the following data (an extension of Problem 11-1)

Family	Savings S	Income Y	Number of Children N
A	\$ 600	\$ 8,000	5
B	1,200	11,000	2
C	1,000	9,000	1
D	700	6,000	3
E	300	6,000	4

(a) Estimate the multiple regression of S on Y and N.
(b) For a family with 5 children and income of $6000, what would you predict savings to be?

*13-3 Combining the data of Problems 13-1 and 13-2, we obtain the following table

Family	Savings S	Income Y	Assets W	Number of Children N
A	$ 600	$ 8,000	$12,000	5
B	1,200	11,000	6,000	2
C	1,000	9,000	6,000	1
D	700	6,000	3,000	3
E	300	6,000	18,000	4

Measuring the independent variables as deviations from the mean, we wish to estimate the regression equation

$$S = \alpha + \beta y + \gamma w + \delta n$$

(a) Generalizing (13-4), use the least squares criterion to derive the system of 4 equations needed to estimate the four parameters.
(b) Using a table such as Table 13-2, calculate the estimates of the four parameters.

13-4 MULTICOLLINEARITY

(a) In Simple Regression

In Figure 12-5*a* it was shown how our estimate b became unreliable if the X_i's were closely bunched, i.e., if the regressor X had little variation. It will be instructive to consider the limiting case, where the X_i's are concentrated on one single value X_0, as in Figure 13-2. Then b is not determined at all. There are any number of differently sloped lines passing through $(\bar{X}, \bar{Y})$ which fit equally well—for each line in Figure 13-2, the sum of squared deviations is the same, since the deviations are measured vertically from $(\bar{X}, \bar{Y})$. This geometric fact has an algebraic counterpart. If all $X_i = \bar{X}$, then all $x_i = 0$, and the term involving b in (11-10) is zero; hence the sum of squares does not depend on b at all. It follows that any b will do equally well in minimizing the sum of squares. An alternative way of looking at the same problem is that

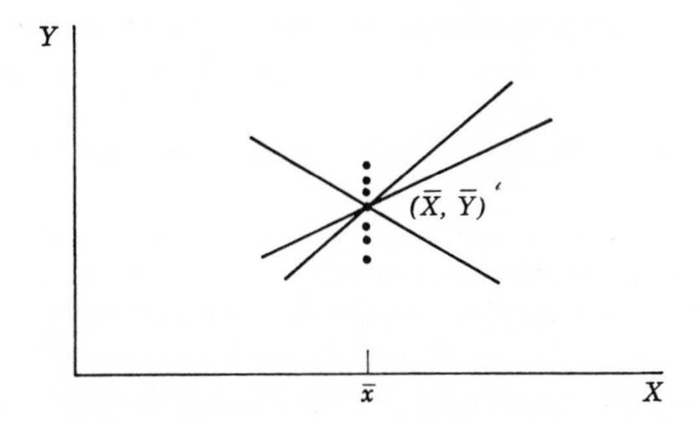

FIG. 13-2 Degenerate regression, because of no spread (variation) in X.

since all x_i are zero, $\sum x_i^2$ in the denominator of (11-16) is zero, and b is not defined.

In conclusion, when the values of X show little or no variation, then the effect of X on Y can no longer be sensibly investigated. But if the problem is *predicting* Y—rather than investigating Y's dependence on X—this bunching of the X values doesn't matter *provided* we stick to this same value of X. All the lines in Figure 13-2 predict Y equally well. The best prediction is $\bar{Y}$, and all these lines give us that result.

(b) In Multiple Regression

Again consider the limiting case where the values of the independent variables X and Z are completely bunched up on a line L, as in Figure 13-3.

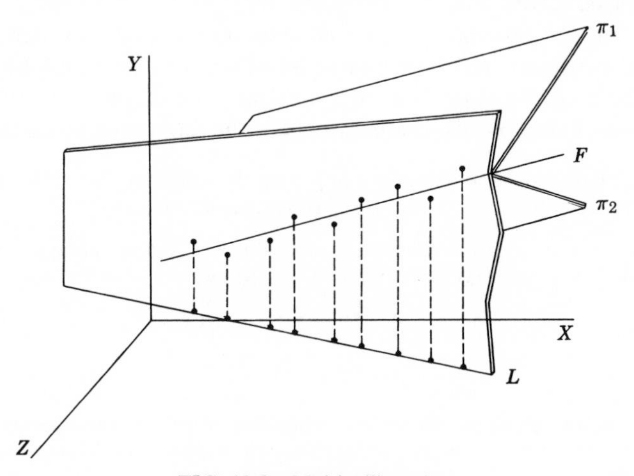

FIG. 13-3 Multicollinearity.

This means that all the observed points in our scatter lie in the vertical plane running up through L. You can think of three-dimensional space as a room in a house; the observations are not scattered throughout this room, but instead lie embedded in an extremely thin pane of glass standing vertically on the floor.

In explaining Y, multicollinearity makes us lose one dimension. In the earlier case of simple regression, our best fit for Y was not a line, but rather a point $(\bar{x}, \bar{Y})$; in this multiple regression case our best fit for Y is not a plane, but rather the line F. To get F, just fit the least squares line through the points on the vertical pane of glass. The problem is identical to the one shown in Figure 11-2; in one case a line is fitted on a flat pane of glass, in the other case, on a flat piece of paper. This regression line F is therefore our best fit for Y. As long as we stick to the same *combination* of X and Z—i.e., so long as we confine ourselves to predicting Y values on that pane of glass—no special problems[3] arise. We can use the regression F on the glass to predict Y in exactly the same way as we did in the simple regression analysis of Chapter 11. But there is no way to examine how X affects Y. Any attempt to define β, the marginal effect of X on Y (holding Z constant), involves moving off that pane of glass, and we have no sample information whatsoever on what the world out there looks like. Or, to put it differently, if we try to explain Y with a plane—rather than a line F—we find there are any number of planes running through F (e.g., π_1 and π_2) which do an equally good job. Since each passes through F, each yields an identical sum of squared deviations; thus each provides an equally good fit. This is confirmed in the algebra in the normal equations (13-4). When X is a linear function of Z (i.e., when x is a linear function of z) it may be shown that the last two equations are not independent, and cannot be solved uniquely for b and c.[4]

Now let's be less extreme in our assumptions and consider the near-limiting case, where z and x are almost on a line, (i.e., where all our observations in the room lie very close to a vertical pane of glass). In this case, a plane may be fitted to our observations, but the estimating procedure is very

[3] In practice, there would be a problem in getting the regression line F, since computer routines typically break down in the face of perfect multicollinearity.

[4] Two equations can usually be solved for two unknowns, but not always. For example, suppose that John's age (X) is twice Harry's (Y). Then we can write

$$X = 2Y$$

or

$$5X = 10Y \tag{13-5}$$

Note that these two equations tell us the same thing. We have two equations with two unknowns, but they don't generate a unique solution, because they don't give us independent information. The second just restates what the first told us.

unstable; it becomes very sensitive to random errors, reflected in large variance of the estimators b and c. Thus, even though X may really affect Y, its statistical significance can't be established because the standard deviation of b is so large. This is analogous to the argument in the simple regression case in Section 12-6.

When the independent variables X and Z are collinear, or nearly so, it is called the problem of multicollinearity. For prediction purposes, it does not hurt provided there is no attempt to predict for values of X and Z removed from their line of collinearity. But structural questions cannot be answered—the *relation* of Y to either X or Z cannot be sensibly investigated.

Example 1

In our wheat yield example, suppose that X is (as before) the amount of fertilizer measured in pounds per acre, and that the statistician makes the incredibly foolish error of defining another independent variable Z as the amount of fertilizer measured in ounces per acre. Since any weight measured in ounces must be sixteen times its measurement in pounds:

$$Z = 16X \tag{13-6}$$

exactly. Thus all combinations of X and Z must fall on this straight line, and we have an example of perfect multicollinearity. Now if we try to fit[5] a regression plane to the observations of yield and fertilizer given in Table 11-1, one possible answer would be our original regression given in (11-18):

$$Y = 32.8 + .068X + 0Z \tag{13-7}$$

But an equally satisfactory solution would follow from substituting (13-6) into (13-7):

$$Y = 32.8 + 0X + .00425Z$$

Another equivalent answer would be to make a partial substitution for X in (13-7) as follows:

$$\begin{aligned} Y &= 32.8 + .068[\lambda X + (1 - \lambda)X] \\ &= 32.8 + .068[\lambda X + (1 - \lambda)(\tfrac{1}{16})Z] \\ Y &= 32.8 + .068\lambda X + .00425(1 - \lambda)Z \end{aligned} \tag{13-8}$$

(13-8) is a whole family of planes depending on the arbitrary value assigned to λ. In fact, all these three-dimensional planes are equivalent expressions for

[5] The computer program would probably "hang up" trying to divide by zero. So we suppose the calculations are handcrafted.

our simple two-dimensional relationship between fertilizer and yield. While all give the same correct prediction of Y, no meaning can be attached to whatever coefficients of X and Z we may come up with.

Example 2

While the previous extreme example may have clarified some of the theoretical issues, no statistician would make that sort of error in model specification. Instead, more subtle difficulties arise. In economics, for example, suppose demand for a group of goods is being related to prices and income, with the overall price index being the first independent variable. Suppose aggregate income measured in money terms is the second independent variable. Since this is real income multiplied by the same price index, the problem of multicollinearity may become a serious one. The solution is to use real income, rather than money income, as the second independent variable. This is a special case of a more general warning: in any multiple regression in which price is one independent variable, beware of other independent variables measured in prices.

The problem of multicollinearity may be solved if there happens to be prior information about the relation of β and γ. For example, if it is known *a priori* that

$$\gamma = 5\beta \tag{13-9}$$

then this information will allow us to uniquely determine the regression plane, even in the case of perfect collinearity. This is evident from the geometry of Figure 13-3. Given a fixed relation between our two slopes (β and γ) there is only one regression plane π which can be fitted to pass through F. This is confirmed algebraically. Using (13-9), our model (13-2) can be written

$$Y_i = \alpha + \beta x_i + 5\beta z_i + e_i \tag{13-10}$$

$$= \alpha + \beta(x_i + 5z_i) + e_i \tag{13-11}$$

It is natural to define a new variable

$$w_i = x_i + 5z_i \tag{13-12}$$

Thus (13-11) becomes

$$Y_i = \alpha + \beta w_i + e_i \tag{13-13}$$

and a regression of Y on w will yield estimates a and b. Finally, if we wish an estimate of γ, it is easily computed using (13-9):

$$c = 5b \tag{13-14}$$

13-5 INTERPRETING AN ESTIMATED REGRESSION

Suppose the multiple regression

$$Y = a + b_1X_1 + b_2X_2 + b_3X_3 + b_4X_4$$

is fitted to 25 observations of Y and the X's. The least squares estimates often are published in the form, for example:

$$\begin{array}{llllll} Y = 10.6 & + 28.4X_1 & + 4.0X_2 & + 12.7X_3 & + .84X_4 & \quad (13\text{-}15) \\ (s_0 = 2.6) & (s_1 = 11.4) & (s_2 = 1.5) & (s_3 = 14.1) & (s_4 = .76) & \\ (t_1 = 4.1) & (t_2 = 2.5) & (t_3 = 2.6) & (t_4 = .9) & (t_5 = 1.1) & \end{array}$$

The bracketed information is used in assessing the reliability of the least squares fit, either in a confidence interval or hypothesis test.

The true effect of X_1 on Y is the unknown population parameter β_1; we estimate it with the sample estimator b_1. While the unknown β_1 is fixed, our estimator b_1 is a random variable, differing from sample to sample. The properties of b_1 may be established, just as the properties of b were established in the previous chapter. Thus b_1 may be shown to be normal—again provided the sample size is large, or the error term is normal. b_1 can also be shown to be unbiased, with its mean β_1. The magnitude of error involved in estimation is reflected in the standard deviation of b_1 which, let us suppose, is estimated to be $s_1 = 11.4$ as given in the first bracket below equation (13-15), and shown in Figure 13-4. When b_1 is standardized with this estimated standard deviation, it will have a t distribution.

To recapitulate: we don't know β_1; all we know is that whatever it may be, our estimator b_1 is distributed around it, as shown in Figure 13-4. This knowledge of how closely b_1 estimates β_1 can, of course, be "turned around" to infer a 95 percent confidence interval for β_1 from our observed sample b_1

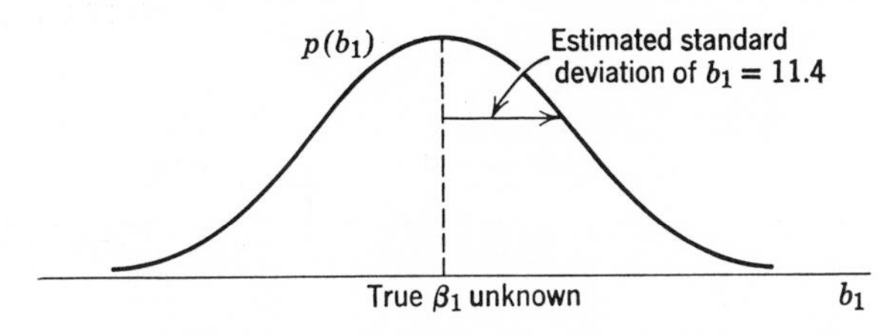

FIG. 13-4 Distribution of the estimator b_1.

as follows:

$$\begin{aligned}\beta_1 &= b_1 \pm t_{.025}s_1 \\ &= 28.4 \pm 2.09(11.4) \\ &= 28.4 \pm 23.8\end{aligned} \tag{13-16}$$

[$n = 25$ is the sample size, $k = 5$ is the number of parameters already estimated in (13-15), and $t_{.025}$ is the critical t value with $n - k$ degrees of freedom.] Similar confidence intervals can be constructed for the other β's.

If we turn to testing hypotheses, extreme care is necessary to avoid very strange conclusions. Suppose it has been concluded on theoretical grounds that X_1 should positively influence Y, and we wish to see if we can statistically confirm this relation. This involves a one-tailed test of the null hypothesis,

$$H_0: \beta_1 = 0 \tag{13-17}$$

against the alternative

$$H_1: \beta_1 > 0 \tag{13-18}$$

If H_0 is true, b_1 will be centered on $\beta_1 = 0$, and there will be only a 5% probability of observing a t value exceeding 1.72; this defines our rejection region in Figure 13-5a. Our observed t value [2.5 as shown below equation (13-15)] falls in this region; hence we reject H_0, thus confirming (at a 5% significance level) that Y is positively related to X_1.

The similar t values [also shown for the other estimators below (13-15)] can be used for testing the null hypothesis on the other β parameters. As we see in Figure 13-5b, the null hypothesis $\beta_2 = 0$ can also be rejected, but a similar conclusion is not warranted for β_3 and β_4. We conclude therefore that

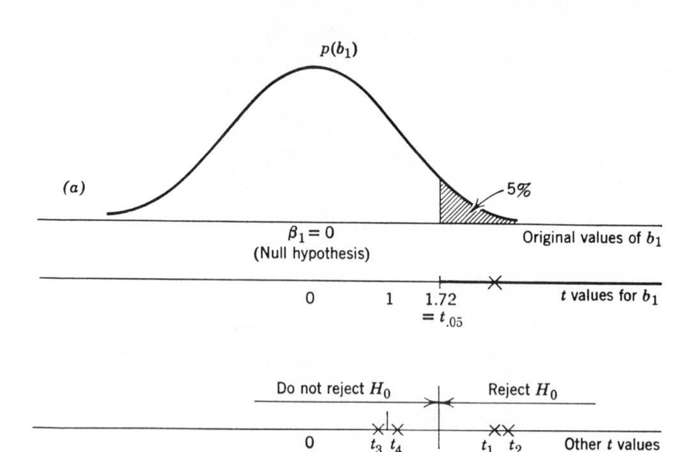

FIG. 13-5 (a) Test of β_1. (b) Test of other β's.

the results are "statistically significant" for X_1 and X_2; the evidence is that Y is related to each. But the results are not statistically significant for X_3 and X_4.

As long as we confine ourselves to rejecting hypotheses—as with β_1 and β_2—we won't encounter too much difficulty. But if we *accept* the null hypothesis about β_3 and β_4, we may run into a lot of trouble of the sort first encountered in Chapter 9. Since this is so important in regression analysis, the argument is reviewed for emphasis.

While it is true, for example, that our t coefficient for X_3 (.9) is not "statistically significant," this does *not* prove there is no relationship between X_3 and Y. It is easy to see why. Suppose that we have strong theoretical grounds for believing that Y is positively related to X_3. In (13-15) this belief is confirmed: Y is related to X_3 by a positive coefficient. Thus our statistical evidence is consistent with our prior belief (even though it is not as strong as we might like it to be).[6] To accept the null hypothesis $\beta_3 = 0$ and conclude that X_3 doesn't affect Y, would be in direct contradiction to both our prior belief and the statistical evidence. We would be reversing a prior belief even though the statistical evidence weakly confirmed it. It would have been better had we not even looked at the evidence. And we note that this remains true for any positive t value, although as t becomes smaller, our statistical confirmation becomes weaker. Only if t is zero or negative, do the statistical results contradict our prior belief.

It follows from this, that if we had strong prior grounds for believing X_3 and X_4 to be positively related to Y, they should not be dropped from the estimating equation (13-15); instead they should be retained, with all the pertinent information on their t values.

It must be emphasized that those who have accepted hypotheses have not *necessarily* erred in this way. But that risk has been run by anyone who has mechanically accepted a null hypothesis because the t value was not statistically significant. The difficulty is especially acute—as in the case we've cited—when the null hypothesis was introduced strictly for convenience (because it was simple), and not because there is any reason to believe it in the first place. It becomes less acute if there is some expectation that H_0 is true—i.e., if there are theoretical grounds for concluding that Y and X are unrelated. Suppose for illustration that we expect a priori that H_0 is true; in such a case, a weak observed relationship (e.g., $t = .6$) would be in some conflict with our prior expectation of no relationship. But it is not a serious conflict, and easily explained by chance. Hence resolving it in favor of our prior expectation and continuing to use H_0 as a working hypothesis might be a reasonable judgment.

---

[6] Perhaps because of too small a sample. Thus 12.7 may be a very accurate description of how Y is related to X_3; but our t value is not statistically significant because our sample is small, and the standard deviation of our estimator ($s_3 = 14.1$) is large as a consequence.

We conclude once again, that classical statistical theory provides incomplete grounds for accepting H_0; acceptance must be based also on extra-statistical judgment, with prior belief playing a key role.

Prior belief plays a less critical role in the rejection of an hypothesis; but it is by no means irrelevant. Suppose, for example that although you believed Y to be related to X_1, X_3, and X_4, you didn't really expect it to be related to X_2; someone had just suggested that you "try on" X_2 at a 5% level of significance. This means that if H_0 (no relation) is true, there is a 5% chance of ringing a false alarm. If this is the *only* variable "tried on," then this is a risk we can live with. However, if many such variables are "tried on" in a multiple regression the chance of a false alarm increases dramatically.[7] Of course, this risk can be kept small by reducing the level of error for each t test from 5 to 1% or less. This has led some authors to suggest a 1% level of significance with the variables just being "tried on," and a 5% level of significance with the other variables expected to affect Y. Using this criterion we would conclude that the relation of Y and X_1 is statistically significant; but the relation of Y to X_2 is *not*—despite its higher t value—because there are no prior grounds for believing it.[8]

To sum up: hypothesis tests require

1. Good judgment, and good prior theoretical understanding of the model being tested;
2. An understanding of the assumptions and limitations of the statistical techniques.

PROBLEMS

13-4 Suppose a multiple regression of Y on three independent variables yields the following estimate, based on a sample of $n = 30$:

$$Y = 25.1 + 1.2X_1 + 1.0X_2 - 0.50X_3$$

Standard deviations	(2.1)	(1.5)	(1.3)	(.060)
t-values	(11.9)	( )	( )	( )
95% confidence limits	(±4.3)	( )	( )	( )

[7] Suppose, for simplicity, that the t tests for the significance of the several variables (say k of them) were independent. Then the probability of no error at all is $(.95)^k$. For $k = 10$, for example, this is .60, making the probability of some error (some false alarm) as high as .40.

[8] Anyone who thinks he would never wish to use such a double standard might suppose that Y is the U.S. price level, X_1 is U.S. wages, and X_2 the number of rabbits in South Australia. With the t values shown in equation (13-15), what would he do?

(a) Fill in the blank spaces in the above estimate.

(b) The following are either true or false. If false, correct.

(1) The coefficient of X_1 is estimated to be 1.2. Other scientists might collect other samples and calculate other estimates. The distribution of these estimates would be centered around the true value of 1.2. Therefore the estimator is called unbiased.

(2) If there were strong prior reasons for believing that X_1 does not influence Y, it is reasonable to reject the null hypothesis $\beta_1 = 0$ at the 5% level of significance.

(3) If there were strong prior reasons for believing that X_2 does influence Y, it is reasonable to use the estimated coefficient 1.0 rather than accept the null hypothesis $\beta_2 = 0$.

13-6 DUMMY VARIABLES

There are two major categories of statistical information: cross section and time series. For example, econometricians estimating the consumption function[9] sometimes use a detailed breakdown of the consumption of individuals at various income levels at one point in time (cross section); sometimes they examine how total consumption is related to national income over a number of time periods (time series); and sometimes they use a combination of the two. In this section we develop a method that is especially useful in analysing time series data; as we shall see, it also has important applications in cross-section studies as well.

(a) Introductory Example

Suppose we wish to investigate how the public purchase of government bonds (B) is related to national income (Y). A hypothetical scatter of annual observations of these two variables is shown for Canada in Figure 13-6, and in Table 13-3. It is immediately evident that the relationship of bonds to income follows two distinct patterns—one applying in wartime (1940–5), the other in peacetime.

The normal relation of B to Y (say L_1) is subject to an upward shift (L_2) during wartime; heavy bond purchases in those years is explained not by Y alone, but also by the patriotic wartime campaign to induce public bond purchases. B therefore should be related to Y *and* another variable—war (W). But this is only a categorical, or indicator variable. It does not have a whole

[9] i.e., how consumption expenditures are related to income.

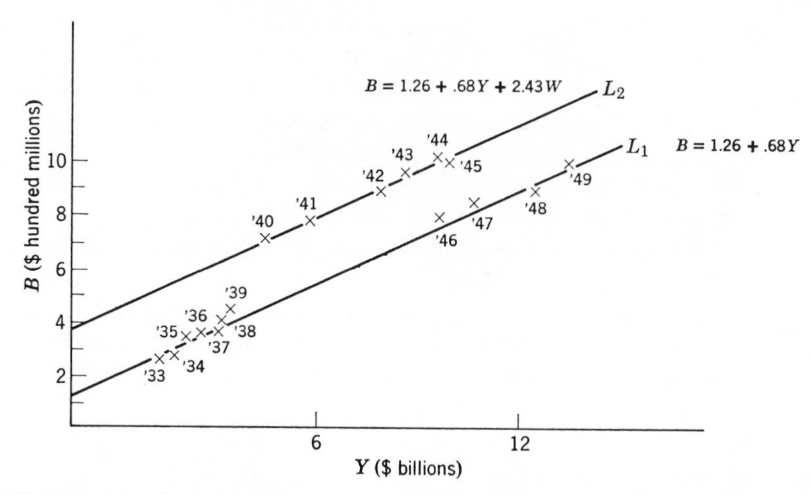

FIG. 13-6 Hypothetical scatter of public purchases of bonds (B) and national income (Y).

range of values, but only two: on the one hand, we arbitrarily set its value at 1 for all wartime years; on the other hand we set its value at 0 for all peacetime years. Since W is either "on" or "off," it is referred to as a "counter" or "dummy" variable. Our model is:

$$B = \alpha + \beta Y + \gamma W + e \tag{13-19}$$

where

$$W = 1 \quad \text{for wartime years,}$$
$$= 0 \quad \text{for peacetime years.}$$

This single equation is seen to be equivalent to the following two equations:

$$B = \alpha + \beta Y + \gamma + e \quad \text{for wartime} \tag{13-20}$$

$$B = \alpha + \beta Y + e \qquad \text{for peacetime} \tag{13-21}$$

W may also be called a "switching" variable. With war and peace, we switch back and forth between (13-20) and (13-21).

We note that γ represents the effect of wartime on bond sales; and β represents the effect of income changes. (The latter is assumed to remain the same in war or peace.) The important point to note is that one multiple regression of B on Y and W as in (13-19) will yield the *two* estimated lines shown in Figure 13-6; L_1 is the estimate of the peacetime function (13-21), and L_2 is the estimate of the wartime function (13-20).

Complete calculations for our example are set out in Table 13-3, and the procedure is interpreted in Figure 13-7. Since all observations are $W = 0$, or $W = 1$, the scatter is spread only in the two vertical planes π_1 and π_2. Estimation involves a multiple (least squares) regression fit of (13-19) to this

TABLE 13-3 Calculations for Regression of B on Y and W, where W is a Dummy Variable.

Year	B	Y	W	$y = Y - \bar{Y}$	$w = W - \bar{W}$	yw	By	Bw	y^2	w^2
1933	2.6	2.4	0	−4.44	−.35	1.55	−11.54	−.91	19.71	.12
1934	3.0	2.8	0	−4.04	−.35	1.41	−12.12	−1.05	16.32	.12
1935	3.6	3.1	0	−3.74	−.35	1.31	−13.46	−1.26	13.99	.12
1936	3.7	3.4	0	−3.44	−.35	1.20	−12.73	−1.29	11.83	.12
1937	3.8	3.9	0	−2.94	−.35	1.03	−11.17	−1.33	8.64	.12
1938	4.1	4.0	0	−2.84	−.35	0.99	−11.64	−1.43	8.07	.12
1939	4.4	4.2	0	−2.64	−.35	0.92	−11.62	−1.54	6.97	.12
1940 (War years)	7.1	5.1	1	−1.74	.65	−1.13	−12.35	4.62	3.03	.42
1941 (War years)	8.0	6.3	1	−.54	.65	−.35	−4.32	5.20	.29	.42
1942 (War years)	8.9	8.1	1	1.26	.65	.82	11.21	5.78	1.59	.42
1943 (War years)	9.7	8.8	1	1.96	.65	1.27	19.01	6.30	3.84	.42
1944 (War years)	10.2	9.6	1	2.76	.65	1.79	28.15	6.63	7.62	.42
1945 (War years)	10.1	9.7	1	2.86	.65	1.86	28.89	6.56	8.18	.42
1946	7.9	9.6	0	2.76	−.35	−.97	21.80	−2.77	7.62	.12
1947	8.7	10.4	0	3.56	−.35	−1.25	30.97	−3.05	12.67	.12
1948	9.1	12.0	0	5.16	−.35	−1.81	46.96	−3.19	26.63	.12
1949	10.1	12.9	0	6.06	−.35	−2.12	61.21	−3.53	36.72	.12
	$\sum B = 115$	$\sum Y = 116.3$	$\sum W = 6$			$\sum yw = 6.52$	$\sum By = 147.25$	$\sum Bw = 13.74$	$\sum y^2 = 193.72$	$\sum w^2 = 3.84$
	$a = \bar{B} = \frac{\sum B}{17} = 6.76$	$\bar{Y} = \frac{\sum Y}{17} = 6.84$	$\bar{W} = \frac{6}{17} = .35$							

Estimating equations (13-4) $\begin{cases} \sum By = b\sum y^2 + c\sum yw \\ \sum Bw = b\sum yw + c\sum w^2 \end{cases}$

or $\begin{cases} 147.25 = b193.72 + c6.52 \\ 13.74 = b6.52 + c3.84 \end{cases}$

Solution: $\begin{cases} b = .68 \\ c = 2.43 \end{cases}$

Thus our estimated regression is: $B = 6.76 + .68y + 2.43w$

Or, expressed in terms of the original variables: $B = 6.76 + .68(Y - \bar{Y}) + 2.43(W - \bar{W})$

$= 6.76 + .68(Y - 6.84) + 2.43(W - .35)$

$$\boxed{B = 1.26 + .68Y + 2.43W}$$

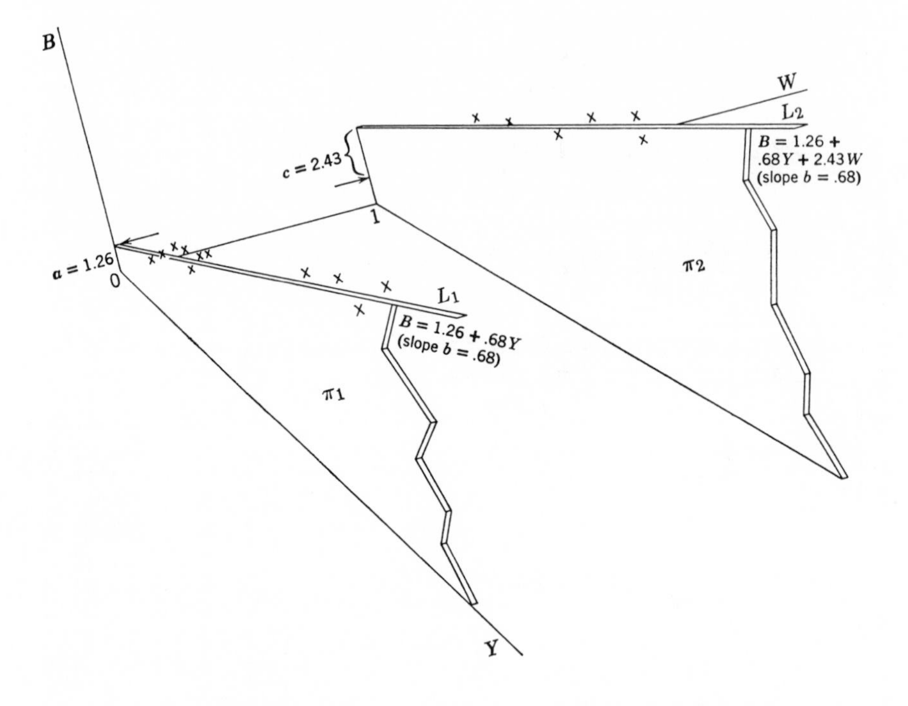

FIG. 13-7 Multiple regression with a dummy variable (W).

scatter. The resulting fitted plane

$$B = a + bY + cW \tag{13-22}$$

can be visualized as a plane resting on its two supporting buttresses π_1 and π_2 The slopes of L_1 and L_2 are (by assumption) equal[10] to the common value b, and c is the estimated wartime shift.

[10] This restriction means that L_1 and L_2 are *not* independently fitted. In other words, our least squares plane (13-22) is fitted first; L_1 and L_2 are simply "read off" this plane. Thus L_1 does *not* represent a least squares fit to the left-hand scatter, nor does L_2 represent a least squares fit to the right-hand scatter.

Thus the dummy variable method of fitting a single multiple regression plane and then reading off L_1 and L_2, can be compared to the alternative method of independently fitting two simple regression lines to the two scatters in Figure 13-7. Our model would be:

$$B = \alpha_1 + \beta_1 Y + e_1 \quad \text{for wartime}$$
$$B = \alpha_2 + \beta_2 Y + e_2 \quad \text{for peacetime,}$$

and the estimated slopes (β_1 and β_2) would generally not be the same.

(cont'd)

In a dummy variable model—as in any regression problem—it is important to understand why *both* variables Y and W must be included. Even if our only interest is in B and Y, their relationship cannot be properly estimated unless W is taken into account. In other words, since experimental control over the "nuisance" variable W is not possible, its effects must explicitly be removed in the regression analysis. To ignore this variable is to invite a bias in our estimators, as well as an increased variance. To see how bias occurs, consider what happens if W is ignored, so that our scatter involves only the two dimensions B and Y. Geometrically this involves projecting the three-dimensional scatter in Figure 13-7 onto the two-dimensional B-Y plane, as in Figure 13-8*a*. This is immediately recognized as the same scatter plotted in Figure 13-6; we also reproduce from that diagram L_1 and L_2, our estimated multiple regression using W as a dummy variable. If we calculate L_3, the simple regression of B on Y, it clearly has too great a slope. This upward bias is due to the fact that war years tended to be high income years: thus on the right-hand side of this scatter, higher bond sales that should be attributed in part to wartime would be (erroneously) attributed to income alone.

A similar error is to be expected in any investigation of B and W which ignores Y. With no Y dimension, our scatter in Figure 13-7 would be projected onto the B-W plane, as in Figure 13-8*b*. In this diagram the only way to estimate the wartime effect is to look at the difference in sample means,[11] which is too large. This upward bias would be due to the same cause: higher bond sales that should be attributed in part to higher income would be (erroneously) attributed to wartime alone.

This example has illustrated the general nature of dummy variables. This can be applied to a wide variety of problems, but one of the most useful applications is in removing seasonal shifts in time series data, as explained next.

Estimates of four parameters are required for this model, rather than the three in the dummy variable model (13-19); thus one advantage of the dummy model is that it conserves one extra degree of freedom. The disadvantage of the dummy model is that it requires an additional prior restriction—that the two slopes are equal. But this is not always a disadvantage. For instance, in our example it may be better to assume the two slopes equal than to independently fit a wartime function to only five observations. The very small wartime sample may yield a very unreliable estimate of slope, and it may make better sense to pool all the data to estimate one slope coefficient.

[11] This is equivalent to a simple regression of B on W. Because of the peculiar scatter involved, this regression line would pass through these two means; thus their difference represents the effect of W on B.

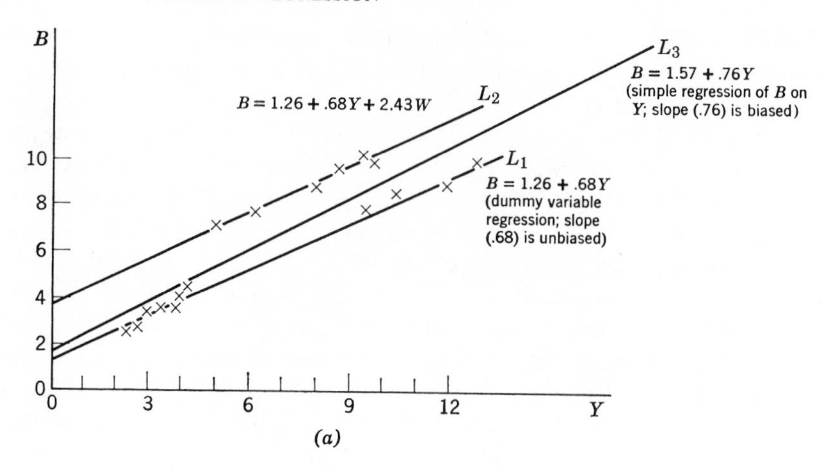

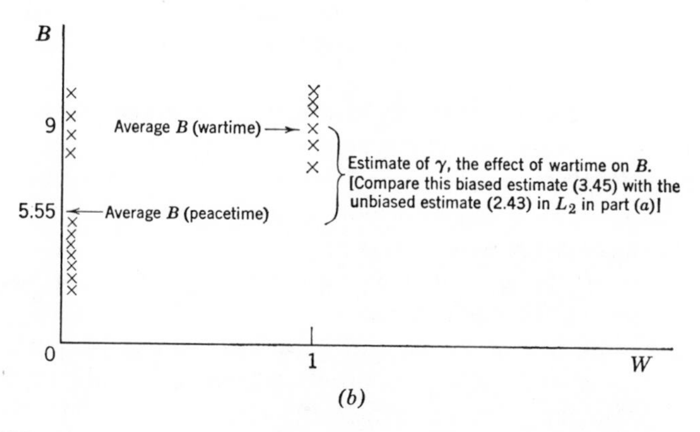

FIG. 13-8 Error when one explanatory variable is ignored. (*a*) Biased estimate of slope (the effect of Y) because the categorical variable W is ignored. (*b*) Biased estimate of the effect of W because the numerical variable Y is ignored.

(b) Seasonal Adjustment

To illustrate, consider a spectacular example from real life. Suppose we wish to examine how department store sales of jewelry increase over time. When we plot quarterly sales (in Table 13-4) against time as in Figure 13-9*a*, we note how sales shoot up every fourth quarter because of Christmas. Since we are interested in the long-term secular increase in sales, these strange Christmas observations should be discounted. This calls for a dummy

TABLE 13-4 Canadian Jewelry Sales (S) and Seasonal Dummies

	T (Quarter Years)	S ($100,000's)	Q_4	Q_3	Q_2
1957	1	24	0	0	0
	2	29	0	0	1
	3	29	0	1	0
	4	50	1	0	0
1958	5	24	0	0	0
	6	30	0	0	1
	7	29	0	1	0
	8	51	1	0	0
1959	9	26	0	0	0
	10	29	0	0	1
	11	30	0	1	0
	12	52	1	0	0
1960	13	25	0	0	0
	14	30	0	0	1
	15	29	0	1	0
	16	50	1	0	0

Source: Dominion Bureau of Statistics, Ottawa.

variable[12] Q_4 (for fourth quarter) so that our model is

$$S = \beta_0 + \beta_1 T + \beta_4 Q_4 + e \tag{13-23}$$

Even this model may not be adequate. If allowance should also be made for shifts in the other quarters, dummies Q_2 and Q_3 should be added. A dummy

[12] There are three points in the analysis at which we might conclude that explicit account should be taken of seasonal swings. We may expect a strong seasonal influence from prior theoretical reasoning. Or, such an influence may be discovered after we plot the scatter. Finally, it may be discovered by examining residuals after the regression is fitted. Clearly those observations indicated by arrows (in Figure 13-9a) have consistently high residuals. To explain this, we look for something they have in common. Their common property is that they all occur in the fourth quarter. Hence the fourth quarter is introduced as a dummy regressor. This technique of "squeezing the residuals till they talk" is important in every kind of regression, not just time series; used with discretion, it indicates which further regressors may be introduced in order to reduce bias and residual variance.

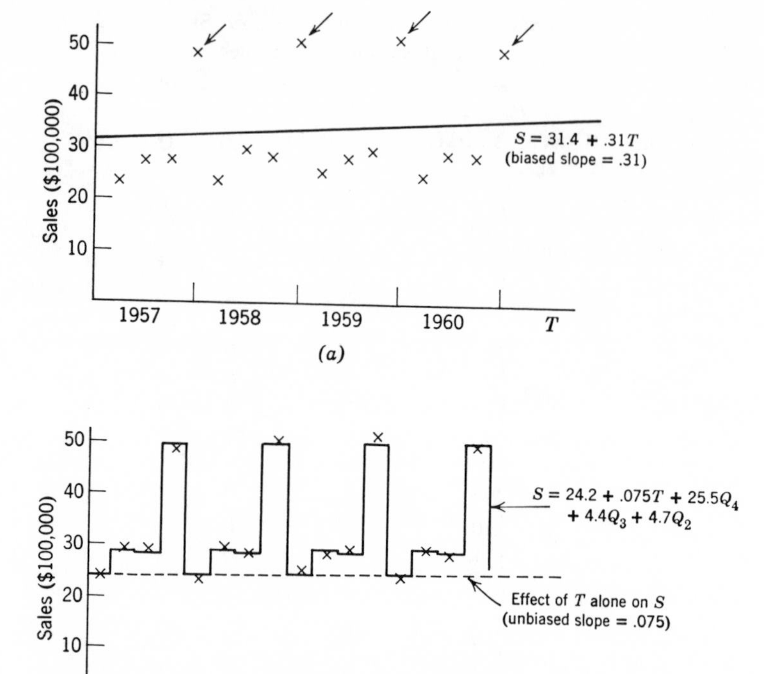

FIG. 13-9 Secular growth in Canadian jewelry sales, with and without seasonal adjustment. (*a*) Inadequate simple regression of *S* on *T* alone. (*b*) Multiple regression of *S* on *T*, including seasonal adjustment.

Q_1 is not needed for the first quarter, because Q_2, Q_3, and Q_4 measure the shift from a first quarter base. (Whether or not to include the various regressors Q_4, Q_3, Q_2, can be decided on statistical grounds, by testing for statistical significance. It is common to include them all in such a test, and reject or accept them as a group. But such a statistical test on data as extreme as ours would be superfluous.) Our modified model is now

$$S = \alpha + \beta_1 T + (\beta_4 Q_4 + \beta_3 Q_3 + \beta_2 Q_2) + e \tag{13-24}$$

The least squares fit[13] is graphed in Figure 13-9*b*. Notice that our seasonal adjustment is exactly the same every year, i.e., each year there is

[13] The least squares fit to this model was calculated by a method similar to that of Table 13-3. Equation system (13-4) was extended to a system of 5 estimating equations for the 5 unknowns.

the same upward shift (b_2) in our fit between the first and second quarters. (These seasonal shift coefficients need not always be positive, as in our example.)

By contrast, the simple regression of S on T without quarterly adjustment is graphed in Figure 13-9a. It is a poor fit, with large residual variance. Even worse, the calculated slope showing the relation of S to T is biased, for the same reasons as in the bond example of part (a).

(c) Seasonal Adjustment without Dummies (Moving Average)

Dummy variables are not the only means of seasonally adjusting data. Another common method is to take a moving average (over a whole year) of the time series, as shown in Table 13-5. Note how the wild seasonal swing at Christmas is ironed out in this averaging process. The desired relation of sales to time can now be estimated by a simple regression of seasonally adjusted S' on T.

It is interesting to compare this method with the dummy variable alternative. An apparent disadvantage is that a total of three observations are lost at the beginning and end of the time series, in order to get the moving average started and finished. However, although it is less evident, the same loss is involved in using dummy variables, since three degrees of freedom are lost in estimating the shift coefficients β_2, β_3, and β_4.

An advantage of the moving average method is that it is not necessary to assume a constant seasonal shift; thus the adjustment for any quarter

TABLE 13-5 Moving Average

Time		S (Unadjusted)	S' (Adjusted by Four Quarter Moving Average)
	1	24	
	2	29	$\frac{1}{4}(24 + 29 + 29 + 50) = 33$
1957	3	29	$\frac{1}{4}(29 + 29 + 50 + 24) = 33$
	4	50	. = 33.25
	1	24	. = 33.25
	2	30	. = 33.5
1958	3	29	.
	4	51	.
	.		.
	.		
	.		

varies from year to year. The advantage of dummy variables is that both seasonal shifts *and* the relation of S to T are estimated simultaneously in the same regression. (A moving average adjustment is only the first stage in a two-step process; only after it is completed can S' be regressed on T.) Another advantage is that the dummy coefficients (β_2, β_3, and β_4) give an index of the average seasonal shift, and tests of significance on them can easily be undertaken using standard procedures.

PROBLEMS

13-5 Referring to the jewelry sales in Figure 13-9, predict the sales S for the next quarter ($T = 17$, the first quarter of 1961)
(a) Using the simple regression of S on T alone;
(b) Using the multiple regression of S on T, including seasonal adjustment. Is this any better than (a)?

13-6 Referring to the two years of jewelry sales in Table 13-5,
(a) Compute the simple regression of S' (adjusted) on T;
(b) Compute the simple regression of S (unadjusted) on T;
(c) Of the 2 slopes in (a) and (b),
(1) Which do you think better shows the time trend of sales?
(2) Which agrees more closely with the slope $b_1 = .075$ estimated by using seasonal dummies?

13-7 Referring to the jewelry sales in Table 13-4, consider the eight quarters from the 4th to the 11th quarter. Supposing this were the only data available:
(a) Fit a simple regression line of S on T, without quarterly adjustment;
(b) Is your slope estimate (time trend) unbiased? Why?

13-8 Referring to Figures 13-6 and 13-8*a*, suppose the last 4 years are missing. If a simple regression of B on Y is calculated (ignoring W), will the bias of the slope be less or greater than before (when all the years were used)? Why?

13-7 REGRESSION, ANALYSIS OF VARIANCE, AND ANALYSIS OF COVARIANCE

(a) Regression with Dummies Equivalent to Analysis of Variance or Analysis of Covariance

If all the independent variables are categorical (dummy) variables, then regression analysis is essentially the familiar analysis of variance (ANOVA). This can be proved in general; but it is more instructive to illustrate it in the simplest case of one independent dummy variable. In Problem 10-3 we applied analysis of variance to the problem of whether the income (Y) of men and women differs. Dummy regression could alternatively have been used, with a model of the form:

$$Y = \alpha + \beta G + e$$

where

$$G = 0 \text{ for men}$$

$$= 1 \text{ for women.} \qquad (13\text{-}25)$$

The data is analyzed in Table 13-6. We find the same value ($b = -8$) for the difference in groups that we found in Problem 10-3 ($\bar{Y}_1 - \bar{Y}_2 = -8$). Note also that in both tests the residual variance (48) is the same; so is the standard error of estimate ($\sqrt{48}\sqrt{1/2}$). Hence the two procedures are seen to be identical.

We referred to our earlier example of explaining bond sales, as a regression on a numerical variable (income) and a dummy variable (wartime). This could alternatively be described as a combination of standard regression analysis and analysis of variance. Technically, this combination is referred to as analysis of covariance (ANOCOVA), although this term is often reserved for cases in which the effect of the dummy variable (wartime) is of prime interest and the other variable (income) is explicitly introduced only to remove its noise effects (i.e., to prevent the sort of error shown in Figure 13-8*b*).

Another application of the analysis of covariance might be a study of the effects of racial discrimination on income; here the major concern would be the effect on income of the dummy variable (negro versus white), with a simultaneous regression on other numerical variables (years of experience, education, etc.) simply a means of keeping these other influences from biasing the result.

(b) Summary

Multiple regression is an extremely useful tool with many broad applications. We define three special cases, distinguished by the nature of the independent variables:

1. "*Standard regression*" is regression on only numerical variables.
2. *ANOVA* is equivalent to regression on only categorical (dummy) variables.
3. *ANOCOVA* (analysis of covariance) is regression on both categorical and numerical variables.

TABLE 13-6 Regression Using Only a Categorical Variable, Being then Equivalent to the Analysis of Variance

Y	G	$g = G - \bar{G}$	Yg	g^2	$\hat{Y} = 56 - 8g$	$Y - \hat{Y}$	$(Y - \hat{Y})^2$
60	0	−1/2	−30	1/4	60	0	0
70	0	−1/2	−35	1/4	60	10	100
62	0	−1/2	−31	1/4	60	2	4
48	0	−1/2	−24	1/4	60	−12	144
48	1	1/2	24	1/4	52	−4	16
56	1	1/2	28	1/4	52	4	16
50	1	1/2	25	1/4	52	−2	4
54	1	1/2	27	1/4	52	2	4
$\bar{Y} = 56 = a$	$\bar{G} = 4/8 = 1/2$		$\sum Yg = -16$	$\sum g^2 = 2$			$\sum (Y - \hat{Y})^2 = 288$

$$b = \frac{\sum Yg}{\sum g^2} = -16/2 = -8$$

$$s^2 = \tfrac{1}{6} \sum (Y - \hat{Y})^2 = 48$$

(This is residual variance, also appearing in solution table in Problem 10-3)

$$t = \frac{b}{s/\sqrt{\sum g^2}} = \frac{-8}{\sqrt{48}/\sqrt{2}} = -1.63 \quad (13\text{-}26)$$

which is less extreme than the critical t value of -2.45, therefore not statistically significant.

These three techniques are compared using the hypothetical data of Figures 13-10 to 13-13, which show the possible ways that mortality may be analyzed.

The hypothetical data in Figure 13-10 shows a sample of observations of the mortality of American men. Applying standard regression, we would reject the hypothesis that the true slope $\beta = 0$; thus we conclude that age does affect the mortality rate. In the process we derive a useful estimate b, of *how* age affects mortality.

If the data is collected into three groups, we come up with the scatter shown in Figure 13-11. Note that this is exactly the same set of mortality (Y) observations as in Figure 13-10. The only difference is that we are no longer as specific about the age (X) variable. Now ANOVA can be applied[14] to this data to test whether the population means of these three scatters are equal. Once again, the conclusion is that age affects mortality. However, ANOVA does not tell us *how* age affects mortality, unless we extend it to multiple comparisons. Moreover, multiple comparisons will yield a whole complicated table, whereas standard regression provides a single descriptive number (b) showing how age affects mortality.

So long as X is numerical, as in Figures 13-10 and 13-11, we conclude that standard regression is generally the preferred technique. But when X is categorical, it cannot be applied. For example, in Figure 13-12 we graph how mortality depends on nationality;[15] our X variable ranges over various categories (American, British, etc.) and there is no natural way of placing these on a numerical scale—or even ordering them. Hence standard regression is out of the question,[16] and ANOVA must be used.

If mortality is dependent on income as well as nationality, the analysis of covariance shown in Figure 13-13 is appropriate. This uses nationality dummies, with the numerical variable income explicitly introduced to eliminate the error that it might otherwise cause. We confirm that this has greatly improved our analysis. Whereas it appeared in Figure 13-12 that a national characteristic of the British was a lower mortality rate than the Chinese, we see in Figure 13-13 that it is not so simple. The height of the fitted planes for China and the United Kingdom are practically the same. The lower U.K. mortality rate is explained solely by higher income.

---

[14] Standard regression could also be applied, with a line fitted to the scatter in Figure 13-11. However, if this technique is to be applied, it is more efficient to use the ungrouped data of Figure 13-10.

[15] In Figures 13-12 and 13-13, all samples are assumed drawn from a single age group; we consider only other factors influencing mortality.

[16] To confirm, the student will note that a standard linear regression line fitted to the scatter in Figure 13-12 will yield $b \approx 0$ and the conclusion that nationality does not matter. Yet if China is graphed last, rather than first, $b \not\approx 0$ and it would be concluded that nationality does matter. Thus, the conclusion depends on the arbitrary ordering of our nationality variable.

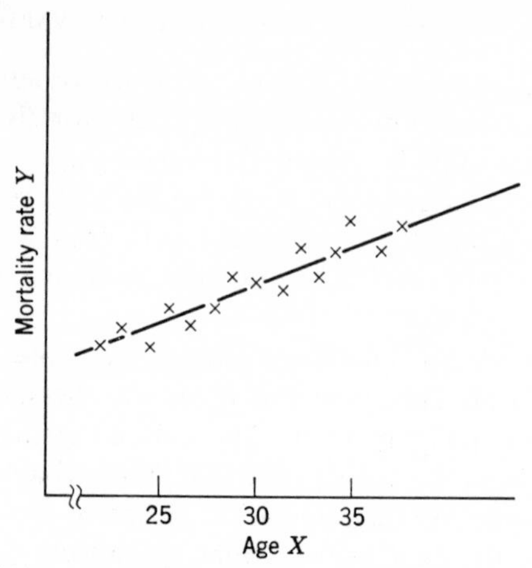

FIG. 13-10 "Standard regression," since X is numerical.

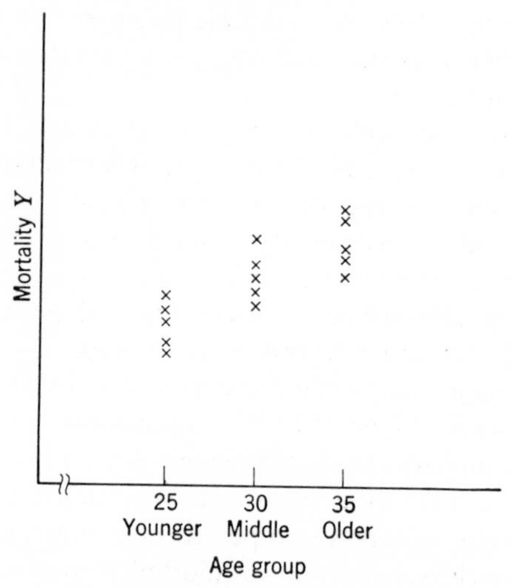

FIG. 13-11 X is grouped into classifications, and ANOVA may be used.

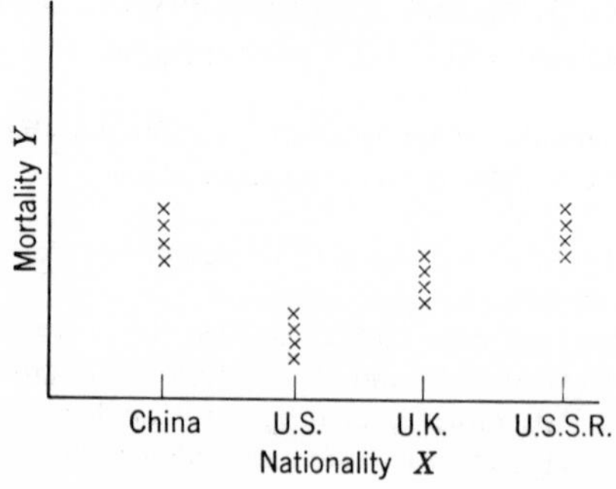

FIG. 13-12 X is categorical, and ANOVA must be used.

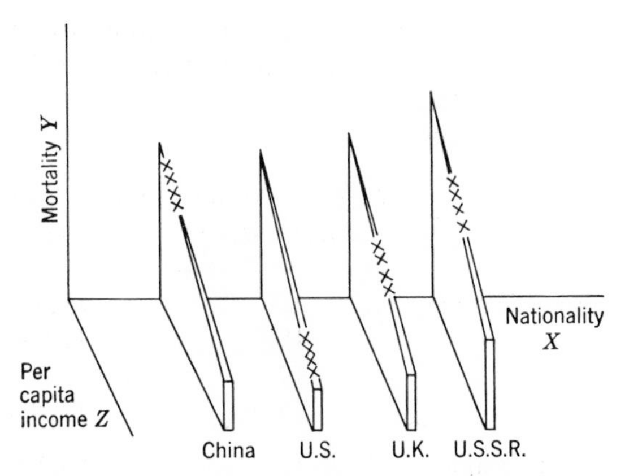

FIG. 13-13 Analysis of covariance for a categorical variable (nationality) and numerical variable (income).

In summary, standard regression is the more powerful tool whenever the independent variable X is numerical and the dependence of Y on X can be described by a simple function. Analysis of variance is appropriate if the independent variable is a set of unordered categories.

PROBLEMS

13-9 Construct a confidence interval for β using the data in Table 13-6. Compare this with the answer to Problem 10-3*b*.

*13-10 Using the data in Problem 10-2, estimate the regression of yield on fertilizer type, using two dummies. Compare with your answer to Problem 10-2.

13-11 The following is the result of a test of gas consumption on a sample of 6 cars

	Miles Per Gallon	Engine Horsepower
Make A	21	210
	18	240
	15	310
Make B	20	220
	18	260
	15	320

(a) Determine the difference in the performance (miles per gallon) of the two makes, allowing for horsepower differences.
(b) Graph your results as in Figure 13-13.

(13-12) (a) Based on the following sample information, use the analysis of covariance to describe how education is related to father's income and place of residence.

(b) Graph your results.

	Years of Formal Education (E)	Father's Income (F)
Urban Sample	15	\$8,000
	18	11,000
	12	9,000
	16	12,000
Rural Sample	13	\$5,000
	10	3,000
	11	6,000
	14	10,000

chapter 14

Correlation

14-1 SIMPLE CORRELATION

Simple regression analysis showed us *how* variables are linearly related; correlation analysis will show us the *degree* to which variables are linearly related. In regression analysis, a whole mathematical function is estimated (the regression equation); but correlation analysis yields only one number—an index designed to give an immediate picture of how closely two variables move together. In correlation analysis, we need not worry about cause and effect relations. Correlation between X and Y can be estimated regardless of whether: (a) X affects Y, or vice versa; (b) both affect each other; or (c) neither directly affects the other, but they move together because some third variable influences both. Although correlation is a less powerful technique than regression, the two are so closely related mathematically that correlation often becomes a useful aid in regression analysis.

(a) The Population Correlation Coefficient ρ (rho)

In equation (5-22) we have already defined a useful index of how two random variables move together: σ_{XY}, the covariance of X and Y. The variables used there were deviations from the mean:

$$\begin{aligned} &X - \mu_X \\ &Y - \mu_Y \end{aligned} \qquad (14\text{-}1)$$

It will be useful to express these deviations in terms of fully standardized units; i.e., define the new variables:

$$\frac{X - \mu_X}{\sigma_X} \qquad \frac{Y - \mu_Y}{\sigma_Y} \tag{14-2}$$

Correlation ρ_{XY} is similar to covariance σ_{XY} in (5-22), the only difference being that the variables in (14-2) replace those in (14-1). Thus

Population correlation

$$\rho_{XY} \triangleq E\left(\frac{X - \mu_X}{\sigma_X}\right)\left(\frac{Y - u_Y}{\sigma_Y}\right) \tag{14-3}$$

This will be interpreted more fully in Section 14-1(c) below; for now we turn our attention to r, the sample correlation coefficient used to estimate this (generally) unknown population ρ.

(b) The Sample Correlation Coefficient r

By analogy with (14-3)

Sample correlation

$$r_{XY} \triangleq \frac{1}{n-1}\sum_{i=1}^{n}\left(\frac{X_i - \bar{X}}{s_X}\right)\left(\frac{Y_i - \bar{Y}}{s_Y}\right) \tag{14-4}$$

Now consider an intuitive development of this index; (because of the similarity of the two concepts, some of this interpretation will closely parallel the development of covariance in Chapter 5-3). As our example, we use the marks on a verbal (Y) and mathematical (X) test scored by a sample of eight college students. Each student's performance is represented by a dot on the scatter shown in Figure 14-1a; this information is set out in the first two columns of Table 14-1.

Since we are after a single measure of how closely these variables are related, our index should be independent of our choice of origin. So we shift *both* axes in Figure 14-1b, with both x and y now defined as deviations from the mean; i.e.,

$$x = X - \bar{X} \qquad \text{and} \qquad y = Y - \bar{Y} \tag{14-5}$$

Values of the translated variables are shown in columns 3 and 4 of Table 14-1.

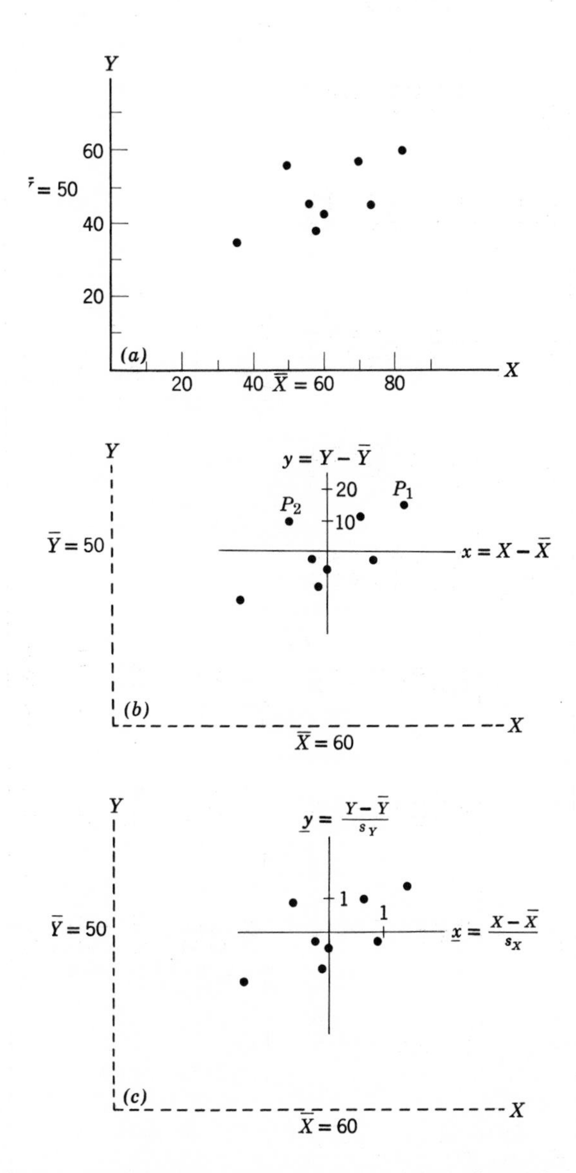

FIG. 14-1 Scatter of math and verbal scores. (*a*) Original observations. (*b*) Shift axes. (*c*) Change scale of axes to standard units.

TABLE 14-1 Math Score (X) and Corresponding Verbal Score (Y) of a Sample of Eight Students Entering College.

(1)	(2)	(3)	(4)	(5)	(6)	(7)	(8)	(9)	(10)	(11)	(12)	(13)
X	Y	$x = X - \bar{X}$	$y = Y - \bar{Y}$	xy	x^2	y^2	$\hat{Y} = 50 + bx$	$Y - \hat{Y}$	$(Y - \hat{Y})^2$	$\hat{X} = 60 + b_* y$	$X - \hat{X}$	$(X - \hat{X})^2$
36	35	−24	−15	360	576	225	37.96	−2.96	8.76	48.27	−12.27	150.55
80	65	20	15	300	400	225	60.03	4.97	24.70	71.73	8.27	68.39
50	60	−10	10	−100	100	100	44.99	15.01	225.30	67.82	−17.82	317.55
58	39	−2	−11	22	4	121	49.00	−10.00	100.00	51.39	6.61	43.69
72	48	12	−2	−24	144	4	56.02	−8.02	64.32	58.44	13.56	183.87
60	44	0	−6	0	0	36	50.00	−6.00	36.00	55.31	4.69	22.00
56	48	−4	−2	8	16	4	47.99	.01	.0001	58.44	−2.44	5.95
68	61	8	11	88	64	121	54.01	6.99	48.86	68.61	−.61	.37
$\Sigma X = 480$	$\Sigma Y = 400$	$\Sigma x = 0$	$\Sigma y = 0$	$\Sigma xy = 654$	$\Sigma x^2 = 1304$	$\Sigma y^2 = 836$			$\Sigma (Y - \hat{Y})^2 = 508$			$\Sigma (X - \hat{X})^2 = 792.37$
$\bar{X} = 60$	$\bar{Y} = 50$			$b = \frac{\Sigma xy}{\Sigma x^2}$	$s_X^2 = \frac{\Sigma x^2}{n - 1}$	$s_Y^2 = \frac{\Sigma y^2}{n - 1}$			$s^2 = \frac{\Sigma (Y - \hat{Y})^2}{n - 2}$			$s_*^2 = \frac{\Sigma (X - \hat{X})^2}{n - 2}$
				$= .5015$	$= \frac{1304}{7}$	$= \frac{836}{7}$			$= \frac{508}{6}$			$= \frac{792}{6}$
				$b_* = \frac{\Sigma xy}{\Sigma y^2}$	$= 186.3$	$= 119.4$			$= 84.7$			$= 132$
				$= .782$	$s_X = 13.65$	$s_Y = 10.93$			$s = 9.20$			$s_* = 11.5$

Suppose we multiply the x and y coordinate values for each student, and sum them all. This ($\sum xy$) gives us a good measure of how math and verbal results tend to move together. Whenever an observation such as P_1 falls in the first quadrant in Figure 14-1*b*, both its x and y coordinates will be positive, and their product xy positive. This also holds true for any observation in the third quadrant, with both coordinates negative. The product is negative only for observations such as P_2 in the second or fourth quadrant, (one coordinate positive, the other negative). If X and Y move together, most observations will fall in the first and third quadrants; consequently most products xy will be positive, as will their sum—a reflection of the positive relationship between X and Y. But if X and Y are negatively related, (i.e., when one rises the other falls), the original scatter will run downhill rather than uphill; most observations will fall in the second and fourth quadrants, yielding a negative value for our $\sum xy$ index. We conclude that as an index of correlation, $\sum xy$ at least carries the right sign. Moreover, when there is no relationship between X and Y, and our observations are distributed evenly over the four quadrants, positive and negative terms will cancel, and this index will be zero.

There are just two ways that $\sum xy$ can be improved. First it depends on the units in which x and y are measured. (Suppose the math test had been marked out of 50 instead of 100; x values and our $\sum xy$ index would be only half as large—even though the degree to which verbal and mathematical performance is related would not have changed.) This difficulty is avoided by measuring both x and y in terms of standard units; i.e., both x and y are divided by their observed standard deviations.

$$\underline{x}_i = \frac{X_i - \bar{X}}{s_X}$$
$$\underline{y}_i = \frac{Y_i - \bar{Y}}{s_Y} \tag{14-6}$$

where, of course,

$$s_X^2 = \frac{1}{n-1} \sum (X_i - \bar{X})^2$$

and

$$s_Y^2 = \frac{1}{n-1} \sum (Y_i - \bar{Y})^2 \tag{14-7}$$

This step is shown in Figure 14-1*c*.

Our new index $\sum \underline{x}_i \underline{y}_i$ has only one remaining flaw: it is dependent on sample size. (Suppose we observed exactly the same sort of scatter from a sample of double the size; our index would also double, even though the

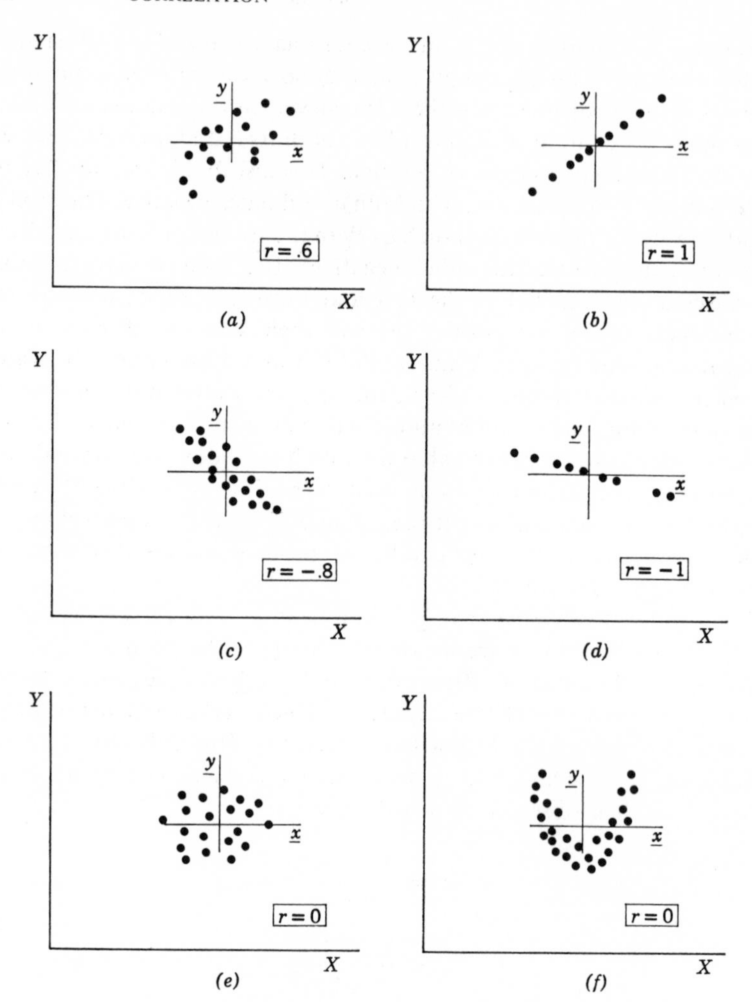

FIG. 14-2 Scatter diagrams and their associated correlation coefficients.

picture of how these variables move together is the same.) To avoid this problem we divide by the sample size n—or rather $n - 1$, the divisor in (14-7). This yields the sample correlation coefficient:

$$r = \frac{1}{n-1} \sum \underline{x}_i \underline{y}_i \tag{14-8}$$

which is recognized to be our definition in (14-4). r may be expressed in terms of the original observations (X_i, Y_i), by substituting s_X and s_Y in

(14-7) into (14-4) and cancelling $(n - 1)$:

$$r = \frac{\sum (X_i - \bar{X})(Y_i - \bar{Y})}{\sqrt{\sum (X_i - \bar{X})^2 \sum (Y_i - \bar{Y})^2}} \tag{14-9}$$

Example

The data in Table 14-1 are applied to (14-9) to calculate the correlation coefficient between the math and verbal scores of our sample of eight students.

$$r = \frac{654}{\sqrt{(1304)(836)}} = .62 \tag{14-10}$$

Some idea of how r behaves is given in Figure 14-2; especially note diagram *b*. When there is a perfect linear association, the product of the coordinates in every case is positive; thus, their sum (and the resulting coefficient of correlation) is as large as possible. The same argument holds for the perfect *inverse* relation of Y and X shown in diagram *d*. This suggests that r has an upper limit of $+1$ and a lower limit of -1. (This is proved in Section (f) below.)

Finally compare diagrams *e* and *f*. Our calculation of r in either case is zero, because positive products of the coordinates are offset by negative ones. Yet when we examine the two scatters, no relation between X and Y is confirmed in *e*—but a strong relation is evident in *f*; in this case a knowledge of X will tell us a great deal about Y. A zero value for r therefore does not imply "no relation"; rather, it means "no linear relation." Thus correlation is a measure of *linear relation* only; it is of no use in describing nonlinear relations. This brings us to the next critical question: "In calculating r, what can we infer about the underlying population ρ?"

(c) Inference from r to ρ

Before we can draw any statistical inference about ρ from our sample statistic r, we must clarify our assumptions about the parent population from which our sample was drawn. In our example, this would be the math and verbal marks scored by *all* college entrants.

This population might appear as in Figure 14-3, except that there would, of course, be many more dots in this scatter, each representing another student. If we subdivide both X and Y into class intervals, the area in our diagram will be divided up in a checkerboard pattern. From the relative

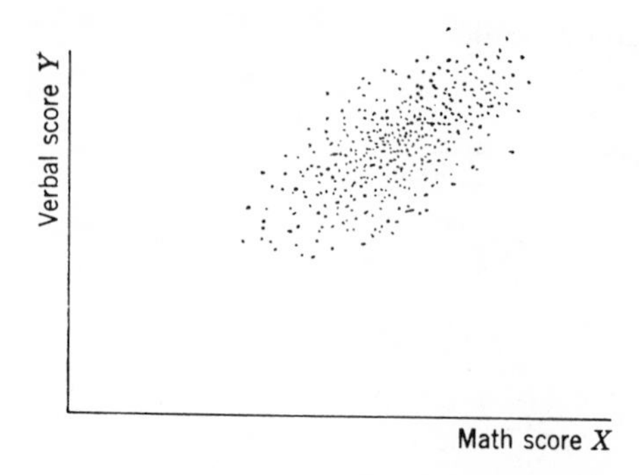

FIG. 14-3 Bivariate population scattergram (math and verbal scores).

frequency (sampling probability) in each of the squares, the histogram in Figure 14-4 is constructed.[1] The histogram would have approximately the shape of the probability density in Figure 14-5. To conclude: in examining a random student, neither his math score X nor his verbal score Y is predetermined; both are random variables. Compare this with our example in Chapter 11, where one variable (fertilizer) was predetermined.

This distribution in Figure 14-5 is called "bivariate normal." This means that the conditional distribution of X or of Y is always normal. Specifically, if we slice the surface at any value of Y, (say Y_0), the shape of the resulting cross section is normal. Similarly, if we select any X value (say X_0) and slice the surface in this other direction, the resulting cross section is also normal.

It is worthwhile pausing briefly to consider the alternative way that the bivariate normal population shown in three dimensions in Figure 14-5 can be graphed in two dimensions. Instead of slicing the surface vertically as we did in that diagram, slice it horizontally as in Figure 14-6. The resulting

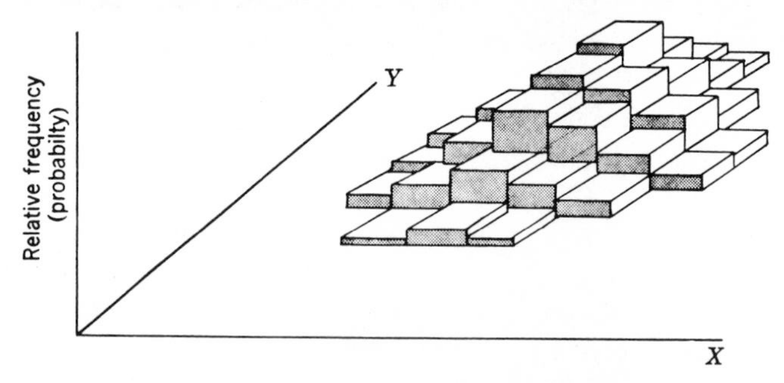

FIG. 14-4 Bivariate population histogram.

[1] Our example is of a finite population, but a similar argument would apply for an infinite population. Moreover, instead of using heights for probabilities, we could use dots of different sizes; see Figure 5-4*a*.

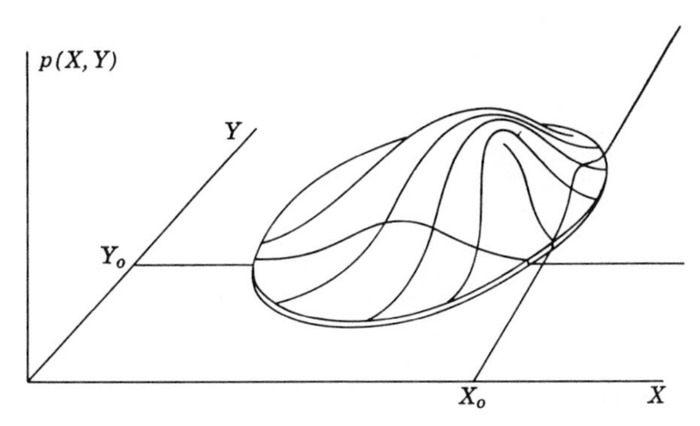

FIG. 14-5 Bivariate normal distribution.

cross section is an ellipse, representing all X, Y combinations with the same probability density. This "isoprobability" curve is marked "c" in the two dimensional X, Y space in Figure 14-7; isoprobability ellipses defined when this surface is sliced horizontally at higher and lower levels are also shown. (Once again, many social scientists will recognize this as the familiar strategy of forcing a three-dimensional function into a two-dimensional space by showing one variable as a set of isoquants, isobars, or whatever.) It will also be useful in Figure 14-7 to mark the major axis (d) common to all these isoprobability ellipses. If the bivariate normal distribution concentrates about its major axis, ρ increases. Several examples of populations, and their associated correlation coefficients ρ are shown in Figure 14-8.

Provided that the parent population is bivariate normal, inferences about the population ρ can easily be made from a sample correlation r. Recall the inferences about π from P in Chapter 8. Using the same reasoning that established Figure 8-4, Figure 14-9 is constructed. Thus from any sample r, a 95% confidence interval for the population ρ can be found. For example, if a sample of 25 students has $r = .80$, the 95% confidence interval

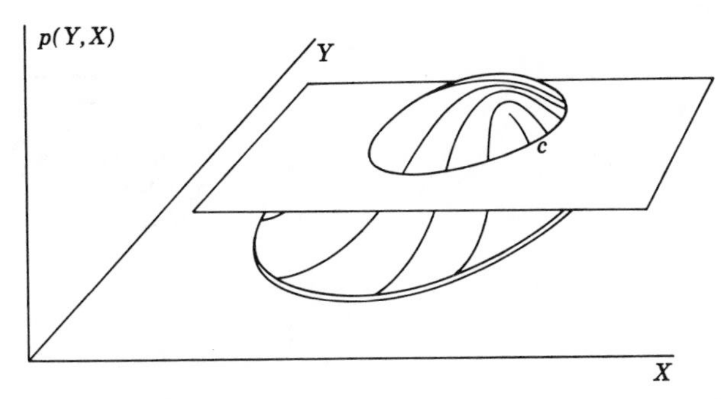

FIG. 14-6 An isoprobability ellipse from a bivariate normal surface.

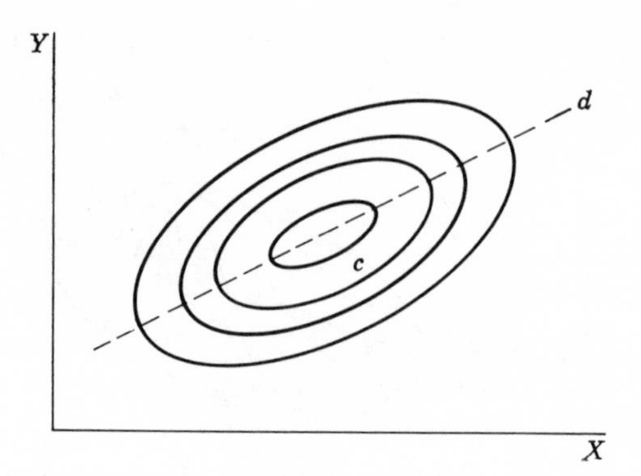

FIG. 14-7 The bivariate normal distribution shown as a set of isoprobability ellipses.

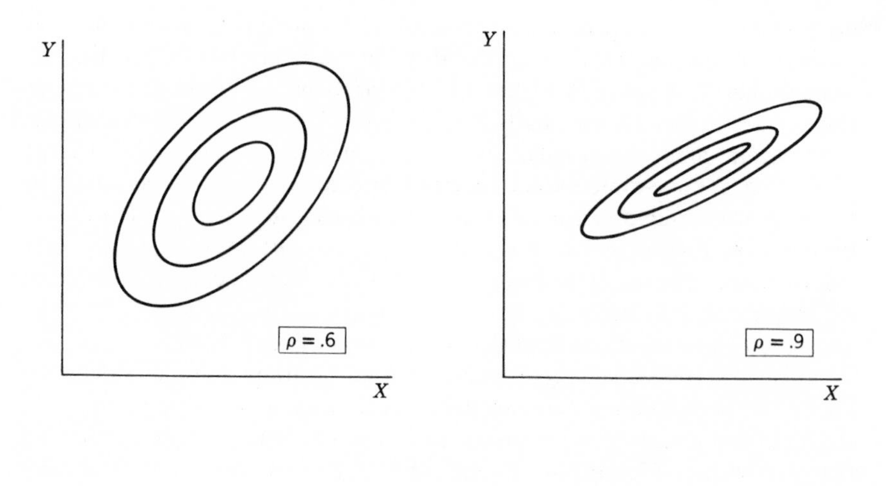

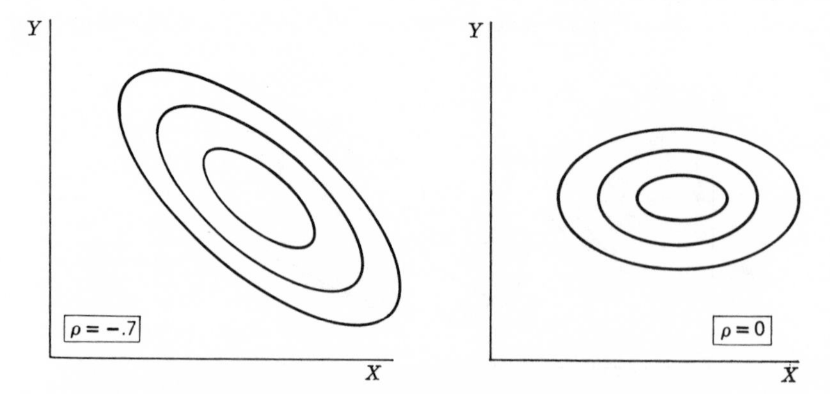

FIG. 14-8 Examples of population correlations.

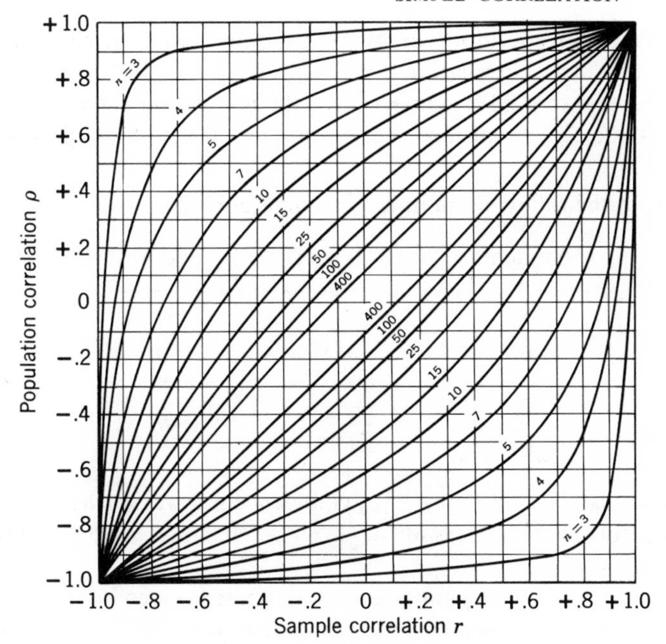

FIG. 14-9 95% confidence bands for correlation ρ in a bivariate normal population, for various sample sizes n. (This chart is reproduced with the permission of Professor E. S. Pearson from F. N. David, *Tables of the Ordinates and Probability Integral of the Distribution of the Correlation Coefficient in Small Samples*, Cambridge University Press, 1938.)

for ρ is read vertically as

$$.58 < \rho < .90 \tag{14-11}$$

Because of space limitations, we shall concentrate in the balance of this chapter on sample correlations, and ignore the corresponding population correlations. But each time a sample correlation is introduced, it should be recognized that an equivalent population correlation is defined similarly, and inferences may be made about it from the sample correlation.

PROBLEMS

14-1

Son's Height (inches)	Father's Height (inches)
68	64
66	66
72	71
73	70
66	69

From the above random sample of 5 son and father heights, find
(a) The sample correlation r;
(b) The 95% confidence interval for the population correlation ρ;
(c) At the 5% significance level, can you reject the hypothesis that $\rho = 0$?

⇒ 14-2 From the following sample of student grades,

Student	First Test X	Second Test Y
A	80	90
B	60	70
C	40	40
D	30	40
E	40	60

(a) Calculate r; and find a 95% confidence interval for ρ;
(b) Calculate the regression of Y on X, and find a 95% confidence interval for β;
(c) Graph the 5 data points and the estimated regression line;
(d) At the 5% significance level, can you reject
(1) The null hypothesis $\rho = 0$?
(2) The null hypothesis $\beta = 0$?

(d) Correlation and Regression

If regression and correlation analysis were both applied to the same scatter of math (X) and verbal (Y) scores, how would they be related? Specifically, consider the relation between the estimated correlation r, and the estimated regression slope b. In Problem 11-4(b) it was confirmed that

$$b = \frac{\sum xy}{\sum x^2} \tag{14-12}$$

and from (14-9) noting that both x and y are defined as deviations

$$r = \frac{\sum xy}{\sqrt{\sum x^2}\sqrt{\sum y^2}} \tag{14-13}$$

When (14-12) is divided by (14-13)

$$\frac{b}{r} = \frac{\sqrt{\sum x^2}\sqrt{\sum y^2}}{\sum x^2} = \sqrt{\frac{\sum y^2}{\sum x^2}} \tag{14-14}$$

If we divide both the numerator and denominator inside the square root sign by $n - 1$

$$\frac{b}{r} = \sqrt{\frac{\sum y^2/(n-1)}{\sum x^2/(n-1)}} = \frac{s_Y}{s_X} \tag{14-15}$$

or

$$b = r\frac{s_Y}{s_X} \tag{14-16}$$

This close correspondence of b and r will play an important role in the argument later. Note that if either r or b is zero, the other will also be zero.

(e) Explained and Unexplained Variation

In Figure 14-10 we reproduce our sample of math (X) and verbal (Y) scores, along with the fitted regression of Y on X, calculated in a straightforward way from the information set out in Table 14-1. Now, if we wished to predict a student's verbal score (Y) without knowing X, then the best prediction would be the average observed value ($\bar{Y}$). At x_i, it is clear from this diagram that we would make a very large error—namely ($Y_i - \bar{Y}$), the deviation in Y_i from its mean. However, once our regression equation has been calculated, we predict Y to be $\hat{Y}_i$. Note how this reduces our error, since ($\hat{Y}_i - \bar{Y}$)—the large part of our deviation—is now "explained." This leaves only a relatively small "unexplained" deviation $Y_i - \hat{Y}_i$. The total

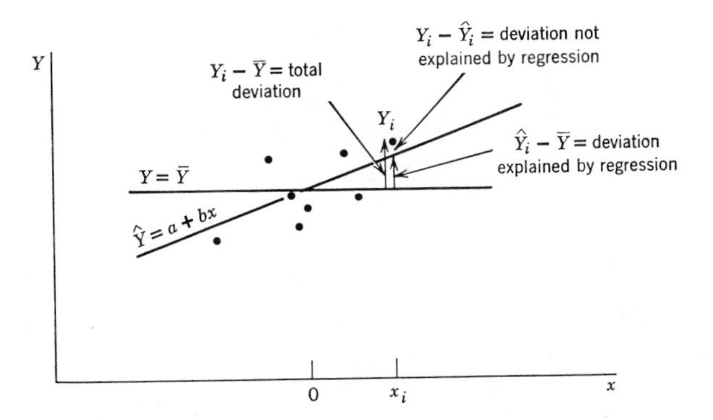

FIG. 14-10 The value of regression in reducing variation in Y.

deviation of Y is the sum:

$$(Y_i - \bar{Y}) = (\hat{Y}_i - \bar{Y}) + (Y_i - \hat{Y}_i), \text{ for any } i \qquad (14\text{-}17)$$

total deviation = explained deviation + unexplained deviation

It follows that

$$\sum (Y_i - \bar{Y}) = \sum (\hat{Y}_i - \bar{Y}) + \sum (Y_i - \hat{Y}_i) \qquad (14\text{-}18)$$

What is surprising is that this same equality holds when these deviations are squared, i.e.

$$\boxed{\sum (Y_i - \bar{Y})^2 = \sum (\hat{Y}_i - \bar{Y})^2 + \sum (Y_i - \hat{Y}_i)^2} \qquad (14\text{-}19)$$

or, total variation = explained variation + unexplained variation

where variation is defined as the sum of squared deviations. Recall a very similar conclusion proved in analysis of variance (10-16); (14-19) can be established in much the same way.[2]

Since we may write, according to Problem 11-4(a),

$$(\hat{Y}_i - \bar{Y}) = \hat{y}_i = bx_i \qquad (14\text{-}21)$$

it is often convenient to rewrite (14-19) as

$$\sum (Y_i - \bar{Y})^2 = b^2 \sum x_i^2 + \sum (Y_i - \hat{Y}_i)^2 \qquad (14\text{-}22)$$

total variation = variation explained by X + unexplained variation

This equation makes explicit the fact that explained variation is that accounted for by the estimated regression coefficient b. This procedure of decomposing total variation and analyzing its components is called "analysis of variance applied to regression." The components of variance are displayed in the ANOVA Table 14-2 similar to Table 10-6, (bearing in mind[3] that we

[2] For proof, square both sides of (14-17), and sum over all values of i:

$$\sum (Y_i - \bar{Y})^2 = \sum [(\hat{Y}_i - \bar{Y}) + (Y_i - \hat{Y}_i)]^2$$
$$= \sum (\hat{Y}_i - \bar{Y})^2 + \sum (Y_i - \hat{Y}_i)^2 + 2\sum (\hat{Y}_i - \bar{Y})(Y_i - \hat{Y}_i) \qquad (14\text{-}20)$$

The last term can be rewritten using (14-21):

$$2b \sum x_i(Y_i - \hat{Y}_i)$$

But this sum vanishes: in fact it was set equal to zero in the normal equation (11-15) used to estimate our regression line. Thus the last term in (14-20) disappears, and (14-19) is proved. This same theorem can similarly be proved in the general case of multiple regression.

A further justification of the least squares technique (not mentioned in Chapter 11) is that it results in this useful relation between explained, unexplained, and total variation.

[3] And also noting that our terminology for degrees of freedom has changed, e.g., the total number of sample observations is now designated simply as n, rather than nr.

TABLE 14-2 ANOVA Table for Linear Regression

(a) General

Source of Variation	Variation	Degrees of Freedom (d.f.)	Variance
Explained (by regression)	$\sum (\hat{Y}_i - \bar{Y})^2$ or $b^2 \sum x_i^2$	1	$\frac{b^2 \sum x_i^2}{1}$
Unexplained (residual)	$\sum (Y_i - \hat{Y}_i)^2$	$n - 2$	$s^2 = \frac{\sum (Y_i - \hat{Y}_i)^2}{n - 2}$
Total	$\sum (Y_i - \bar{Y})^2$	$n - 1$	

(b) For Sample of Verbal and Math Scores (Table 14-1)

Sources of Variation	Variation	Degrees of Freedom (d.f.)	Variance	F
Explained (by regression)	328	1	328	3.87
Unexplained (residual)	508	6	84.7	
Total	836 √	7 √		

are now explaining Y, rather than X). From this, a null hypothesis test on β may be constructed; as before, the question is whether the ratio of the explained variance to unexplained variance is sufficiently greater than 1 to reject the hypothesis that Y is unrelated to X. Specifically, a test of the hypothesis

$$H_0: \beta = 0, \tag{14-23}$$

involves forming the ratio

$$F = \frac{\text{variance explained by regression}}{\text{unexplained variance}}$$

$$= \frac{b^2 \sum x_i^2}{s^2} \tag{14-24}$$

A 5% significance test involves finding the critical F value which leaves 5% of the distribution in the right-hand tail. If the sample F value calculated from (14-24) exceeds this, reject the hypothesis.

We must emphasize that this is just an alternate way of testing the null hypothesis (14-23). The first method—using the t distribution to find the confidence interval for β (as in Section 12-7)—is usually preferable.

Note that the F and t distributions are related, in general, by

$$F = t^2$$

where there is one degree of freedom in the numerator of F. Since the F calculated in (14-24) is just the t^2 of (12-36), the ANOVA F-test of this section is justified.

Example

In Table 14-2(b) the ANOVA calculations are presented for our verbal and math score example. (The necessary computational details are shown on the bottom of Table 14-1.) To test $\beta = 0$, (14-24) is evaluated to be:

$$F = \frac{328}{84.7} = 3.87 \qquad (14\text{-}25)$$

Since this falls short of 5.99, the critical 5% point of F, we do not reject the null hypothesis.

The same test of $\beta = 0$ could be equivalently done using (12-36):

$$t = \frac{b}{s/\sqrt{\sum x_i^2}} = \frac{.50}{9.2/\sqrt{1304}} = 1.97$$

Since this falls short of 2.45, (the critical value leaving a total of 5% in both tails of the t distribution), the null hypothesis is not rejected. Since $t^2 = F$, (both for the calculated and for the critical values), the same conclusion must follow from both tests.

Alternatively, a 95% confidence interval for β could be constructed from (12-30):

$$\beta = .50 \pm (2.45)\,.254$$
$$= .50 \pm .62$$

This includes the value $\beta = 0$, once more confirming that H_0 cannot be rejected. (Of course, this inconclusive result may be partly due to the smallness of the sample.)

(f) Interpretation of Correlation

These variations in Y are now related to r. It follows from (14-14) that

$$b = r\sqrt{\frac{\sum y_i^2}{\sum x_i^2}} \qquad (14\text{-}26)$$

Substituting this value for b in (14-22)

$$\sum (Y_i - \bar{Y})^2 = r^2 \sum y_i^2 + \sum (Y_i - \hat{Y}_i)^2 \tag{14-27}$$

Noting that $\sum y_i^2$ is by definition $\sum (Y_i - \bar{Y})^2$, the solution for r^2 is:

$$\frac{\sum (Y_i - \bar{Y})^2 - \sum (Y_i - \hat{Y}_i)^2}{\sum (Y_i - \bar{Y})^2} = r^2 \tag{14-28}$$

Finally, we can reexpress the numerator by noting (14-19). Thus

$$\boxed{r^2 = \frac{\sum (\hat{Y}_i - \bar{Y})^2}{\sum (Y_i - \bar{Y})^2} = \frac{\text{explained variation of } Y}{\text{total variation of } Y}} \tag{14-29}$$

This equation provides a clear intuitive interpretation of r^2. (Note that this is the *square* of the correlation coefficient r, often called the coefficient of determination.) *It is the proportion of the total variation in Y explained by fitting the regression.* Since the numerator cannot exceed the denominator, the maximum value of the right-hand side of (14-29) is 1. Since the maximum value of r^2 is 1, the limits on r are ± 1. These two limits were illustrated in Figure 14-2: in part (b), $r = 1$ and all observations lie on a straight line running uphill; in part (d), $r = -1$ and this perfect inverse correlation reflects the fact that all observations lie on a straight line running downhill. In either case, a regression fit will explain all the variation in Y.

When $r = 0$ (and $r^2 = 0$) the explained variation of Y is zero and a regression line explains nothing; i.e., the regression line will be parallel to the X-axis, with $b = 0$. Thus $r = 0$ and $b = 0$ are seen to be equivalent ways of formally stating "no observed linear relation between X and Y."

(g) Regression Analysis Applied to a Bivariate Normal Population

In Table 14-1 a regression was calculated for sample values assumed taken from a bivariate normal population. We now ask: "Is the b we calculated an estimator of a population β, or does β even exist? For a bivariate normal population, does there exist a true regression line of Y on X?" It will now be shown that the answer is yes.

Our assumed bivariate normal population is shown in Figure 14-11 as a set of isoprobability ellipses, with major axis d. Now consider the straight line $Y = \alpha + \beta X$, defined by joining points of vertical tangency such as P_1 and P_3. Each of these vertical tangents defines a cross section slice of Y which is normal. Concentrating on the slice through P_1Q_1, for example, we

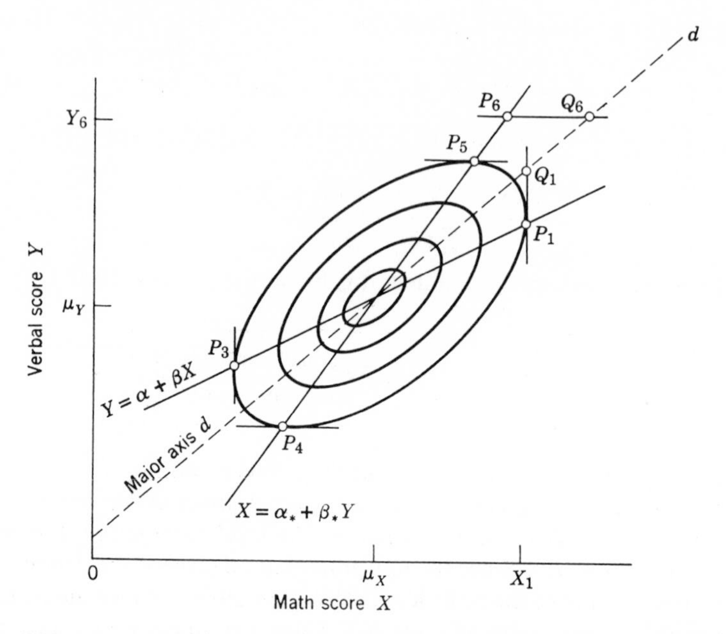

FIG. 14-11 Two regression lines found from the isoprobability ellipses.

see that the mean of these Y values occurs at the point of tangency P_1; at this point our vertical line touches its highest isoprobability ellipse, and the highest point on any normal curve is at the mean. Thus we see that the means of the Y populations lie on the straight line $Y = \alpha + \beta X$. Next, the variance of the Y populations can be shown to be constant.[4] Thus the assumptions of the *regression* model (12-2) are satisfied by a bivariate normal (*correlation*) population. The line $Y = \alpha + \beta X$ may therefore be regarded as a true linear regression of Y on X.

Thus if we know a student's math score and we wish to predict his verbal score, this regression line would be appropriate, (e.g., if his math score were X_1, we would predict his verbal score to be P_1). It is important to fully

[4] This may seem like a curious conclusion, since in Figure 14-5 the size of each cross section slice differs depending on the value of X_0. However each slice $p(X_0, Y)$ must be adjusted by division by $p(X_0)$ in order to define the conditional distribution of Y. Thus recalling the argument in Section 5-1(c), and in particular equation (5-10), the conditional distribution is

$$p(Y/X_0) = \frac{p(X_0, Y)}{p(X_0)}$$

In fact, this adjustment makes all the conditional distributions of Y "look alike," and thus have the same variance.

understand why we would *not* predict Q_1; i.e., we do *not* use the major axis of the ellipse (line d) for prediction, even though this represents "equivalent" performance on the two tests. Since this student is far above average in mathematics, an equivalent verbal score seems too optimistic a prediction. Recall that there is a large random element involved in performance. There are a lot of students who will do well in one exam, but poorly in the other; technically, ρ is less than 1 for this population. Therefore, instead of predicting at Q_1, we are more moderate and predict at P_1—a sort of average[5] of "equivalent" performance Q_1 and "average" performance μ_Y.

This is the origin of the term regression. Whatever a student's score in math, there will be a tendency for his verbal score to "regress" toward mediocrity (i.e., the average).[6] It is evident from Figure 14-11 that this is equally true for a student with a math score below average; in this case the predicted verbal score regresses upward toward the average.

Another interesting observation is that the correlation coefficient between X and Y is unique (i.e., ρ_{XY} is identically ρ_{YX}); but there are two regressions, the regression of Y on X *and* the regression of X on Y. This is immediately evident if we ask how we would predict a student's math score (X) if we knew his verbal score (e.g., Y_6).

Exactly the same argument holds. Equivalent performance (point Q_6 on line d) is a bad predictor; since he has done very well in the verbal test, we would expect him to do less well in math, although still better than average. Thus, the best prediction is P_6 on the line $X = \alpha_* + \beta_* Y$, the regression of X (math) on Y (verbal). This is the direct analogue to our regression of Y on X, but in this case our regression is defined by joining points (P_5, P_4, etc.) of *horizontal*, rather than vertical tangency. Each of these horizontal tangents defines a normal conditional distribution of X, given Y; each of these distributions has the same variance, with its mean lying on this regression line thus satisfying our conditions of a true regression of X on Y; hence least squares values a_* and b_* are used to estimate α_* and β_*.

---

[5] P_1 is in fact a weighted average of Q_1 and μ_Y, with weights depending on ρ. Thus in the limiting case in which $\rho = 1$, X and Y are perfectly correlated, and we would predict Y at Q_1. At the other limit, in which $\rho = 0$, we can learn nothing about likely performance on one test from the result of the other, and we would predict Y at μ_Y. But for all cases between these two limits, we predict using both Q_1 and μ_Y; and the greater the ρ, the more heavily Q_1 is weighted.

[6] A classical case, encountered by Pearson & Lee (Biometrika, 1903), involved trying to predict a son's height from his father's height. If the father is a giant the son is likely to be tall; but there are good reasons for expecting him to be shorter than his father. (For example, how tall was his mother? And his grandparents? An so on.) So the prediction for the son was derived by "regressing" his father's height towards the population average.

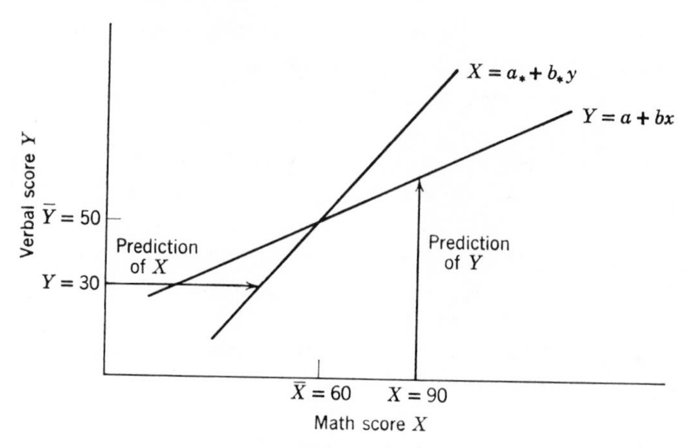

FIG. 14-12 Regressions estimated from a sample of verbal and math scores.

Example

Our sample of eight student's scores shown in Figure 14-1 and Table 14-1 was, by assumption, drawn from a bivariate normal population as shown in Figure 14-11. We have already estimated ρ with

$$r = .62 \quad \text{(14-10) repeated}$$

And from Table 14-1, we estimated $Y = \alpha + \beta X$ with

$$Y = 50 + .50x \quad (14\text{-}30)$$

$$= 20 + .50X \quad (14\text{-}31)$$

We now estimate $X = \alpha_* + \beta_* Y$. The coefficients in this simple regression of X on Y are calculated in Table 14-1; this involves using the estimating equations (11-13) and (11-16), taking care to interchange X and Y throughout.

Thus

$$X = 60 + .78y$$

$$= 60 + .78(Y - \bar{Y}) \quad (14\text{-}32)$$

$$= 21 + .78Y \quad (14\text{-}33)$$

The two estimated regressions (14-31) and (14-33) are shown in Figure 14-12. Thus, for example, the predicted verbal score of a student with a math result of 90 is 65; and the predicted math score of a student with a verbal result of 30 is 44.4.

(h) When Correlation, When Regression?

Both the standard regression and correlation models require that Y be a random variable. But the two models differ in the assumptions made about X. The regression model makes few assumptions about X, but the more restrictive correlation model of this chapter requires that X be a random variable, having with Y a bivariate normal distribution. We therefore conclude that the standard regression model has wider application. Regression may be used for example to describe the fertilizer-yield problem in Chapter 11 where X was fixed, or the bivariate normal population of X and Y in this chapter. However, the standard correlation model describes only the latter. (It is true that r^2 can be *calculated* even when X is fixed as an indication of how effectively regression reduces variation; but r cannot be used for inferences about ρ in Figure 14-9.)

In addition, regression answers more interesting questions. Like correlation, it indicates if two variables move together; but it also estimates how. Moreover, it can be shown that a key issue in correlation analysis—the test of the null hypothesis

$$H_0 : \rho = 0 \tag{14-34}$$

can be answered directly from regression analysis by testing the equivalent null hypothesis

$$H_0 : \beta = 0 \tag{14-35}$$

Thus rejection of $\beta = 0$ implies rejection of $\rho = 0$, and the conclusion that correlation does not exist between X and Y. If this is the only correlation question, then it can be answered by the regression test of (14-35), and there is no need to introduce correlation analysis at all.

Since regression answers a broader and more interesting set of questions, (and some correlation questions as well), it becomes the preferred technique; correlation is useful primarily as an aid to understanding regression, and as an auxiliary tool.

(i) "Nonsense" Correlations

In interpreting correlation, one must keep firmly in mind that absolutely no claim is made that this necessarily indicates cause and effect. For example, suppose that the correlation of teachers' salaries and the consumption of liquor over a period of years turns out to be .98. This would not prove that teachers drink; nor would it prove that liquor sales increase teachers' salaries. Instead, both variables moved together, because both are influenced by a third variable—long-run growth in national income. If only third factors of this kind could be kept constant—or their effects fully discounted—then correlation would become more meaningful. This is the objective of *partial correlation* in the next section.

Correlations such as the above are often called "nonsense" correlations. It would be more accurate to say that the observed mathematical correlation is real enough, but any naive inference of cause and effect is nonsense. Moreover, it should be recognized that the same charge can also be leveled at the conclusions sometimes drawn from regression analysis. For example, a regression applied to teachers' salaries and liquor sales would also yield a statistically significant b coefficient. Any inference of cause and effect from this would still be nonsense.

Although correlation and regression cannot be used as *proof* of cause and effect, these techniques are very useful in two ways. First, they may provide *further confirmation* of a relation that theory tells us should exist (e.g., prices depend on wages). Second, they are often helpful in *suggesting* causal relations that were not previously suspected. For example, when cigarette smoking was found to be highly correlated with lung cancer, possible links between the two were investigated further. This included more correlation studies in which third factors were more rigidly controlled, as well as extra-statistical studies such as experiments with animals, and chemical theories.

PROBLEMS

14-3 For the following random sample of 5 shoes, find

(a) The proportion of the variation in Y explained by regression on X.

(b) The proportion unexplained.

(c) Whether Y depends on X, at the 5% significance level. Answer this in three alternate ways—using the F test, t test, and a 95% confidence interval.

X = Cost of Shoe	Y = Months of Wear
10	8
15	10
10	6
20	12
20	9

14-4 Suppose a bivariate normal distribution of scores is perfectly symmetric, with $\rho = .50$ and with isoprobability ellipses as follows:

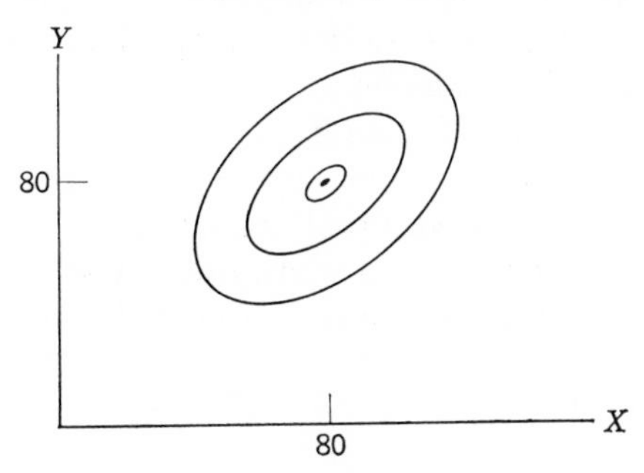

True or False? If false, correct it.

(a) The regression curve of Y on X is

$$Y = 80 + .5(X - 80)$$

(b) The regression line of Y on X has graph as follows:

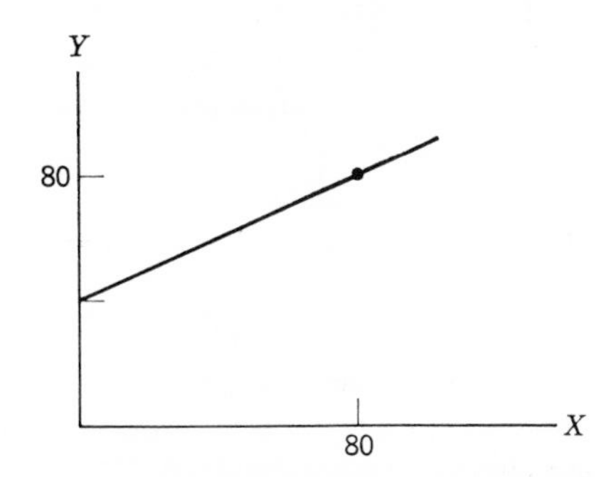

(c) The variance of Y is 1/4 the variance of X.

(d) The proportion of the Y variation explained by X is only 1/4.

(e) Thus the residual Y values (after fitting X) would have 3/4 the variance of the original Y values.

14-5 Let b and b_* be the sample regression slopes of Y on X, and X on Y, for any given scatter of points.
True or False? If false, correct it.

(a) $b = r \dfrac{s_Y}{s_X}$.

(b) $b_* = r \dfrac{s_X}{s_Y}$.

(c) $bb_* = r^2$.

(d) If $b > 1$, then $b_* < 1$ necessarily.

(e) If $b < 1$, then $b_* > 1$ necessarily.

14-6 In the following graph of 4 students' marks find geometrically (without doing any algebraic calculations):

(a) The regression line of Y on X.

(b) The regression line of X on Y.

(c) The correlation r (*Hint.* Problem 14-5c).

(d) The predicted Y-score of a student with X-score of 70.

(e) The predicted X-score of a student with Y-score of 70.

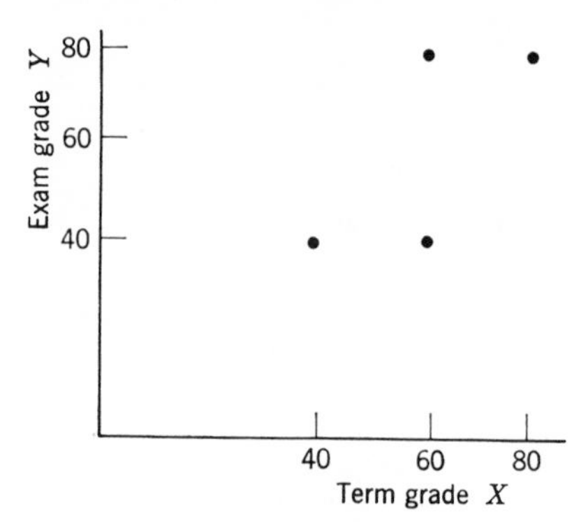

14-2 PARTIAL CORRELATION

As soon as we move from the simple two-variable case to relations which involve more than two variables, complications arise. To illustrate, consider a simple three-variable example: suppose that yield of hay (Y) depends on spring temperature (X) and rainfall (Z).

Following the techniques of Chapter 13 we could fit the following regression plane to a scatter of observations of Y, X, and Z:

$$Y = a + bX + cZ \tag{14-36}$$

Recall how we interpreted the multiple regression coefficient: b estimates how Y is related to X if Z were constant. The partial correlation coefficient $r_{XY.Z}$ is a similar concept. It estimates how X and Y move together if Z were held constant. (For convenience variables Y, X, and Z are often defined as variables 1, 2, and 3; thus $r_{YX.Z}$ becomes $r_{12.3}$, the partial correlation of the first two variables, when the third is assumed constant.)

While the previous sections of this chapter correspond to the simple regression analysis of Chapter 12, the partial correlation analysis in this section corresponds to the multiple regression analysis of Chapter 13. Thus we could embark here on a whole chapter on partial correlation, and a long one at that. However, since we have argued in the previous section that correlation is relatively less important, we confine ourselves to a brief intuitive introduction to this concept, and how it may be used.

The following assumptions are generally made about the parent population. The distribution of X, Y, and Z is multivariate normal. This implies that for any value of Z, the conditional distribution of Y and X is bivariate normal as shown in Figure 14-5. $\rho_{YX.Z}$ is defined as the simple correlation of this conditional joint distribution of X and Y.

In computing its estimator $r_{YX.Z}$ a problem arises. Since Z is a random variable, it is simply not possible to fix a single value Z_0 and sample the corresponding conditional distribution of X and Y. Thus, unless the sample is extremely large, it is unlikely that more than a single Y, X, Z_0 combination involving Z_0 will be observed. The alternative is to compute $r_{YX.Z}$ as the correlation of Y and X after the influence of Z has been removed from each.[7]

The resulting partial correlation $r_{XY.Z}$ can, after considerable manipulation, be expressed as the simple correlation of Y and X (r_{YX}), adjusted by applying the two simple correlations involving Z (namely r_{XZ} and r_{YZ}) as follows:

$$r_{YX.Z} = \frac{r_{YX} - r_{YZ}r_{XZ}}{\sqrt{1 - r_{XZ}^2}\sqrt{1 - r_{YZ}^2}} \tag{14-40}$$

This formula shows explicitly that there need be no close correspondence between the partial and simple correlation coefficient; however, in the special case that both X and Y are completely uncorrelated with Z (i.e., $r_{XZ} = r_{YZ} = 0$), then (14-40) reduces to:

$$r_{YX.Z} = r_{YX} \tag{14-41}$$

and, as we would expect, the partial and simple correlation coefficients are the same.

It is instructive to note what happens at the other extreme when X becomes perfectly correlated with Z. In this case $r_{YX.Z}$ cannot be calculated

[7] By the "influence" of Z on Y we mean the fitted regression of Y on Z:

$$\hat{Y} = a + bZ \tag{14-37}$$

By "removing the influence," we mean subtracting the fitted from the observed Y value, obtaining the residual deviation:

$$\hat{u} = Y - \hat{Y} = Y - a - bZ \tag{14-38}$$

which is recognized to be that part of Y not explained by Z. Similarly, we obtain $\hat{v}$, the residual deviation of X from its fitted value on Z. The partial correlation coefficient $r_{XY.Z}$ is the simple correlation of $\hat{u}$ and $\hat{v}$, thus:

$$r_{XY.Z} = r_{\hat{u}\hat{v}} \tag{14-39}$$

since $r_{XZ} = 1$ and the denominator of (14-40) becomes zero as a consequence. This is recognized as the multicollinearity problem of Chapter 13, where the corresponding multiple regression estimate b could not be defined.

The parallel statistical properties of b and $r_{YX.Z}$ can be extended further: rejection of the hypothesis that $\beta = 0$ in Chapter 13 is equivalent to rejecting the null hypothesis that $\rho_{YX.Z} = 0$. Again, one reason for emphasizing regression analysis is confirmed: multiple regression will not only answer its own set of regression questions, but also partial correlation questions as well.

14-3 MULTIPLE CORRELATION

A partial correlation coefficient may be computed for each independent variable in a multiple regression. In addition, one single overall index of value of fitting the multiple regression equation can be defined: the *multiple correlation coefficient, R, is the simple correlation coefficient of the observed Y and the corresponding fitted* $\hat{Y}$. Thus, if our estimated regression is:

$$\hat{Y} = a + bX + cZ \tag{14-42}$$

then

$$\boxed{R \triangleq r_{Y\hat{Y}}} \tag{14-43}$$

This has all the nice algebraic properties of any simple correlation. In particular, we note (14-29) which takes the form

$$R^2 = \frac{\sum (\hat{Y}_i - \bar{Y})^2}{\sum (Y_i - \bar{Y})^2} = \frac{\text{explained variation of } Y}{\text{total variation of } Y} \tag{14-44}$$

Note that this is identical to r^2 if there is only one regressor (independent variable). If there is more than one regressor, then the numerator represents the variation of Y explained by all of them [with Y estimated from the full multiple regression (e.g., (14-42)]. Thus, as we add additional explanatory variables to our model, by watching how fast R^2 increases we can immediately see in (14-44) how helpful these variables are in improving our explanation of Y. Our conclusion is the same as in simple correlation: one of the major values of calculating R^2 is to clarify how successfully our regression explains the variation in Y.

It remains, finally, to relate this to our t-test of multiple regression coefficients, using our example in (13-15). We could extend (14-22) to

$$\begin{aligned}\text{Total variation} &= \text{variation explained by } (X_1 \cdots X_3) + \text{additional} \\ &\quad \text{variation explained by } X_4 + \text{unexplained variation}\end{aligned} \tag{14-45}$$

We could set this up in an ANOVA table like Table 10-10, and construct the ratio

$$F = \frac{\text{additional variance explained by } X_4}{\text{unexplained variance}} \tag{14-46}$$

just as the ratio (10-30) was constructed. A test of significance of this observed value of F is thus seen to be a test of the significance of the (last included) regressor X_4. Similarly, we could construct an observed F ratio for each of the other regressors in turn. These F values are translated into t values[8] that appear under equation (13-15), and lend themselves so easily to tests of significance on each regressor.

PROBLEMS

14-7 For the data of Problem 13-1 relating savings S to income Y and assets W, find

(a) r_{SY}, the simple correlation of S and Y.

(b) $r_{SY.W}$, the partial correlation of S and Y, holding W fixed.

(c) R, the multiple correlation of S on Y and W.

(d) The proportion of the variation of S which is explained by

(1) Y alone;

(2) By Y and W.

(e) Comparing (a) and (c), is R larger than r in this problem? Is R necessarily larger than r always?

(f) Is r_{SY} or $r_{SY.W}$ a better measure of "how S and Y are related, other things being equal"?

14-8 Repeat Problem 14-7, using the data of Problem 13-2 and substituting N for W throughout.

*14-9 Following Problem 14-7(d), find

(a) The proportion of the variation explained by the addition of W as a regressor.

(b) The proportion of the variation which is unexplained after regression of S on Y and W.

(c) How many degrees of freedom are there for the two components of variation in (a) and (b)?

(d) Using parts (a), (b), and (c), calculate the variance ratio F, to test the statistical significance of adding W to the regression model.

(e) Calculate $t = -\sqrt{F}$. This is one way the t-values could be found in an equation such as (13-15).

*14-10 Repeat the steps of Problem 14-9 to find the t-value to test the statistical significance of adding Y as a regressor after S is regressed on W.

[8] Using, of course, $t = \pm\sqrt{F}$, with 1 degree of freedom in the numerator.

chapter 15

Decision Theory

This chapter is devoted to making decisions in the face of uncertainty. A large part of the discussion involves Bayesian methods, which are not only useful for their own sake, but also sharpen our understanding of the limitations of classical statistics.

15-1 PRIOR AND POSTERIOR DISTRIBUTIONS

Problem 3-24b on Bayes' theorem is important enough to repeat, in a slightly altered form. If we were to predict tomorrow's weather before consulting a barometer, we would use Table 15-1:

TABLE 15-1 Prior Probabilities

State θ	Rain (θ_1)	Shine (θ_2)
Prior probability $p(\theta)$	.40	.60

But we can do better, by using a barometer characterized by Table 15-2:

TABLE 15-2 Conditional Probabilities $p(x/\theta)$

Prediction x \ State θ	Rain (θ_1)	Shine (θ_2)
"Rain" (x_1)	.90	.20
"Shine" (x_2)	.10	.80
Σ	1.00	1.00

TABLE 15-3 Posterior Probabilities, $p(\theta/x)$

State θ	Rain (θ_1)	Shine (θ_2)
Posterior probability $p(\theta/\text{"rain"})$	.75	.25

After it is observed that the barometer's prediction is "rain," the probabilities of Table 15-1 are no longer relevant, and should be replaced by Table 15-3.

We recall that this was derived, diagrammatically, by combining Tables 15-1 and 15-2 into Figure 15-1. The new sample space is "rain" with the relative size of these two hatched areas explaining the two posterior probabilities in Table 15-3.

Since this is so important, we now write down its full formal confirmation. We use (5-10) to express the probability of rain and "rain" as

$$p(\theta_1, x_1) = p(\theta_1) \cdot p(x_1/\theta_1) \tag{15-1}$$

$$= (.4)(.9) = .36 \tag{15-2}$$

Similarly the probability of the state shine and the prediction "rain" is

$$p(\theta_2, x_1) = p(\theta_2) \cdot p(x_1/\theta_2) \tag{15-3}$$

$$= (.6)(.2) = .12 \tag{15-4}$$

These two calculations define the hatched areas in Figure 15-1. Comparing areas we conclude that it is three times as likely for a "rain" prediction to be associated with rain, as with shine. Formally, the hatched area in Figure 15-1 becomes the new sample space, within which we calculate the new (conditional) probabilities.

To do this, we note that

$$p(\text{"rain"}) = p(x_1) = .36 + .12 = .48 \tag{15-5}$$

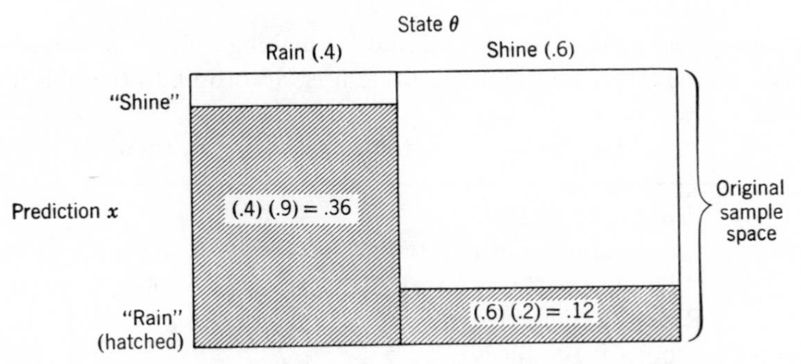

FIG. 15-1 How posterior probabilities are determined.

Using (5-10) again

$$p(\theta_1/x_1) = \frac{p(\theta_1, x_1)}{p(x_1)} = \frac{.36}{.48} = .75 \tag{15-6}$$

Similarly

$$p(\theta_2/x_1) = \frac{p(\theta_2, x_1)}{p(x_1)} = \frac{.12}{.48} = .25 \tag{15-7}$$

When this new (hatched) sample space has its probabilities blown up in this way by using the divisor $p(x_1)$, the result is the posterior probability distribution in Table 15-3. This is often written in the more convenient and general form

$$p(\theta/x) = \frac{p(\theta, x)}{p(x)} = \frac{p(\theta)\, p(x/\theta)}{p(x)} \tag{15-8}$$

To keep the mathematical manipulations in perspective, we repeat the physical interpretation for emphasis. Before the evidence (barometer) is seen, the prior probabilities $p(\theta)$ give the proper betting odds on the weather. But after the evidence is in we can do better; the posterior probabilities $p(\theta/x)$ now give the proper betting odds. (This may be intuitively grasped by appealing to the relative frequency interpretation. Of all the times the barometer registers "rain," in what proportion will rain actually occur? The answer is 75%.) As a simple summary, we note that the prior probability distribution is adjusted by the empirical evidence to yield the posterior distribution. Schematically:

Prior probabilities	and	Probability of empirical evidence	yields →	Posterior probabilities	(15-9)
$p(\theta)$		$p(x/\theta)$		$p(\theta/x)$	

PROBLEMS

15-1 A factory has 3 machines (θ_1, θ_2, and θ_3) making bolts. The newer the machine, the larger and more accurate it is, according to the following table:

Machine →	θ_1 (oldest)	θ_2	θ_3 (newest)
Proportion of total output produced by this machine	10%	40%	50%
Rate of defective bolts it produces	5%	2%	1%

Thus, for example, θ_3 produces half of the factory's output, and of all the bolts θ_3 produces, 1% are defective.
(a) Suppose a bolt is selected at random; *before* it is examined, what is the chance it was produced by machine θ_1? By θ_2? By θ_3?
(b) Suppose the bolt is examined and found defective; *after* this examination, what is the chance it was produced by machine θ_1? By θ_2? By θ_3?

⇒ 15-2 Suppose a man is drawn at random from a roomful of ten people, whose heights θ have the following distribution:

θ (inches)	$p(\theta)$
70	.1
71	.3
72	.2
73	.2
74	.1
75	.1

(a) Graph this (prior) distribution of θ.
(b) Suppose also that a crude measuring device is available, that makes errors with the following distribution:

e (error in inches)	$p(e)$
−2	.1
−1	.2
0	.4
1	.2
2	.1

Surely this can help us to be more accurate in estimating the man's height. For example, suppose his measured height using this crude device is $x = 74$ inches. We now have further information about θ; i.e., this measurement changes the probabilities for θ from the prior distribution $p(\theta)$ to a posterior distribution $p(\theta/x = 74)$. Calculate and graph this posterior distribution.

15-2 OPTIMAL DECISIONS

(a) Example

Suppose a salesman regularly sells umbrellas or lemonade on Saturday afternoons at football games. To keep matters simple, suppose he has just three possible options (actions, a_i):

$$a_1 = \text{sell only umbrellas;}$$
$$a_2 = \text{sell some umbrellas, some lemonade;}$$
$$a_3 = \text{sell only lemonade.}$$

If he chooses a_1 and it rains, his profit is \$20; but if it shines, he loses \$10. It will be more convenient to describe everything as a loss (negative profit); thus his losses will be -20 or $+10$ respectively.

If he chooses action a_2 or a_3, there will also be certain losses. All this information may be assembled conveniently in the following loss table:

TABLE 15-4 Loss Function $l(a, \theta)$

State θ / Action a	Rain (θ_1)	Shine (θ_2)
a_1	-20	10
a_2	5	5
a_3	25	-7

Suppose further that the probability distribution (long-run relative frequency) of the weather is as shown in Table 15-5.

TABLE 15-5 Probability Distribution of θ

State θ	Rain	Shine
Probability $p(\theta)$	.20	.80

What is the best action for the salesman to take? (You are urged to work this out, before reading on; it will be easier that way.)

Solution. If he chooses a_1, what could he expect his loss to be, on the average? Intuitively, we calculate the expected loss if he chooses a_1:

$$L(a_1) = -20(.20) + 10(.80) = 4 \tag{15-10}$$

We recognize this as the concept of expected value,[1] as given by (4-17):

$$L(a_1) = l(a_1, \theta_1)\, p(\theta_1) + l(a_1, \theta_2)\, p(\theta_2) = \sum_\theta l(a_1, \theta)\, p(\theta) \quad (15\text{-}11)$$

Similarly, we evaluate

$$L(a_2) = 5(.20) + 5(.80) = 5 \quad (15\text{-}12)$$

and

$$L(a_3) = 25(.20) - 7(.80) = -.6 \quad (15\text{-}13)$$

In general

$$L(a) = \sum_\theta l(a, \theta)\, p(\theta) \quad (15\text{-}14)$$

The optimal action is seen to be a_3, which minimizes the expected loss; in fact, this is the only option that allows any expected profit. To summarize, we assemble all our information and calculations in Table 15-6:

TABLE 15-6 Calculation of the Optimal Action a

$p(\theta)$	.20	.80	
a \ θ	θ_1	θ_2	$L(a)$ = expected loss
a_1	−20	10	4
a_2	5	5	5
a_3	25	−7	−.6 ← minimum
	Loss function $l(a, \theta)$		

(b) Generalization

It hardly seems necessary to state that this problem can be generalized to any number of states θ or actions a (even an infinite number, as in the next section). The objective remains the same: to minimize expected loss. We now pause to consider:

1. The probabilities $p(\theta)$, and
2. The loss function $l(a, \theta)$.

[1] For those who wish to review, we give an alternative intuitive calculation. In, say, 100 days he would get about 20 rainy days at \$−20 each, yielding \$−400; and about 80 shiny days, at \$+10 each, yielding \$+800—for a sum of about \$+400 in 100 days, or an average of \$4 per day.

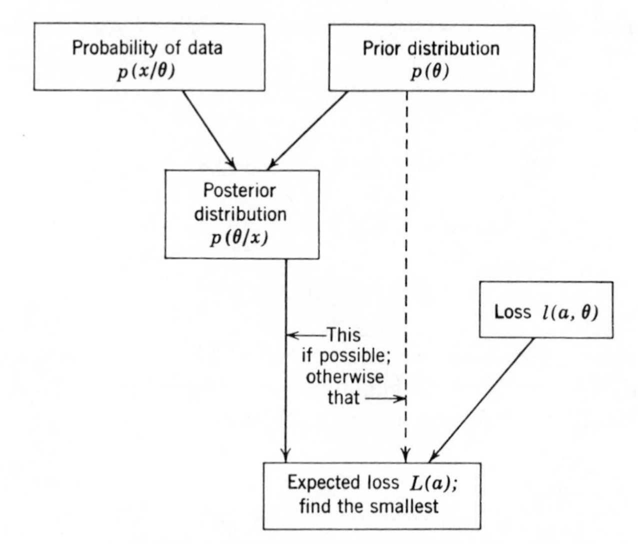

FIG. 15-2 The logic of Bayesian decisions to minimize expected loss.

The probabilities $p(\theta)$. These of course should represent the best possible intelligence on the subject. For example, suppose the salesman moves to another state, with weather probabilities as given in Table 15-1. If he has no barometer, he will have to use the (prior) probabilities in this table. But if he can consult the barometer (described in Table 15-2), then of course the posterior probabilities $p(\theta/x)$ in Table 15-3 should be used. (See Problem 15-3.)

The logic of Bayesian inference is laid out in the block diagram, Figure 15-2. Incidentally, in the calculation of the average loss $L(a)$ in (15-14) it would not hurt to use $kp(\theta)$ instead of $p(\theta)$ as weights, where k is any constant (independent of θ and a). For $kp(\theta)$ would generate losses $kL(a)$, which would rank in the same order as the true losses $L(a)$, and hence point to the same correct optimizing action. This is a very useful observation. Thus, for example, our umbrella salesman need not undertake the last step in calculating the posterior probabilities of rain $p(\theta/x_1)$ in (15-6) and (15-7); he can forget about the denominator $p(x_1)$, and use (15-2) and (15-4) instead—without affecting his decision.[2]

The loss function, $l(a, \theta)$. In our example, we assumed that monetary loss is the appropriate consideration. This may be valid enough if the decision is made ("game is played") over and over again: whatever minimizes the expected loss in each game will minimize total expected loss in the long run.

[2] i.e., attaching weights of .36 and .12 to his losses would yield the same result as weights of .75 and .25.

Yet there are some decisions that are made only once, and then expected monetary loss may not be the right criterion. To illustrate: suppose you were offered (tax-free) a choice between

(a) $100,000 for sure, or
(b) a 1/2 chance (lottery ticket) on a $210,000 prize. (15-15)

Most people would prefer choice (a), even though its expected monetary value

$$\left.\begin{array}{l}\$100{,}000\ (1) = \$100{,}000 \\ \text{is less than that of choice (b):} \\ \$210{,}000\ (1/2) = \$105{,}000\end{array}\right\} \qquad (15\text{-}16)$$

The reason is that most people value the first hundred thousand much more than the second. (The student should speculate on how he would spend the first hundred thousand. Once these purchases have been made, less exciting opportunities would be available for spending the second hundred thousand; the sports car has already been bought, and so on.) Such a decision should be based not on money itself as in (15-16), but rather on a subjective valuation of money, or the "utility" of money. As an illustration, Figure 15-3 shows one author's subjective evaluation[3] $U(M)$. Since utility is the more appropriate measure, the decision should be based on *expected utility*, rather than expected money. Using Figure 15-3, the expected utilities of the two choices are:

$$\begin{aligned}&(a)\ u_1(1) = u_1 \\ &(b)\ u_2(\tfrac{1}{2}) = 1.4u_1(\tfrac{1}{2}) = .7u_1\end{aligned} \qquad (15\text{-}17)$$

which is a clear victory for choice (a). In decision situations, a *loss-of-utility* function of this kind should typically be used as our loss function $l(a, \theta)$; hereafter we shall interpret losses in this way.

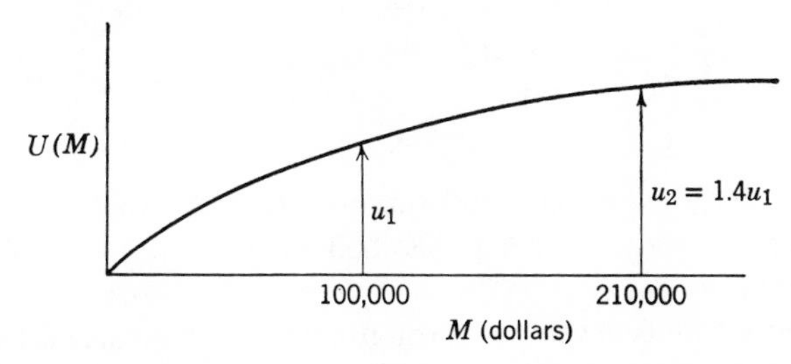

FIG. 15-3 Author's subjective evaluation of money.

[3] This utility curve is highly personal, and temporary. It is defined empirically for an individual by asking him which bets he prefers. In other words, many bets like (15-15) are used to define utility, rather than vice-versa.

PROBLEMS

15-3 Using the losses of Table 15-4, calculate the optimal action if

(a) The only available probabilities are the prior probabilities of Table 15-1.

(b) The barometer reads "rain" (so that the posterior probabilities of Table 15-3 are relevant).

(c) The barometer reads "shine."

(d) Is the following a true or false summary of questions (a) to (c) above? If false correct it.

If the salesman must choose his action (order his merchandise) before consulting a barometer, then a_2 (umbrellas and lemonade) is best. However, if the barometer can be consulted first, then the salesman should

Choose a_1 (umbrellas) if the barometer predicts "rain."

Choose a_3 (lemonade) if the barometer predicts "shine."

But a bright salesman could have seen this obvious solution without going to all the trouble of learning about Bayesian decisions.

15-4 A farmer has to decide whether to sell his corn for use A or use B. His losses depend on its water content, (determined by the mill during processing, after the farmer's decision has been made) according to the following table.

Action a \ State θ	Dry	Wet
Use A	−10	30
Use B	20	10

(a) If his only additional information is that, through long past experience his corn has been classified as dry one third of the time, what should his decision be?

(b) Suppose he has developed a rough-and-ready means of determining whether it is wet or dry—a method which is correct 3/4 of the time regardless of the state of nature. If this indicates that his corn is "dry" what should his decision be? How much is this method worth, i.e., how much does it reduce his expected loss?

⇒ 15-5 A school is to be built to serve 125 students, who live along a single road.

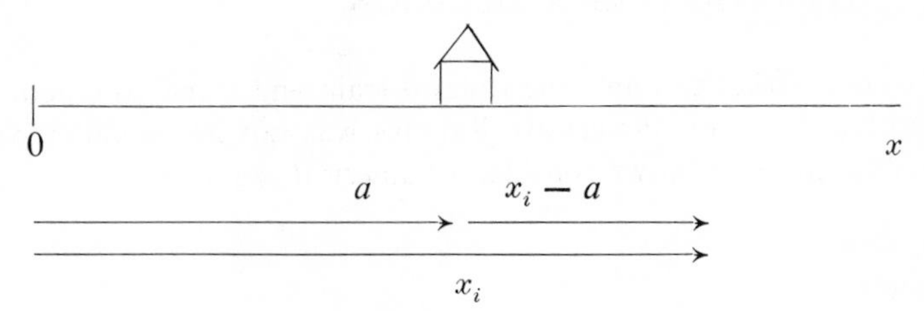

Let x_i = distance student i lives from origin
a = distance of school from origin

Thus

$$(x_i - a) = \text{distance of student } i \text{ from school.}$$

(a) Where is the optimum place (mean, median, mode, midrange?) to build the school in order to

(1) Minimize the distance that the farthest student has to walk.

(2) Minimize the total walking done, i.e., minimize the sum of the absolute deviations:

$$\sum |x_i - a|$$

*(3) Minimize the sum of the squared deviations:

$$\sum (x_i - a)^2$$

(*Hint*. Calculus suggests differentiating with respect to a, setting the result equal to zero.)

(4) Maximize the number of students who live where the school is built, and do not have to walk at all.

(b) Does the following accurately reflect your conclusions in question (a) above? If not, correct it.

In (2) we are concerned only about the total walking done; walking is considered a loss, no matter who does it. In (1), on the other hand, only the walking done by the two extreme people is considered a loss; walking done by any others is of no concern whatsoever. (3) is a compromise; we imply that although all walking is some kind of loss, the more a student has walked, the greater his loss in walking one more mile. Thus the person who walks 3 miles ($x_i - a = 3$) contributes 9 to the loss function, whereas the person who walks 1 mile contributes only 1.

15-3 ESTIMATION AS A DECISION

In our earlier example the states θ (rain and shine) and actions a were categorical (i.e., nonnumerical). But this was not an essential part of the theory; in this section we consider a numerical example.

Example

Suppose the judge at a beauty contest is asked to guess the height θ of the first contestant, whom he has never seen. Yet he is not in complete ignorance; suppose he knows that the heights of contestants follow the probability distribution $p(\theta)$ shown in Figure 15-4.

θ (inches)	$p(\theta)$
64	.1
65	.1
66	.2
67	.2
68	.3
69	.1

FIG. 15-4 Prior distribution of heights θ.

(i) Suppose, in order to encourage an intelligent guess, the judge is to be fined $1 if he makes a mistake (no matter how large or small); "a miss is as good as a mile." What should the rational judge guess?

(ii) Suppose the rules become more severe, by fining the judge $\$x$ for an error of x inches; the greater his error, the greater his loss. What is his rational guess?

(iii) Suppose the rules are made even more severe, by fining the judge $\$x^2$ for an error of x inches; this is the same as (b), except that the loss becomes more severe as his error increase. What is his rational guess now?

Solution. (i) The most likely (modal) value 68.

(ii) The median value 67.

(iii) The mean value 66.8.

Thus (i), (ii), and (iii) are like (4), (2), and (3) in the schoolhouse Problem 15-5, with the same solution.

To translate this into the familiar language of decision theory, the girl's height is the state of nature θ, and the guessed height (estimate) is the action

a to be taken. The fine the judge must pay is the loss function $l(a, \theta)$; since a and θ are numerical, the loss function is most conveniently given by a formula, rather than a table. Each of the 3 loss functions, along with its corresponding optimal estimator, is shown in Table 15-7.

TABLE 15-7 How the Optimal Estimator of θ Depends on the Loss Function

If the Loss Function $l(a, \theta)$ is:	Then the Corresponding Optimal Estimator a is:
(i) 0 if $a = \theta$ exactly, 1 otherwise ("the 0–1 loss function")	Mode of $p(\theta)$
(ii) $\lvert a - \theta \rvert$	Median
(iii) $(a - \theta)^2$	Mean

The "quadratic" loss function (iii) is the one that is usually used in decision theory. It is graphed in Figure 15-5. It is justified not only by its intuitive appeal, but also by its attractive mathematical properties. For example, it is easily differentiated (an important requirement in minimization problems); on the other hand (i) obviously cannot be differentiated, nor can (ii), since it is an absolute value function.

We reemphasize that the probability distribution $p(\theta)$ used in the decision process ought to reflect the best available information. Thus we may be forced to use the prior distribution $p(\theta)$ if we have not yet collected any data, but after data is collected, the posterior distribution $p(\theta/x)$ is appropriate.

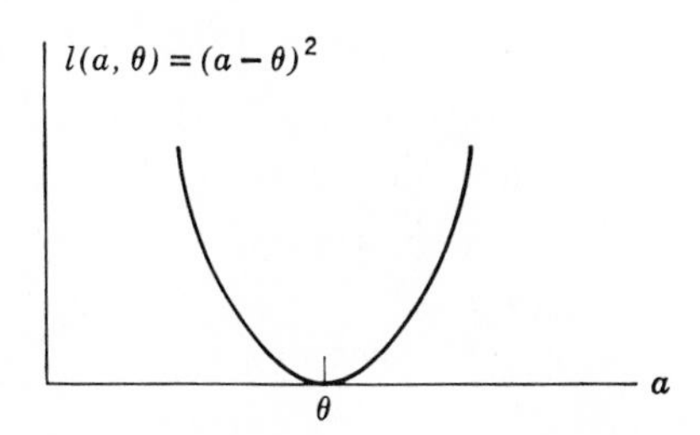

FIG. 15-5 The quadratic loss function.

PROBLEMS

These are extensions of Problem 15-2.

15-6 Suppose you have to guess the height of the man drawn in Problem 15-2, with only the prior distribution $p(\theta)$ known. Find the optimal estimate of θ

(a) Assuming

$$l(a, \theta) = 0 \quad \text{if } a = \theta$$
$$= 1 \quad \text{otherwise.} \tag{15-18}$$

(b) Assuming $l(a, \theta) = |a - \theta|$. (15-19)

(c) Assuming $l(a, \theta) = (a - \theta)^2$. (15-20)

15-7 Repeat Problem 15-6, *after* the man's height has been crudely measured as $x = 74$, so that the posterior distribution $p(\theta/x)$ is relevant.

15-4 ESTIMATION: BAYESIAN VERSUS CLASSICAL

This comparison is best shown with an extended example, illustrated in Figure 15-6; from this we shall draw conclusions later.

(a) Example

Suppose it is essential to estimate the length θ of a beetle accidentally caught in a delicate piece of machinery. A measurement x is possible, using a device which is subject to some error; suppose x is normally distributed about the true value θ, with $\sigma = 1$. Suppose x turns out to be 20 mm.

Question (a). What is the classical 95% confidence interval for θ?

Solution. Our information on the sampling distribution of x, i.e.,

$$p(x/\theta) = N(\theta, \sigma^2), \quad \text{specifically } N(\theta, 1) \tag{15-21}$$

can be "turned around" to construct the following confidence interval for θ:

$$\theta = 20 \pm 1.96(1)$$
$$= 20 \pm 1.96 \tag{15-22}$$

and, of course:

$$\text{point estimate of } \theta = 20 \tag{15-23}$$

Question (b). Suppose we take the effort to find out from a biologist that the population of all beetles has a normally-distributed length, with mean $\theta_0 = 25$ mm and variance $\sigma_0^2 = 4$. How can this be used to define a posterior distribution of θ?

Solution. It will be useful to develop a general formula applying for any θ_0, σ_0, etc., and then solve it for our specific example. Since our prior distribution is

$$p(\theta) = N(\theta_0, \sigma_0^2) \tag{15-24}$$

and the distribution of our empirical evidence x is:

$$p(x/\theta) = N(\theta, \sigma^2) \tag{15-21}$$

repeated

it can be shown[4] that the posterior distribution is also normal; specifically:

$$p(\theta/x) = N(ab, a) \tag{15-36}$$

---

[4] (15-24) and (15-21) may be written:

$$p(\theta) = K_1 e^{-(1/2\sigma_0^2)(\theta-\theta_0)^2} \tag{15-25}$$

$$p(x/\theta) = K_2 e^{-(1/2\sigma^2)(x-\theta)^2} \tag{15-26}$$

where K_1 and K_2 and other similar constants introduced in this footnote are of a form not critical to the argument. Since

$$p(x, \theta) = p(\theta)\, p(x/\theta) \tag{15-27}$$

we can use (15-25) and (15-26) to write

$$p(x, \theta) = K_1 K_2 e^{-(1/2)[(1/\sigma_0^2)(\theta^2-2\theta\theta_0+\theta_0^2)+(1/\sigma^2)(x^2-2x\theta+\theta^2)]} \tag{15-28}$$

Now consider only the exponent, which may be rearranged to

$$-\tfrac{1}{2}\left[\theta^2\left(\frac{1}{\sigma_0^2} + \frac{1}{\sigma^2}\right) - 2\theta\left(\frac{\theta_0}{\sigma_0^2} + \frac{x}{\sigma^2}\right) + K_3\right] \tag{15-29}$$

Let

$$\frac{1}{\sigma_0^2} + \frac{1}{\sigma^2} = \frac{1}{a} \tag{15-30}$$

$$\frac{\theta_0}{\sigma_0^2} + \frac{x}{\sigma^2} = b \tag{15-31}$$

Using these definitions, the exponent (15-29) can be written

$$-\frac{1}{2a}[\theta^2 - 2ab\theta + K_4] \tag{15-32}$$

$$= -\frac{1}{2a}[(\theta - ab)^2 + K_5] \tag{15-33}$$

Finally we use this to write (15-28) as

$$p(x, \theta) = K_6\, e^{-(1/2a)(\theta-ab)^2} \tag{15-34}$$

and

$$p(\theta/x) = \frac{p(x, \theta)}{p(x)} = K_7\, e^{-(1/2a)(\theta-ab)^2} \tag{15-35}$$

This means that θ, given x, is a normal variable with mean ab and variance a, provided a appears appropriately in K_7. But it must, since $p(\theta/x)$ is a bona fide probability function (integrating to 1), and K_7 is just the scale factor necessary to ensure this.

where

$$\frac{1}{a} = \frac{1}{\sigma_0^2} + \frac{1}{\sigma^2} \tag{15-37}$$

$$b = \frac{\theta_0}{\sigma_0^2} + \frac{x}{\sigma^2} \tag{15-38}$$

Now apply this to our example. Since

$$\sigma_0^2 = 4$$

$$\sigma^2 = 1$$

$$\theta_0 = 25$$

$$x = 20$$

it follows that

$$\frac{1}{a} = \frac{1}{4} + \frac{1}{1} = \frac{5}{4}$$

and

$$b = \frac{25}{4} + \frac{20}{1} = \frac{105}{4}$$

Thus:

$$\text{mean} = ab = 21.0$$

$$\text{variance} = a = .8$$

Hence the posterior distribution may be formally written

$$p(\theta/x = 20) = N(21, .8) \tag{15-39}$$

compared with the prior:

$$p(\theta) = N(25, 4) \tag{15-40}$$

The Bayesian logic is shown in Figure 15-6. A prior distribution is adjusted to take account of observed data (x), with the weight attached to the observed x depending on its probability $p(x/\theta)$. The result is the posterior distribution, with mean (21) falling, as expected, between the prior mean (25) and the observed value (20). (As a bonus, variance is reduced in the posterior distribution. Although this does not always happen, it is evident that it must happen for normal distributions; for (15-37) shows that the posterior variance a is less than σ_0^2, and also less than σ^2 incidentally.)

***Question* (*c*).** With the posterior distribution (15-39) now in hand, defining a Bayesian estimate of θ requires only a loss function. Suppose this is the quadratic loss function; what is the Bayesian point estimator of θ? Find also the 95% probability interval for θ.

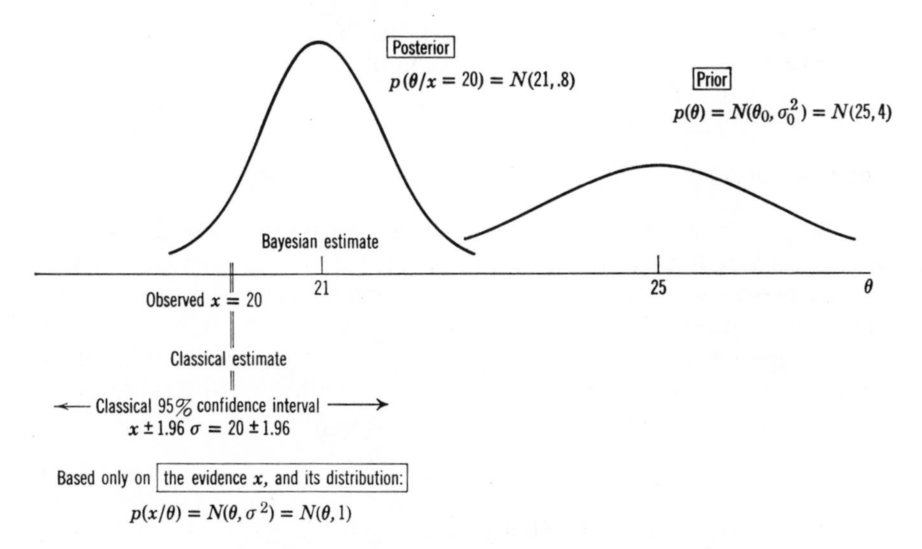

FIG 15-6 Bayesian versus classical estimation.

Solution. For the quadratic loss function, the posterior mean (21) is the optimum estimator. (Note that because $p(\theta/x)$ is normal, this is also the posterior median and mode, so that all the loss functions in Table 15-7 yield the same answer. This is reassuring, and frequently happens in practice.)

To construct a 95% probability interval, we know from (15-39) that, given the observation $x = 20$, there is a 95% probability that θ will fall in the interval

$$\begin{aligned} &21 \pm 1.96\sqrt{.8} \\ &= 21 \pm 1.76 \end{aligned}$$

Note that this is narrower (more precise) than the classical interval (15-22), reflecting the value of the prior information $p(\theta)$.

PROBLEMS

15-8 As our means of measuring (beetles) becomes more and more precise ($\sigma^2 \to 0$), show that in the posterior distribution $p(\theta/x)$,

$$\begin{aligned} \text{the mean} &\to x \\ \text{variance} &\to 0 \end{aligned} \tag{15-41}$$

In other words, if we use an errorless measuring device, we can be certain that the true θ will be its measured value x.

⇒ 15-9 Using Figure 15-6, what would you expect intuitively of the posterior mean if two independent measurements of the beetle had yielded an average of 20 mm? (For an extension of your answer, see the section immediately following.)

(b) Generalization

Suppose that a sample of n independent measurements $x_1, x_2, \ldots, x_n$ can be taken rather than just a single x. Using the sample mean $\bar{x}$, what now is the Bayesian estimate of θ? In particular, what happens as we get more and more observations ($n \to \infty$)?

This problem may be solved, using (15-36) to (15-38) with one important change. Since our data now is $\bar{x}$ instead of x we must make this substitution in (15-38), and also substitute

$$\sigma_{\bar{x}}^2 = \frac{\sigma^2}{n}, \tag{15-42}$$

for σ^2 in (15-37) and (15-38). [Of course, (15-42) is just the variance of a sample mean when σ^2 is the variance of a single observation.] Thus, our generalized definition of a and b in (15-36) is:

$$\frac{1}{a} = \frac{1}{\sigma_0^2} + \frac{n}{\sigma^2} \qquad \text{for } n = 1, \text{ this reduces to (15-37)} \tag{15-43}$$

$$b = \frac{\theta_0}{\sigma_0^2} + \frac{n\bar{x}}{\sigma^2} \qquad \text{for } n = 1, \text{ this reduces to (15-38)} \tag{15-44}$$

In the limit, as sample size $n \to \infty$:

$$\frac{1}{a} \simeq \frac{n}{\sigma^2} \tag{15-45}$$

$$b \simeq \frac{n\bar{x}}{\sigma^2} \tag{15-46}$$

Incidentally, exactly these same results follow, whether $n \to \infty$, or

$$\sigma_0^2 \to \infty \tag{15-47}$$

Thus, evaluating (15-36):

$$\text{posterior mean} = ab \simeq \bar{x} \tag{15-48}$$

$$\text{posterior variance} = a \simeq \frac{\sigma^2}{n} \tag{15-49}$$

Again the normality of this posterior distribution ensures that its mean, mode, and median coincide. Hence, regardless of which loss function we may use:

$$\text{Bayesian estimator of } \theta \simeq \bar{x} \tag{15-50}$$

$$95\% \text{ probability interval} \simeq \bar{x} \pm 1.96 \frac{\sigma}{\sqrt{n}} \tag{15-51}$$

We conclude that, as $n \to \infty$, Bayesian estimation approaches the classical. This is exactly as it should be: as more and more data are collected, less and

TABLE 15-8 Relation of Classical and Bayesian Estimation. (Although Normality is Assumed, Results are Instructive for Other Cases Too)

Procedure to Estimate θ	Point and Interval Estimates	Requires, Along With Observed x	And Gets the Answer: In Our Example ($n = 1$)	In the Limit, as $n \to \infty$ or $\sigma_0^2 \to \infty$
Classical	Point estimate		20	$\bar{x}$
	Confidence interval	$p(x/\theta)$	20 ± 1.96	$\bar{x} \pm 1.96 \frac{\sigma}{\sqrt{n}}$
Bayesian	Point estimate	$p(x/\theta)$, $p(\theta)$ and loss function	21	Same as classical
	Probability interval	$p(x/\theta)$, $p(\theta)$	21 ± 1.76	Same as classical

less weight need be attached to prior information; and with an unlimited sample, prior information is completely disregarded, as in classical estimation. The classical and Bayesian approaches are compared in more detail in Table 15-8.

We now turn to the other condition that leads to the same result. Bayes estimators also approach the classical if prior information is very vague (i.e., if $\sigma_0^2 \to \infty$, as stated in (15-47). Thus the less the prior distribution tells us, the less weight we attach to it. To sum up, the two reasons for completely disregarding prior information are (1) if present data is in unlimited supply, or (2) if prior information is useless.

(c) Is θ Fixed or Variable?

In this chapter we regard the target to be estimated as a random variable—for example, the beetle's length θ in Figure 15-6. Yet in all preceding chapters, we have regarded the target as a fixed parameter—for example, the average height μ of American men. Nevertheless, we may often find it useful to think of μ as having a *subjective* probability distribution—with this being a description of the betting odds we would give that μ is bracketed by any two given values (see the description of subjective probability in Section 3-6). In the problem of men's heights it may be helpful to boil down our best prior knowledge of μ into a prior subjective distribution of μ. Then the posterior subjective distribution of μ would reflect how the sampling data changed the betting odds.

PROBLEMS

15-10 Following the beetle example in Section 15-4(b), suppose that:

$$\sigma_0^2 = 100$$
$$\theta_0 = 25$$
$$\sigma^2 = 1$$

and a sample of 4 independent observations on the trapped beetle yields an average length $\bar{x}$ of 20 mm.
(a) Calculate the Bayesian point estimate for θ, the length of the beetle. For two reasons this estimate is closer to the observed value of 20 than the Bayesian estimate (21) in Figure 15-6. Explain.
(b) Calculate the Bayesian 95% probability interval for θ.

⇒ 15-11 Suppose that, in a random sample of 10 students on an American college campus, you find only one is a Democrat. Which would you rather quote as your "best estimate" of the proportion π of Democrats in the population (whole campus):
(a) The classical estimate,

$$P = \frac{x}{n} = \frac{1}{10} = .10,$$

or
(b) The Bayesian estimate which, assuming this subjective prior distribution:

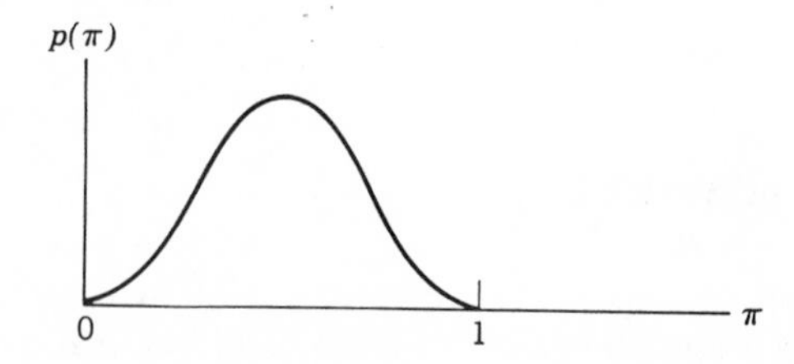

and a quadratic loss function, yields[5] the estimate

$$\frac{x + 3}{n + 6} = \frac{4}{16} = .25$$

15-5 CRITIQUE OF BAYESIAN METHODS

(a) Strength

Bayesian inference is the optimal statistical method (in the sense of minimizing loss of utility) if there is a known prior distribution $p(\theta)$ and loss function $l(a, \theta)$. Compared to classical methods, Bayesian methods often yield shorter interval estimates (e.g., Table 15-8), more credible point estimates (e.g., Problem 15-11), and more appropriate hypothesis tests (e.g., Problem 15-13 below). Bayesian methods are particularly useful in the social sciences and business, where sample size is often very small, and Bayesian methods differ considerably from the classical methods.

(b) Weakness

The major criticism of Bayesian estimation is that it is highly subjective. The prior $p(\theta)$ and loss function $l(a, \theta)$ are usually not known[6]—nor is there often any hope at all of specifying them exactly. For example, what is the loss function for an economist measuring a population's unemployment rate, with inevitable statistical error? We have already seen that this is not as serious a difficulty as it seems at first glance, since in many problems any of the three loss functions of Table 15-7 lead to the same Bayes estimator. Then selecting the "wrong" loss function would still lead to the right estimator.

The other information required—the prior distribution $p(\theta)$—usually remains unknown too. Moreover, there are often difficulties in interpreting θ as a random variable; an economist cannot regard the unemployment rate θ as a random variable (as though it is drawn from a bowlful of chips). Instead he must think of $p(\theta)$ as a subjective distribution reflecting his prior betting odds on θ. But he may not view even this as entirely satisfactory.

Since Bayesian techniques require a rough-and-ready specification of these unknown functions, they do indeed involve subjective judgments. The

---

[5] For proof, see for example, Lindgren, B. W., *Statistical Theory*, 2nd Ed. New York: Macmillan, 1967.

[6] The other required information for Bayesian inference is $p(x/\theta)$, the distribution of sample data x. But this can often be borrowed from classical statistics. [For example, recall how we borrowed a classical deduction in (15-42).]

interesting observation however, is that classical methods which require no such *explicit* specifications, are by no means free of the same subjective elements. One of the major contributions of the Bayesian method has been to lay bare the assumptions implicit in classical techniques. As we shall see in the next section, some of these fare badly when exposed; in extreme cases *any* intelligent guess is substantially better.[7]

*(c) Classical Methods as Bayesian Methods in Disguise

Suppose a Bayesian wishes to estimate θ with no prior knowledge. In desperation he might use the "equiprobable" prior:

$$p(\theta) = c, \text{ a constant} \tag{15-52}$$

Further suppose that, rather than using the familiar and attractive quadratic loss function, he opts for the 0-1 loss function. He thus will estimate θ with the mode of the posterior distribution:

$$p(\theta/x) = \frac{p(\theta)\, p(x/\theta)}{p(x)} \tag{15-53}$$

(15-8) repeated

But because of (15-52):

$$= \left[\frac{c}{p(x)}\right] p(x/\theta) \tag{15-54}$$

To find the mode, he finds the value of θ which makes $p(\theta/x)$ largest. But since the bracketed term $[c/p(x)]$ doesn't depend on θ, he only needs to find:

$$\text{the value of } \theta \text{ which makes } p(x/\theta) \text{ largest} \tag{15-55}$$

But this statement is recognized as just the definition of the classical MLE.[8]

From this, we conclude that a classical statistician who uses MLE is getting the same result as a Bayesian using the 0-1 loss function and an "equiprobable" prior. This seems a very unflattering description of MLE, since neither this prior nor this loss function is easy to justify. But in many cases, MLE is not nearly this restrictive. If $p(\theta/x)$ is unimodal and symmetric, as it often is, then its mean, mode and median coincide; in such circumstances MLE is equivalent to Bayesian estimation using *any* of our three loss functions.

As if the discussion of MLE above has not been damaging enough, we consider an even more questionable application. Suppose we are estimating

[7] A further criticism of Bayesian methods is that there is too great a cost of computing Bayesian estimates (not to mention learning about them); but this criticism is being weakened with the advent of better computer programs.

[8] Note that in developing MLE in Section 7-3, the notation $p(x;\theta)$ was used, equivalent to $p(x/\theta)$ used here.

a population proportion π (as in Problem 15-11). It has been proved[9] that a classical statistician using MLE will arrive at the same result (estimating π with x/n) as a Bayesian using the quadratic loss function and the prior distribution shown in Figure 15-7.

This prior distribution is obviously hopeless, the worst we have yet encountered. (It means that a huge majority of students are Republican, or a huge majority are Democratic.) We recall that we may have been uncomfortable about the prior distribution graphed in Problem 15-11; but it

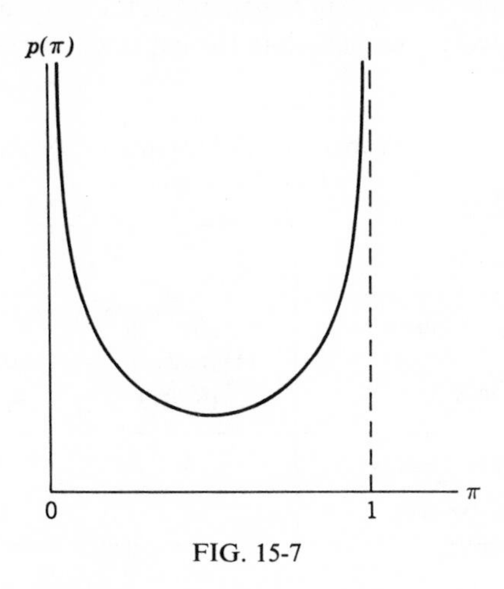

FIG. 15-7

was vastly better than this. This explains why MLE can occasionally give a very strange result in a small sample; our intuition was correct in leading us to reject it in Problem 15-11.

In conclusion, although MLE has many attractive characteristics [see Section 7-3(e)], these are large sample properties; in small sample estimation, it should be used with great caution.

*15-6 HYPOTHESIS TESTING AS A BAYESIAN DECISION

(a) Example

Suppose there are two species of beetle. Species S_0 is harmless, while species S_1 is a serious pest, requiring an expensive insecticide. A beetle is sighted in a new, as yet uninfested territory; but this sighting provides no

[9] Again, see for example, Lindgren, B. W., *Statistical Theory*, 2nd Ed., New York: Macmillan 1967.

information useful in establishing whether the beetle was S_0 or S_1. Should insecticide be used or not?

To answer this question, we need to know the costs $l(a, \theta)$ of a wrong decision, and the probabilities $p(\theta)$ of it being one species or the other; these are given in Table 15-9. Obviously action a_0 (don't spray) is appropriate if the state of nature is S_0 (harmless beetle) while a_1 is appropriate if the state is S_1.

Question (a). Should we spray, or not?

Solution. It will be convenient to generalize the loss table, calling $l(a_i, \theta_j) = l_{ij}$, for short. As always, we calculate the expected losses $L(a)$, by weighting

TABLE 15-9 Probabilities of States of Nature, and Loss Function

$p(\theta)$	.7	.3
State θ / Action a	S_0 (Harmless Species)	S_1 (Harmful Species)
a_0 (don't spray)	5	100
a_1 (spray)	15	15

elements in each row of this table by their appropriate probabilities:

$$\begin{aligned} L(a_0) &= p(\theta_0)l_{00} + p(\theta_1)l_{01} \\ &= (.7)5 + (.3)100 = 33.5 \end{aligned} \tag{15-56}$$

and

$$L(a_1) = (.7)15 + (.3)15 = 15 \leftarrow \text{min}$$

Thus the optimal action is a_1: spray.

We see that this problem may be expressed in terms of hypothesis testing: action a_0 (don't spray) may be interpreted as accepting H_0 (harmless beetle), while action a_1 (spray) may be interpreted as accepting H_1 (harmful beetle).

Question (b). Suppose that prior information about the beetles is that species S_0 is 9 times as common as S_1. Given this new information about $p(\theta)$, what is the optimum action?

Solution. Don't spray, as shown in Table 15–10.

In this case the harmful species is so rare, that it is better to "take the risk," i.e., assume the beetle is harmless as our working hypothesis.

TABLE 15-10 Calculation of Optimal Action, a priori

$p(\theta)$	.9	.1	
Action a \ State θ	S_0 (H_0)	S_1 (H_1)	$L(a)$
a_0 (Don't spray)	5	100	14.5 ← min
a_1 (Spray)	15	15	15

***Question* (*c*).** So far we have assumed no statistical information on the beetle that has been sighted. Now suppose it has been captured, with its length measured as 27 mm. Suppose further that the two species are distinguishable by their lengths, which are normal random variables with $\sigma = 4$, and means $\theta_0 = 25$ and $\theta_1 = 30$ respectively. What now is the best action, a posteriori? [Assume $p(\theta)$ and losses given in Table 15-9.]

Solution. It will be most instructive to develop a general solution, leaving substitution of particulars to the end. Losses are calculated as in (15-56), substituting the appropriate posterior probabilities $p(\theta/x)$ for $p(\theta)$:

$$L(a_0) = p(\theta_0/x)l_{00} + p(\theta_1/x)l_{01} \tag{15-57}$$

Similarly

$$L(a_1) = p(\theta_0/x)l_{10} + p(\theta_1/x)l_{11} \tag{15-58}$$

We choose action a_0 if and only if

$$L(a_0) < L(a_1) \tag{15-59}$$

Substituting (15-57) and (15-58) into (15-59), and collecting like terms, we obtain the criterion: choose a_0 iff

$$p(\theta_1/x)[l_{01} - l_{11}] < p(\theta_0/x)[l_{10} - l_{00}] \tag{15-60}$$

The bracketed quantities

$$r_0 \triangleq l_{10} - l_{00} \tag{15-61}$$

and

$$r_1 \triangleq l_{01} - l_{11} \tag{15-62}$$

are called regrets. It is easy to see why: the regret if the beetle is harmless (r_0) is the extra loss incurred if we used the wrong action—i.e., sprayed (a_1), rather than not sprayed (a_0). Evaluating (15-61) we see that r_0 is $15 - 5 = 10$, the difference in column elements in Table (15-10). Our much larger regret

$r_1 = 100 - 15$ represents our net loss if we employ the wrong action (don't spray) on a beetle that turns out to be harmful.

Returning to (15-60), it may now be written in terms of regrets:

$$p(\theta_1/x)r_1 < p(\theta_0/x)\, r_0 \tag{15-63}$$

i.e.,

$$\frac{p(\theta_1/x)}{p(\theta_0/x)} < \frac{r_0}{r_1} \tag{15-64}$$

The posterior probabilities in this equation can now be expressed in full using (15-8), and noting that $p(x)$ cancels,

$$\frac{p(\theta_1)\, p(x/\theta_1)}{p(\theta_0)\, p(x/\theta_0)} < \frac{r_0}{r_1} \tag{15-65}$$

Recall that this is our criterion for action a_0 (don't spray), interpreted as acceptance of H_0: (beetle harmless, $\theta = \theta_0$). An appropriate cross-multiplication of (15-65) leads us to an important theorem, called the

Bayesian Likelihood-Ratio Criterion:
Accept H_0 iff

$$\frac{p(x/\theta_1)}{p(x/\theta_0)} < \frac{r_0\, p(\theta_0)}{r_1\, p(\theta_1)} \tag{15-66}$$

where r_i is the regret if θ_i is true, $p(\theta_i)$ is the prior distribution, and $p(x/\theta_i)$ is the distribution of the observed data.

As stated earlier, $p(x/\theta_i)$ is often borrowed directly from classical deduction, and is the distribution of the estimator x, given the parameter θ_i. Specifically, it appeared in maximum likelihood estimation in Section 7-3 as the likelihood function. Thus the left-hand side of (15-66) is called the "likelihood ratio."

This criterion is certainly reasonable. If θ_1 is a sufficiently implausible explanation of the data [i.e., $p(x/\theta_1)$ is sufficiently less than $p(x/\theta_0)$], then the likelihood ratio will be small enough to satisfy this inequality. Thus H_0 will be accepted, as it should be.

To illustrate further, consider the very simple case in which the regrets (penalties for error) are assumed equal, and the prior probabilities $p(\theta_0)$ and $p(\theta_1)$ are also assumed equal. The right-hand side of (15-66) becomes 1; thus H_0 is accepted if the likelihood of θ_0 generating the sample [$p(x/\theta_0)$] is greater than the likelihood of θ_1 generating the sample [$p(x/\theta_1)$]. Otherwise, the alternative H_1 is accepted. In simplest terms: we select the hypothesis which is more likely to generate the observed x. In this sense, this could be viewed as hypothesis testing, within a maximum likelihood context, shown in Figure 15-8*a*. In *b* we make the further assumption that the two likelihood functions (centered on θ_0 and θ_1 respectively) have the same normal[10] distribution. Then criterion (15-66) reduces to

$$\text{Accept } H_0 \text{ iff } x \text{ is observed closer to } \theta_0 \text{ than } \theta_1 \qquad (15\text{-}67)$$

Again, a very reasonable result.

Evaluating (15-66) when $r_0 \neq r_1$ or $p(\theta_0) \neq p(\theta_1)$ is obviously a more complicated matter. To keep things simple, we assume that $\theta_0 < \theta_1$, and that $p(x/\theta_0)$ and $p(x/\theta_1)$ are normal with a common σ. Then (15-66)—our criterion

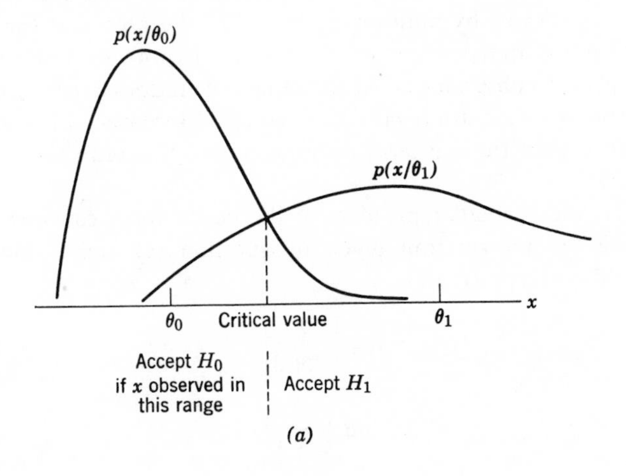

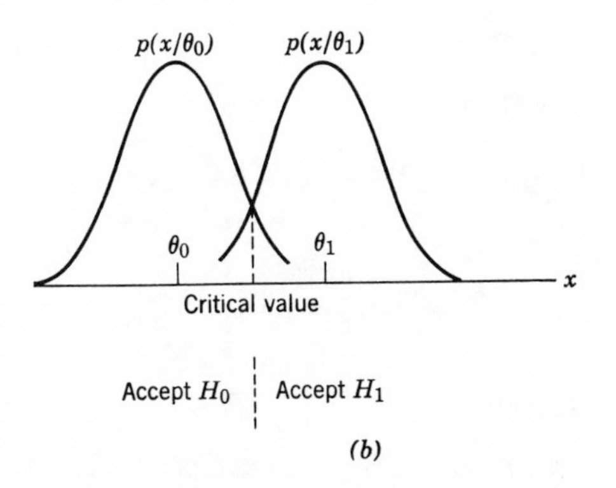

FIG. 15-8 Hypothesis testing, using the Bayesian likelihood ratio [special case when $r_0 = r_1$ and $p(\theta_2) = p(\theta_1)$]. (*a*) For any $p(x/\theta_i)$. (*b*) If $p(x/\theta_i) = N(\theta_i, \sigma)$.

[10] In fact, normality is not required; the two distributions need only be unimodal and symmetric.

for accepting H_0—becomes

$$\frac{e^{-(1/2\sigma^2)(x-\theta_1)^2}}{e^{-(1/2\sigma^2)(x-\theta_0)^2}} < \frac{r_0\, p(\theta_0)}{r_1\, p(\theta_1)} \tag{15-68}$$

This may be reduced[11] to: accept H_0 iff

$$x < \frac{\sigma^2}{\theta_1 - \theta_0} \log\left[\frac{r_0\, p(\theta_0)}{r_1\, p(\theta_1)}\right] + \frac{\theta_1 + \theta_0}{2} \tag{15-69}$$

(The logarithms used throughout this section are natural logarithms, to the base e. The common logarithms of appendix Table VIII can be converted to natural logarithms by multiplying by 2.30.) We note that the right-hand side of (15-69) is independent of x; as in all hypothesis tests, this can be evaluated prior to observing x. At the same time it does depend, as expected, on background information $p(\theta)$ and regrets. Moreover, when $r_0 = r_1$ and $p(\theta_0) = p(\theta_1)$, then the log term disappears and this reduces to the special case (15-67).

Finally, the particular problem of the beetle spray can now be solved. Substituting the information given in question (c) and Table 15-9 into (15-69) yields: accept H_0 iff

$$x < \frac{16}{5} \log\left[\frac{10(.7)}{85(.3)}\right] + 27.5 \tag{15-75}$$

$$< 3.2 \log\left(\frac{7}{25.5}\right) + 27.5 \tag{15-76}$$

$$< 3.2(-1.29) + 27.5 \tag{15-77}$$

$$< 23.4 \tag{15-78}$$

---

[11] Details: taking logarithms of (15-68):

$$-\frac{1}{2\sigma^2}(x - \theta_1)^2 + \frac{1}{2\sigma^2}(x - \theta_0)^2 < K \tag{15-70}$$

where

$$K = \log\left[\frac{r_0\, p(\theta_0)}{r_1\, p(\theta_1)}\right] \tag{15-71}$$

Rearranging (15-70):

$$\frac{1}{2\sigma^2}(2\theta_1 x - 2\theta_0 x - \theta_1^2 + \theta_0^2) < K \tag{15-72}$$

$$2(\theta_1 - \theta_0)x - (\theta_1^2 - \theta_0^2) < 2\sigma^2 K \tag{15-73}$$

i.e., accept H_0 iff:

$$x < \frac{\sigma^2}{\theta_1 - \theta_0} K + \frac{(\theta_1^2 - \theta_0^2)}{2(\theta_1 - \theta_0)} \tag{15-74}$$

Using the definition of K in (15-71), (15-74) may be written as (15-69).

Since we observed a 27 mm beetle, this condition is violated, and we reject H_0. But what does seem strange is the critical value in (15-78): even if the beetle were 25 mm—exactly θ_0, the length we would expect of a *harmless* beetle—we would still spray. With further thought we see that this answer is, after all, reasonable. The heavy damage involved if the beetle turns out to be harmful induces us to spray to avoid this risk. [From (15-75) we confirm that it is in fact the relative size of the two regrets that explains this result.]

(b) Comparison with Classical Methods

The Bayesian hypothesis testing described here involves only two competing hypotheses H_0 and H_1 (two states of nature θ_0 and θ_1), one of which must be chosen. This analysis is of limited scope, since hypothesis testing often involves a composite H_1. Thus we have covered only that material paralleling the first section of Chapter 9. In recalling that classical test, we note that it had the advantage of being far simpler; but it was also less satisfactory. It used only the probability function $p(x/\theta)$, while the Bayesian method also exploits the prior distribution $p(\theta)$ and regrets (the loss function); we have seen in the last section how important both these can be in setting up an appropriate test. Restated, the classical method sets $\alpha = 5$ or 1%—sometimes arbitrarily, sometimes with implicit reference to vague considerations of loss and prior belief. Bayesians would argue that these considerations should be explicitly introduced—with all the assumptions exposed, and open to criticism and improvement.

PROBLEMS

15-12 Using $p(\theta)$ and losses given in Table 15-10:

(a) Reconstruct the hypothesis test of question (c) above. With your measurement of 27 mm, what would you do? Why does our argument in the last paragraph (spray even if beetle is 25 mm) no longer hold?

(b) Suppose that species S_0 and S_1 were equally frequent. Would that alter your decision?

(c) How frequent would species S_0 have to be in order to alter your decision?

15-13 Suppose a psychiatrist has to classify people as sick or well (hospitalized or not) on the basis of a psychological test. The test scores are normally distributed, with $\sigma = 8$, and mean $\theta_0 = 100$ if they are well or $\theta_1 = 120$ if they are sick. The losses (regrets) of a wrong classification are obvious: if a healthy person is hospitalized, resources are wasted and the person himself may even be hurt by the treatment. Yet the other loss is even worse: if a sick person is not hospitalized, he may do damage, conceivably fatal. Suppose this second loss is considered roughly five times as serious. From past records it has been found that of the people taking the test, 60% are sick and 40% are healthy.

(a) (1) What should be the critical score above which the person is classified as sick? Then
(2) What is α? (Probability of type I error).
(3) What is β? (Probability of type II error).

(b) (1) If a classical test is used, arbitrarily setting $\alpha = 5\%$, what then will be the critical score? Then
(2) What is β?
(3) By how much has the average loss increased by using this less-than-optimal method?

(c) What would we have to assume the ratio of the two regrets to be in order to arrive at a Bayesian test having $\alpha = 5\%$? Do you think it is reasonable?

*15-7 GAME THEORY

At this point we leave the general argument of this book to consider a rather interesting branch of decision theory. Recall that the concept of probability was developed in Chapter 3 as a groundwork for the statistical deduction and induction that followed. Game theory is not part of this statistical theory; rather, it illustrates a quite different application of the concept of probability.

Game theory is a way of analyzing conflict situations. These may arise, for example, in poker, business, or politics; thus our conflicting parties might be card players playing for insignificant stakes, oligopolists playing to remain in business, or military leaders engaged in a desperate set of moves and countermoves.

(a) Strictly Determined Games

The players employ strategies. Because a player can choose his strategy, he has some control over the outcome of the game. But he is not in complete control; the outcome will also depend on the strategy of his opponent.

The way the outcome of the game is related to the strategy of both players is shown in Table 15-11; this is called the "payoff matrix," and defines

TABLE 15-11 An Example of a Payoff Matrix for A (Loss Function for B)

		B's Strategies		
		1	2	3
A's Strategies	1	25	6	11
	2	20	10	18
	3	11	7	5

the payoff going to player A. Thus if A selects strategy 2 and B strategy 1, A receives 20. There might also be a payoff matrix for B, similarly dependent on the strategies selected by the two players. However, to keep the discussion simple, suppose that this is a "zero-sum" game—i.e., what A gains, B loses. Thus, Table 15-11 defines not only the gain matrix for A, but also the loss matrix (or loss function) for B. A should be selecting a strategy to make the outcome as large as possible, while B should be trying to keep the outcome as small as possible.

Obviously B will have no interest in playing the game shown in Table 15-11 since he can do nothing but lose. So we might think of a payoff matrix normally involving some positive elements (where B pays A) and some negative ones (where A pays B). Alternatively, in order to induce B to play the game shown in Table 15-11, A might bribe B \$12 for each time he plays. This is the assumption we now make, in order to keep our payoff matrix all-positive for easier geometric interpretation. The question is "With this \$12 side payment, is it in B's interest to play this game?"

If a player can select his strategy after he knows how his opponent has committed himself, his appropriate strategy is obvious and the game becomes a trivial one. For example, if it is known that B has chosen strategy 3, A will just scan column 3, select the largest payoff (18) and then play that strategy 2. The essence of game theory, however, is that each player must commit himself without knowledge of his opponent's decision; he only knows the payoff matrix. We further assume that the game is repeated many times. The only clues a player has about his opponent's strategy must come from observing his past pattern of play.

In these circumstances A finds the continuous play of strategy 1 unattractive. It is true that this row has the largest possible payoff (\$25). But this requires B's cooperation in playing his strategy 1, and it is clearly not in B's interest to cooperate. Indeed if B observes that A is continuously playing strategy 1, he will select strategy 2, thus keeping the payoff down to 6. A finds strategy 3 similarly unattractive; B will counter with strategy 3, reducing A's payoff to only 5. A chooses strategy 2; the very best play by B

will still yield A a payoff of 10. Now review why A chose strategy 2. He calculated the minimum value in each row—and then selected the largest of these minimum values. This *maxi*mum of the row *min*ima is called the "maximin."

Now consider the problem from B's point of view. Recall that he wants to keep the payoff as low as possible. Strategy 1 is ill-advised; when A observes

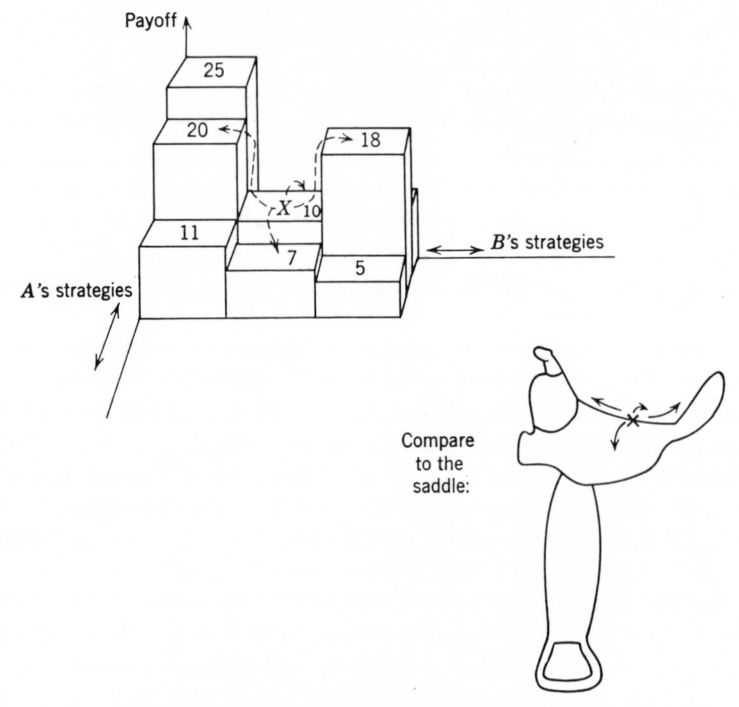

FIG. 15-9 Payoff matrix in Table 15-11.

him playing this, he will counter with 1, leaving B with a loss of \$25. Strategy 3 is also rejected—it may cost him \$18. He selects strategy 2; the most it can cost him is \$10. Note that B calculates the maximum value of each column, and then selects the smallest of these. This *mini*mum of the column *max*ima is called the "minimax." Note that in this special case minimax occurs at the same point as maximin, with a payoff of 10. In this game, A will play his strategy 2, and B will play his strategy 2; this is called a "strictly determined" game—because minimax and maximin coincide.

This is illustrated in Figure 15-9, a diagram of the payoff matrix with each payoff measured vertically. At X we note a "saddle point," which is both the largest element in its column and the smallest element in its row. When such a saddle point exists, it is both maximin and minimax.

Summary. In this strictly determined game, both *A* and *B* will play strategy 2. The payoff (from *B* to *A*) is always 10, so that it is clearly a game *B* will wish to play if he is bribed $12 to do so.

PROBLEMS

15-14 What is the appropriate strategy for each opponent, in the following games; in each case decide which player the game favors.

(a)

		B 1	2	3	4
	1	3	−1	0	−2
A	2	2	2	1	3
	3	−2	0	−1	−3

(b)

		B 1	2	3
A	1	10	−2	10
	2	20	−1	5

(c)

		B 1	2
	1	−20	−4
A	2	−6	−2
	3	1	0

(b) Mixed Strategies

Let us now try to apply the theory of part (a) to the following game:

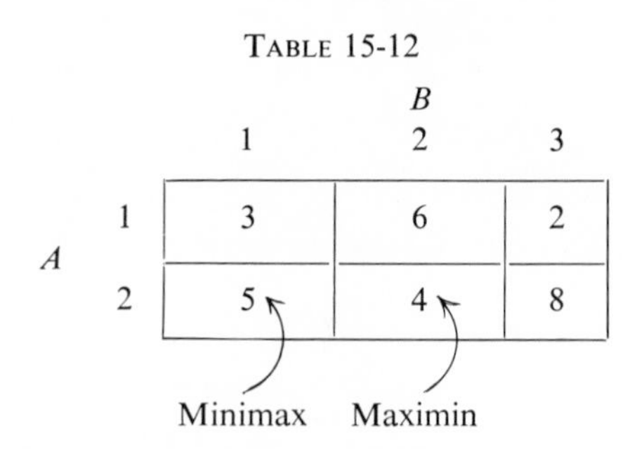
TABLE 15-12

		B 1	B 2	B 3
A	1	3	6	2
A	2	5	4	8

A would select his strategy 2; this is the row with the largest minimum value (maximin = 4); at the same time *B* would select his strategy 1; this is the column with the smallest maximum value (minimax = 5). But now problems arise; because minimax and maximin do not coincide, there is no saddle point. Such a game is not strictly determined, with each playing only one strategy; it is easy to see why. *B* begins by playing column 1, while *A* plays row 2; the payoff is 5. Now *B* observes, that as long as *A* is playing row 2, he can do better by playing column 2, thus reducing the payoff to 4. But when *B* switches to column 2, it is now in *A*'s interest to switch to row 1, raising the payoff to 6. As an exercise the student should confirm that a whole series of such moves and countermoves are set into play—eventually drawing the players in a circle around to the initial position. Then a new cycle begins.

This will continue until the players recognize a fundamental point. Once a player allows his strategy to be predicted, he will be hurt. Thus, for example, when *A*'s strategy becomes clear, he can be hurt by *B*. What is his defense?

A's best plan is to keep *B* guessing. Thus if *A* determines his strategy by a chance process, *B* will be unable to predict what he will do. For example, *A* might toss a coin, playing row 1 if heads, or row 2 if tails. He is using a "mixed strategy," weighting each row with a probability of .50. Now *B* doesn't know what to expect; the only question left for *A* is whether this 50/50 mix is the best set of odds to use.

The best mix of strategies for *A* is determined in Figure 15-10. Along the horizontal axis we consider various probabilities that *A* may attach to playing row 1. This is all *A* has to select; once this is determined (e.g., if

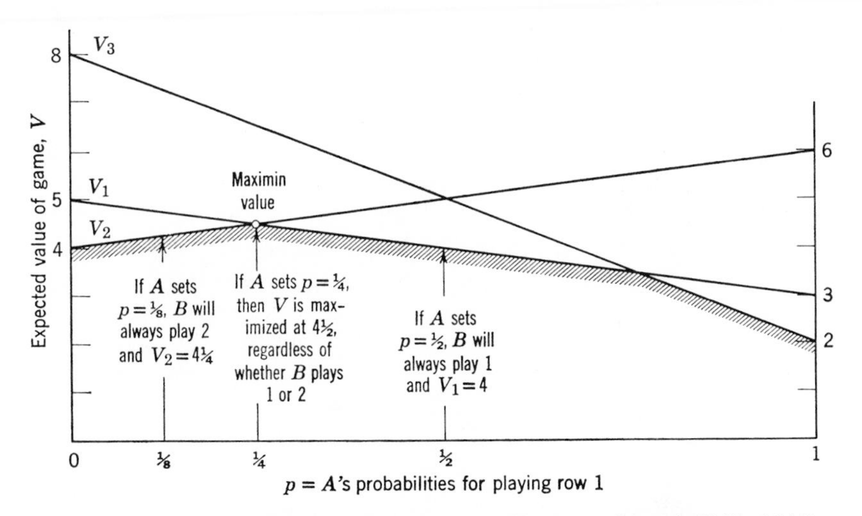

FIG. 15-10 Determination of A's mixed strategy (for game shown in Table 15-12).

A sets $p = 1/3$), then the probability attached to playing row 2 is also determined ($1 - 1/3 = 2/3$).

Vertically, we plot the expected value of the game—which, of course, not only depends on the probabilities A may select, but also on what B may do. If B plays *only* column 1, then the expected value of the game is a function only of the probability A may select; this appears in this diagram as the line V_1. It is worth examining in detail.

At the extreme left, if A sets $p = 0$ (i.e., never plays row 1, but always plays row 2) then the value of the game V_1 is 5. On the other hand, at the extreme right, if A sets $p = 1$ (and always plays row 1), then $V_1 = 3$. Or if A sets $p = 1/2$, then

$$V_1 = 3(1/2) + 5(1/2) = 4 \tag{15-79}$$

Generally, for any probability p that A may select:

$$V_1 = 3(p) + 5(1 - p) = 5 - 2p \tag{15-80}$$

The form of (15-80) confirms that V_1 is a straight line function of p, the probability A selects. Similarly, if B plays only column 2, then

$$V_2 = 6p + 4(1 - p) = 4 + 2p \tag{15-81}$$

Or, if B plays only column 3, then

$$V_3 = 2p + 8(1 - p) = 8 - 6p \tag{15-82}$$

These last two equations are also graphed in Figure 15-10.

The game is now laid out for A to analyze, with his problem being to select p. If he selects $p = 1/8$, his opponent will counter by always playing 2, and keep the expected value of the game at $4\frac{1}{4}$. [This is shown geometrically, and confirmed by evaluating (15-81) setting $p = 1/8$.] Or if A selects $p = 1/2$, B will counter with 1, thus keeping the expected value of the game down to 4. Since A is dealing with an opponent who will be selecting strategies to keep V *low*, the expected value of the game from A's point of view is shown as the hatched line in Figure 15-10. The best A can do, therefore, is to select $p = 1/4$. This guarantees $V = 4\frac{1}{2}$; moreover, note that this is the intersection of V_1 and V_2. Thus this is the value of the game regardless of whether B plays 1 or 2. This geometric solution may be read from Figure 15-10, or determined algebraically by setting $V_1 = V_2$; using (15-80) and (15-81):

$$5 - 2p = 4 + 2p \tag{15-83}$$

$$p = \tfrac{1}{4} \tag{15-84}$$

Finally, this value of p is substituted back into (15-81) for the value of the game:

$$V_2 = 4 + 2(\tfrac{1}{4}) = 4\tfrac{1}{2} \tag{15-85}$$

Thus A decides to attach a probability of 1/4 to playing row 1. How does he put this into practice? There are several possibilities; for example, he might toss 2 coins. If they both come up heads (probability 1/4), then he plays row 1; if not, he plays row 2. If this game is repeated many times, A will insure that he receives an average payoff which will tend towards $4\frac{1}{2}$—and there is nothing B can do to reduce this. All B can hope for is that A has bad luck; (e.g., by the luck of the toss, A plays row 1 when B is playing column 1). This sort of bad luck can reduce A's average winning below $4\frac{1}{2}$ if the game is played only a few times (or A's good luck can raise his average winnings above $4\frac{1}{2}$); but as the game is played over and over, the element of bad luck tends to fade out.

PROBLEMS

⇒ 15-15 (a) Let's play a variation on matching coins. Each of us will choose heads or tails, independently and secretly. I'll pay you \$30 if I show tails and you show heads. I'll pay you \$10 if I show heads and you show tails. Finally, to make it fair, you pay me \$20 if we match (i.e., both show heads, or both show tails). Do you want to play? Why?
(b) What are the optimal strategies of the two players in an *ordinary* game of matching pennies? (Recall that in this game, one player gets the pennies if they match, the other gets them if they don't match.)

Would you still toss your penny in such a game, rather than secretly selecting a head or tail? Why?

15-16 You find yourself on a long sea voyage. You wish to match pennies, but your companion wants to play cards. He therefore suggests a compromise. You choose heads or tails while he selects an ace. If you select a head he pays you \$15, \$4, \$−5, and \$1 respectively, depending on whether he's chosen the spade, heart, diamond or club ace. If you select a tail, he pays you \$−10, \$−2, \$1, and \$−5, again depending on which ace he's chosen.
(a) Do you agree to play? Why? What strategies?
(b) If you were to play this game five times and found you had won \$5, what would you conclude?
(c) Are there any two lines in your diagram that do *not* intersect? From both the diagram and the payoff matrix, show that, no matter what the circumstances, it is always preferable for him to select the club ace instead of the heart ace (i.e., the heart strategy is "dominated by" the club strategy). By initially examining the payoff matrix, couldn't he have dropped the heart strategy from all further consideration?

(c) Conclusions

In solving for the best game strategy, the first step is to test whether maximin and minimax coincide. If they do, this is a strictly determined game, and the single strategy to be used by each player is determined.

If minimax and maximin do not coincide, the game is not strictly determined. Mixed strategies are called for, and are determined in simple cases geometrically or algebraically as we have illustrated. In more complex cases, more advanced mathematical techniques are required; but rather than extending the mechanical solution, it is more important to consider the fundamental philosophy and assumptions underlying game theory:

1. A player using his best mixed strategy can guarantee a certain expected value for the game, regardless of what his opponent may do. However, this is only the value towards which the average of many games will tend. If the game is only played a few times, luck may raise or lower this payoff.

2. Once the optimal mixed strategy is determined (e.g., $p = 1/2$), the play is dictated by a random process (tossing a coin). It is simply not good enough to decide to play each strategy half the time—for example, alternating 1, then 2, then 1, then 2, and so on. Once the opponent observes this pattern,

he can predict your next play, and hurt you. Note in the simplest game of matching pennies (Problem 5-15b), how badly a player would be hurt if he interchanged heads and tails rather than tossed the coin. Once an intelligent opponent observed this pattern, he could win every time. Each player must be unpredictable, by deciding his play be chance.

3. The theory of games is a very conservative strategy. It is appropriate if a game is being replayed many times against an intelligent strategist, who is out to get you, knows the payoff matrix, and can observe your strategy mix. If these conditions are not met, chances are you can find a better strategy than game play. To illustrate, consider an extreme example. Your payoff is:

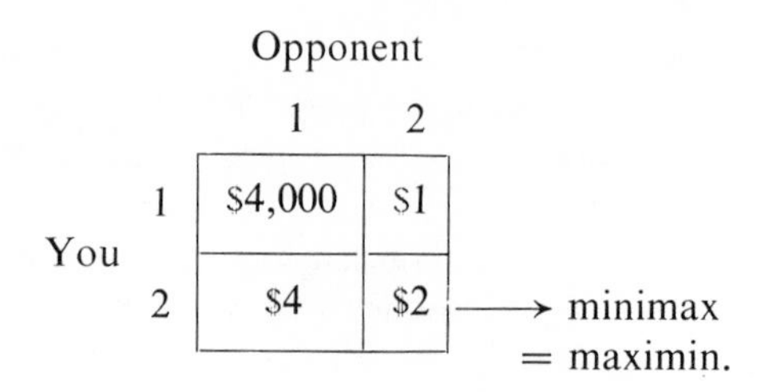

		Opponent 1	Opponent 2
You	1	\$4,000	\$1
	2	\$4	\$2 → minimax = maximin.

Since maximin and minimax coincide, both you and your opponent should play strategy 2 every time. But on the first play, your opponent plays strategy 1! This means either that he is a fool, or unaware of the payoff matrix (and that \$4,000 debacle that he faces). It doesn't matter which; in these circumstances you drop game strategy, play row 1 and punish your opponent for his stupidity or ignorance.

Game theory also should not be used in games against nature. As an example, suppose you are trying to decide whether to hold a picnic indoors or outdoors.

		Nature: Rain 1	Nature: Not Rain 2
You hold picnic:	Indoors 1	100	0
	Outdoors 2	0	1,000

Your profit depends on both where the picnic is held, and the weather. You can easily confirm that game theory means selecting $p = 10/11$, with an expected profit of just over \$90. These odds mean that you will probably hold the picnic indoors.

Something clearly has gone wrong. An intuitive glance at the payoff matrix suggests you should go outdoors, providing there is a reasonable expectation that it won't rain. Game strategy is inappropriate because it is based on the false premise that nature is an opponent—determining the weather with the sole objective of ruining your picnic (i.e., minimizing V). Instead, nature's odds are determined independently; and let us suppose that the probability is 4/5 that it will *not* rain. With these odds, you should be holding the picnic outdoors, with an expected profit of:

$$0(\tfrac{1}{5}) + 1000(\tfrac{4}{5}) = 800 \qquad (15\text{-}86)$$

The more complicated game solution is dead wrong in this case, because one of the key game theory assumptions (nature is intelligent and out to get you) simply does not hold. The student will immediately see that the simpler solution (15-86) is required; and this of course is the Bayesian, or expected value, solution outlined in Section 15-2.

In conclusion, if a prior distribution $p(\theta)$ does not exist independently, but rather is determined by a hostile opponent, then game theory is appropriate; but even under these conditions it may be too conservative a line of play, unless the opponent is highly intelligent and informed. On the other hand, if $p(\theta)$ is determined independently (e.g., rain versus shine), then Bayesian methods are required.

Appendix

LIST OF TABLES

TABLE I Squares and Square Roots

N	N^2	$\sqrt{N}$	$\sqrt{10N}$
1.00	1.0000	1.00000	3.16228
1.01	1.0201	1.00499	3.17805
1.02	1.0404	1.00995	3.19374
1.03	1.0609	1.01489	3.20936
1.04	1.0816	1.01980	3.22490
1.05	1.1025	1.02470	3.24037
1.06	1.1236	1.02956	3.25576
1.07	1.1449	1.03441	3.27109
1.08	1.1664	1.03923	3.28634
1.09	1.1881	1.04403	3.30151
1.10	1.2100	1.04881	3.31662
1.11	1.2321	1.05357	3.33167
1.12	1.2544	1.05830	3.34664
1.13	1.2769	1.06301	3.36155
1.14	1.2996	1.06771	3.37639
1.15	1.3225	1.07238	3.39116
1.16	1.3456	1.07703	3.40588
1.17	1.3689	1.08167	3.42053
1.18	1.3924	1.08628	3.43511
1.19	1.4161	1.09087	3.44964
1.20	1.4400	1.09545	3.46410
1.21	1.4641	1.10000	3.47851
1.22	1.4884	1.10454	3.49285
1.23	1.5129	1.10905	3.50714
1.24	1.5376	1.11355	3.52136
1.25	1.5625	1.11803	3.53553
1.26	1.5876	1.12250	3.54965
1.27	1.6129	1.12694	3.56371
1.28	1.6384	1.13137	3.57771
1.29	1.6641	1.13578	3.59166
1.30	1.6900	1.14018	3.60555
1.31	1.7161	1.14455	3.61939
1.32	1.7424	1.14891	3.63318
1.33	1.7689	1.15326	3.64692
1.34	1.7956	1.15758	3.66060
1.35	1.8225	1.16190	3.67423
1.36	1.8496	1.16619	3.68782
1.37	1.8769	1.17047	3.70135
1.38	1.9044	1.17473	3.71484
1.39	1.9321	1.17898	3.72827
1.40	1.9600	1.18322	3.74166
1.41	1.9881	1.18743	3.75500
1.42	2.0164	1.19164	3.76829
1.43	2.0449	1.19583	3.78153
1.44	2.0736	1.20000	3.79473
1.45	2.1025	1.20416	3.80789
1.46	2.1316	1.20830	3.82099
1.47	2.1609	1.21244	3.83406
1.48	2.1904	1.21655	3.84708
1.49	2.2201	1.22066	3.86005
1.50	2.2500	1.22474	3.87298
N	N^2	$\sqrt{N}$	$\sqrt{10N}$

N	N^2	$\sqrt{N}$	$\sqrt{10N}$
1.50	2.2500	1.22474	3.87298
1.51	2.2801	1.22882	3.88587
1.52	2.3104	1.23288	3.89872
1.53	2.3409	1.23693	3.91162
1.54	2.3716	1.24097	3.92428
1.55	2.4025	1.24499	3.93700
1.56	2.4336	1.24900	3.94968
1.57	2.4649	1.25300	3.96232
1.58	2.4964	1.25698	3.97492
1.59	2.5281	1.26095	3.98748
1.60	2.5600	1.26491	4.00000
1.61	2.5921	1.26886	4.01248
1.62	2.6244	1.27279	4.02492
1.63	2.6569	1.27671	4.03733
1.64	2.6896	1.28062	4.04969
1.65	2.7225	1.28452	4.06202
1.66	2.7556	1.28841	4.07431
1.67	2.7889	1.29228	4.08656
1.68	2.8224	1.29615	4.09878
1.69	2.8561	1.30000	4.11096
1.70	2.8900	1.30384	4.12311
1.71	2.9241	1.30767	4.13521
1.72	2.9584	1.31149	4.14729
1.73	2.9929	1.31529	4.15933
1.74	3.0276	1.31909	4.17133
1.75	3.0625	1.32288	4.18330
1.76	3.0976	1.32665	4.19524
1.77	3.1329	1.33041	4.20714
1.78	3.1684	1.33417	4.21900
1.79	3.2041	1.33791	4.23084
1.80	3.2400	1.34164	4.24264
1.81	3.2761	1.34536	4.25441
1.82	3.3124	1.34907	4.26615
1.83	3.3489	1.35277	4.27785
1.84	3.3856	1.35647	4.28952
1.85	3.4225	1.36015	4.30116
1.86	3.4596	1.36382	4.31277
1.87	3.4969	1.36748	4.32435
1.88	3.5344	1.37113	4.33590
1.89	3.5721	1.37477	4.34741
1.90	3.6100	1.37840	4.35890
1.91	3.6481	1.38203	4.37035
1.92	3.6864	1.38564	4.38178
1.93	3.7249	1.38924	4.39318
1.94	3.7636	1.39284	4.40454
1.95	3.8025	1.39642	4.41588
1.96	3.8416	1.40000	4.42719
1.97	3.8809	1.40357	4.43847
1.98	3.9204	1.40712	4.44972
1.99	3.9601	1.41067	4.46094
2.00	4.0000	1.41421	4.47214
N	N^2	$\sqrt{N}$	$\sqrt{10N}$

Table I (Continued)

N	N^2	$\sqrt{N}$	$\sqrt{10N}$
2.00	4.0000	1.41421	4.47214
2.01	4.0401	1.41774	4.48330
2.02	4.0804	1.42127	4.49444
2.03	4.1209	1.42478	4.50555
2.04	4.1616	1.42829	4.51664
2.05	4.2025	1.43178	4.52769
2.06	4.2436	1.43527	4.53872
2.07	4.2849	1.43875	4.54973
2.08	4.3264	1.44222	4.56070
2.09	4.3681	1.44568	4.57165
2.10	4.4100	1.44914	4.58258
2.11	4.4521	1.45258	4.59347
2.12	4.4944	1.45602	4.60435
2.13	4.5369	1.45945	4.61519
2.14	4.5796	1.46287	4.62601
2.15	4.6225	1.46629	4.63681
2.16	4.6656	1.46969	4.64758
2.17	4.7089	1.47309	4.65833
2.18	4.7524	1.47648	4.66905
2.19	4.7961	1.47986	4.67974
2.20	4.8400	1.48324	4.69042
2.21	4.8841	1.48661	4.70106
2.22	4.9284	1.48997	4.71169
2.23	4.9729	1.49332	4.72229
2.24	5.0176	1.49666	4.73286
2.25	5.0625	1.50000	4.74342
2.26	5.1076	1.50333	4.75395
2.27	5.1529	1.50665	4.76445
2.28	5.1984	1.50997	4.77493
2.29	5.2441	1.51327	4.78539
2.30	5.2900	1.51658	4.79583
2.31	5.3361	1.51987	4.80625
2.32	5.3824	1.52315	4.81664
2.33	5.4289	1.52643	4.82701
2.34	5.4756	1.52971	4.83735
2.35	5.5225	1.53297	4.84768
2.36	5.5696	1.53623	4.85798
2.37	5.6169	1.53948	4.86826
2.38	5.6644	1.54272	4.87852
2.39	5.7121	1.54596	4.88876
2.40	5.7600	1.54919	4.89898
2.41	5.8081	1.55242	4.90918
2.42	5.8564	1.55563	4.91935
2.43	5.9049	1.55885	4.92950
2.44	5.9536	1.56205	4.93964
2.45	6.0025	1.56525	4.94975
2.46	6.0516	1.56844	4.95984
2.47	6.1009	1.57162	4.96991
2.48	6.1504	1.57480	4.97996
2.49	6.2001	1.57797	4.98999
2.50	6.2500	1.58114	5.00000
N	N^2	$\sqrt{N}$	$\sqrt{10N}$

N	N^2	$\sqrt{N}$	$\sqrt{10N}$
2.50	6.2500	1.58114	5.00000
2.51	6.3001	1.58430	5.00999
2.52	6.3504	1.58745	5.01996
2.53	6.4009	1.59060	5.02991
2.54	6.4516	1.59374	5.03984
2.55	6.5025	1.59687	5.04975
2.56	6.5536	1.60000	5.05964
2.57	6.6049	1.60312	5.06952
2.58	6.6564	1.60624	5.07937
2.59	6.7081	1.60935	5.08920
2.60	6.7600	1.61245	5.09902
2.61	6.8121	1.61555	5.10882
2.62	6.8644	1.61864	5.11859
2.63	6.9169	1.62173	5.12835
2.64	6.9696	1.62481	5.13809
2.65	7.0225	1.62788	5.14782
2.66	7.0756	1.63095	5.15752
2.67	7.1289	1.63401	5.16720
2.68	7.1824	1.63707	5.17687
2.69	7.2361	1.64012	5.18652
2.70	7.2900	1.64317	5.19615
2.71	7.3441	1.64621	5.20577
2.72	7.3984	1.64924	5.21536
2.73	7.4529	1.65227	5.22494
2.74	7.5076	1.65529	5.23450
2.75	7.5625	1.65831	5.24404
2.76	7.6176	1.66132	5.25357
2.77	7.6729	1.66433	5.26308
2.78	7.7284	1.66733	5.27257
2.79	7.7841	1.67033	5.28205
2.80	7.8400	1.67332	5.29150
2.81	7.8961	1.67631	5.30094
2.82	7.9524	1.67929	5.31037
2.83	8.0089	1.68226	5.31977
2.84	8.0656	1.68523	5.32917
2.85	8.1225	1.68819	5.33854
2.86	8.1796	1.69115	5.34790
2.87	8.2369	1.69411	5.35724
2.88	8.2944	1.69706	5.36656
2.89	8.3521	1.70000	5.37587
2.90	8.4100	1.70294	5.38516
2.91	8.4681	1.70587	5.39444
2.92	8.5264	1.70880	5.40370
2.93	8.5849	1.71172	5.41295
2.94	8.6436	1.71464	5.42218
2.95	8.7025	1.71756	5.43139
2.96	8.7616	1.72047	5.44059
2.97	8.8209	1.72337	5.44977
2.98	8.8804	1.72627	5.45894
2.99	8.9401	1.72916	5.46809
3.00	9.0000	1.73205	5.47723
N	N^2	$\sqrt{N}$	$\sqrt{10N}$

TABLE I (Continued)

N	N²	√N	√10N
3.00	9.0000	1.73205	5.47723
3.01	9.0601	1.73494	5.48635
3.02	9.1204	1.73781	5.49545
3.03	9.1809	1.74069	5.50454
3.04	9.2416	1.74356	5.51362
3.05	9.3025	1.74642	5.52268
3.06	9.3636	1.74929	5.53173
3.07	9.4249	1.75214	5.54076
3.08	9.4864	1.75499	5.54977
3.09	9.5481	1.75784	5.55878
3.10	9.6100	1.76068	5.56776
3.11	9.6721	1.76352	5.57674
3.12	9.7344	1.76635	5.58570
3.13	9.7969	1.76918	5.59464
3.14	9.8596	1.77200	5.60357
3.15	9.9225	1.77482	5.61249
3.16	9.9856	1.77764	5.62139
3.17	10.0489	1.78045	5.63028
3.18	10.1124	1.78326	5.63915
3.19	10.1761	1.78606	5.64801
3.20	10.2400	1.78885	5.65685
3.21	10.3041	1.79165	5.66569
3.22	10.3684	1.79444	5.67450
3.23	10.4329	1.79722	5.68331
3.24	10.4976	1.80000	5.69210
3.25	10.5625	1.80278	5.70088
3.26	10.6276	1.80555	5.70964
3.27	10.6929	1.80831	5.71839
3.28	10.7584	1.81108	5.72713
3.29	10.8241	1.81384	5.73585
3.30	10.8900	1.81659	5.74456
3.31	10.9561	1.81934	5.75326
3.32	11.0224	1.82209	5.76194
3.33	11.0889	1.82483	5.77062
3.34	11.1556	1.82757	5.77927
3.35	11.2225	1.83030	5.78792
3.36	11.2896	1.83303	5.79655
3.37	11.3569	1.83576	5.80517
3.38	11.4244	1.83848	5.81378
3.39	11.4921	1.84120	5.82237
3.40	11.5600	1.84391	5.83095
3.41	11.6281	1.84662	5.83952
3.42	11.6964	1.84932	5.84808
3.43	11.7649	1.85203	5.85662
3.44	11.8336	1.85472	5.86515
3.45	11.9025	1.85742	5.87367
3.46	11.9716	1.86011	5.88218
3.47	12.0409	1.86279	5.89067
3.48	12.1104	1.86548	5.89915
3.49	12.1801	1.86815	5.90762
3.50	12.2500	1.87083	5.91608
N	N²	√N	√10N

N	N²	√N	√10N
3.50	12.2500	1.87083	5.91608
3.51	12.3201	1.87350	5.92453
3.52	12.3904	1.87617	5.93296
3.53	12.4609	1.87883	5.94138
3.54	12.5316	1.88149	5.94979
3.55	12.6025	1.88414	5.95819
3.56	12.6736	1.88680	5.96657
3.57	12.7449	1.88944	5.97495
3.58	12.8164	1.89209	5.98331
3.59	12.8881	1.89473	5.99166
3.60	12.9600	1.89737	6.00000
3.61	13.0321	1.90000	6.00833
3.62	13.1044	1.90263	6.01664
3.63	13.1769	1.90526	6.02495
3.64	13.2496	1.90788	6.03324
3.65	13.3225	1.91050	6.04152
3.66	13.3956	1.91311	6.04979
3.67	13.4689	1.91572	6.05805
3.68	13.5424	1.91833	6.06630
3.69	13.6161	1.92094	6.07454
3.70	13.6900	1.92354	6.08276
3.71	13.7641	1.92614	6.09098
3.72	13.8384	1.92873	6.09918
3.73	13.9129	1.93132	6.10737
3.74	13.9876	1.93391	6.11555
3.75	14.0625	1.93649	6.12372
3.76	14.1376	1.93907	6.13188
3.77	14.2129	1.94165	6.14003
3.78	14.2884	1.94422	6.14817
3.79	14.3641	1.94679	6.15630
3.80	14.4400	1.94936	6.16441
3.81	14.5161	1.95192	6.17252
3.82	14.5924	1.95448	6.18061
3.83	14.6689	1.95704	6.18870
3.84	14.7456	1.95959	6.19677
3.85	14.8225	1.96214	6.20484
3.86	14.8996	1.96469	6.21289
3.87	14.9769	1.96723	6.22093
3.88	15.0544	1.96977	6.22896
3.89	15.1321	1.97231	6.23699
3.90	15.2100	1.97484	6.24500
3.91	15.2881	1.97737	6.25300
3.92	15.3664	1.97990	6.26099
3.93	15.4449	1.98242	6.26897
3.94	15.5236	1.98494	6.27694
3.95	15.6025	1.98746	6.28490
3.96	15.6816	1.98997	6.29285
3.97	15.7609	1.99249	6.30079
3.98	15.8404	1.99499	6.30872
3.99	15.9201	1.99750	6.31664
4.00	16.0000	2.00000	6.32456
N	N²	√N	√10N

TABLE I (Continued)

N	N^2	$\sqrt{N}$	$\sqrt{10N}$
4.00	16.0000	2.00000	6.32456
4.01	16.0801	2.00250	6.33246
4.02	16.1604	2.00499	6.34035
4.03	16.2409	2.00749	6.34823
4.04	16.3216	2.00998	6.35610
4.05	16.4025	2.01246	6.36396
4.06	16.4836	2.01494	6.37181
4.07	16.5649	2.01742	6.37966
4.08	16.6464	2.01990	6.38749
4.09	16.7281	2.02237	6.39531
4.10	16.8100	2.02485	6.40312
4.11	16.8921	2.02731	6.41093
4.12	16.9744	2.02978	6.41872
4.13	17.0569	2.03224	6.42651
4.14	17.1396	2.03470	6.43428
4.15	17.2225	2.03715	6.44205
4.16	17.3056	2.03961	6.44981
4.17	17.3889	2.04206	6.45755
4.18	17.4724	2.04450	6.46529
4.19	17.5561	2.04695	6.47302
4.20	17.6400	2.04939	6.48074
4.21	17.7241	2.05183	6.48845
4.22	17.8084	2.05426	6.49615
4.23	17.8929	2.05670	6.50384
4.24	17.9776	2.05913	6.51153
4.25	18.0625	2.06155	6.51920
4.26	18.1476	2.06398	6.52687
4.27	18.2329	2.06640	6.53452
4.28	18.3184	2.06882	6.54217
4.29	18.4041	2.07123	6.54981
4.30	18.4900	2.07364	6.55744
4.31	18.5761	2.07605	6.56506
4.32	18.6624	2.07846	6.57267
4.33	18.7489	2.08087	6.58027
4.34	18.8356	2.08327	6.58787
4.35	18.9225	2.08567	6.59545
4.36	19.0096	2.08806	6.60303
4.37	19.0969	2.09045	6.61060
4.38	19.1844	2.09284	6.61816
4.39	19.2721	2.09523	6.62571
4.40	19.3600	2.09762	6.63325
4.41	19.4481	2.10000	6.64078
4.42	19.5364	2.10238	6.64831
4.43	19.6249	2.10476	6.65582
4.44	19.7136	2.10713	6.66333
4.45	19.8025	2.10950	6.67083
4.46	19.8916	2.11187	6.67832
4.47	19.9809	2.11424	6.68581
4.48	20.0704	2.11660	6.69328
4.49	20.1601	2.11896	6.70075
4.50	20.2500	2.12132	6.70820
N	N^2	$\sqrt{N}$	$\sqrt{10N}$

N	N^2	$\sqrt{N}$	$\sqrt{10N}$
4.50	20.2500	2.12132	6.70820
4.51	20.3401	2.12368	6.71565
4.52	20.4304	2.12603	6.72309
4.53	20.5209	2.12838	6.73053
4.54	20.6116	2.13073	6.73795
4.55	20.7025	2.13307	6.74537
4.56	20.7936	2.13542	6.75278
4.57	20.8849	2.13776	6.76018
4.58	20.9764	2.14009	6.76757
4.59	21.0681	2.14243	6.77495
4.60	21.1600	2.14476	6.78233
4.61	21.2521	2.14709	6.78970
4.62	21.3444	2.14942	6.79706
4.63	21.4369	2.15174	6.80441
4.64	21.5296	2.15407	6.81175
4.65	21.6225	2.15639	6.81909
4.66	21.7156	2.15870	6.82642
4.67	21.8089	2.16102	6.83374
4.68	21.9024	2.16333	6.84105
4.69	21.9961	2.16564	6.84836
4.70	22.0900	2.16795	6.85565
4.71	22.1841	2.17025	6.86294
4.72	22.2784	2.17256	6.87023
4.73	22.3729	2.17486	6.87750
4.74	22.4676	2.17715	6.88477
4.75	22.5625	2.17945	6.89202
4.76	22.6576	2.18174	6.89928
4.77	22.7529	2.18403	6.90652
4.78	22.8484	2.18632	6.91375
4.79	22.9441	2.18861	6.92098
4.80	23.0400	2.19089	6.92820
4.81	23.1361	2.19317	6.93542
4.82	23.2324	2.19545	6.94262
4.83	23.3289	2.19773	6.94982
4.84	23.4256	2.20000	6.95701
4.85	23.5225	2.20227	6.96419
4.86	23.6196	2.20454	6.97137
4.87	23.7169	2.20681	6.97854
4.88	23.8144	2.20907	6.98570
4.89	23.9121	2.21133	6.99285
4.90	24.0100	2.21359	7.00000
4.91	24.1081	2.21585	7.00714
4.92	24.2064	2.21811	7.01427
4.93	24.3049	2.22036	7.02140
4.94	24.4036	2.22261	7.02851
4.95	24.5025	2.22486	7.03562
4.96	24.6016	2.22711	7.04273
4.97	24.7009	2.22935	7.04982
4.98	24.8004	2.23159	7.05691
4.99	24.9001	2.23383	7.06399
5.00	25.0000	2.23607	7.07107
N	N^2	$\sqrt{N}$	$\sqrt{10N}$

TABLE I (Continued)

N	N^2	$\sqrt{N}$	$\sqrt{10N}$	N	N^2	$\sqrt{N}$	$\sqrt{10N}$
5.00	25.0000	2.23607	7.07107	**5.50**	30.2500	2.34521	7.41620
5.01	25.1001	2.23830	7.07814	5.51	30.3601	2.34734	7.42294
5.02	25.2004	2.24054	7.08520	5.52	30.4704	2.34947	7.42967
5.03	25.3009	2.24277	7.09225	5.53	30.5809	2.35160	7.43640
5.04	25.4016	2.24499	7.09930	5.54	30.6916	2.35372	7.44312
5.05	25.5025	2.24722	7.10634	5.55	30.8025	2.35584	7.44983
5.06	25.6036	2.24944	7.11337	5.56	30.9136	2.35797	7.45654
5.07	25.7049	2.25167	7.12039	5.57	31.0249	2.36008	7.46324
5.08	25.8064	2.25389	7.12741	5.58	31.1364	2.36220	7.46994
5.09	25.9081	2.25610	7.13442	5.59	31.2481	2.36432	7.47663
5.10	26.0100	2.25832	7.14143	**5.60**	31.3600	2.36643	7.48331
5.11	26.1121	2.26053	7.14843	5.61	31.4721	2.36854	7.48999
5.12	26.2144	2.26274	7.15542	5.62	31.5844	2.37065	7.49667
5.13	26.3169	2.26495	7.16240	5.63	31.6969	2.37276	7.50333
5.14	26.4196	2.26716	7.16938	5.64	31.8096	2.37487	7.50999
5.15	26.5225	2.26936	7.17635	5.65	31.9225	2.37697	7.51665
5.16	26.6256	2.27156	7.18331	5.66	32.0356	2.37908	7.52330
5.17	26.7289	2.27376	7.19027	5.67	32.1489	2.38118	7.52994
5.18	26.8324	2.27596	7.19722	5.68	32.2624	2.38328	7.53658
5.19	26.9361	2.27816	7.20417	5.69	32.3761	2.38537	7.54321
5.20	27.0400	2.28035	7.21110	**5.70**	32.4900	2.38747	7.54983
5.21	27.1441	2.28254	7.21803	5.71	32.6041	2.38956	7.55645
5.22	27.2484	2.28473	7.22496	5.72	32.7184	2.39165	7.56307
5.23	27.3529	2.28692	7.23187	5.73	32.8329	2.39374	7.56968
5.24	27.4576	2.28910	7.23878	5.74	32.9476	2.39583	7.57628
5.25	27.5625	2.29129	7.24569	5.75	33.0625	2.39792	7.58288
5.26	27.6676	2.29347	7.25259	5.76	33.1776	2.40000	7.58947
5.27	27.7729	2.29565	7.25948	5.77	33.2929	2.40208	7.59605
5.28	27.8784	2.29783	7.26636	5.78	33.4084	2.40416	7.60263
5.29	27.9841	2.30000	7.27324	5.79	33.5241	2.40624	7.60920
5.30	28.0900	2.30217	7.28011	**5.80**	33.6400	2.40832	7.61577
5.31	28.1961	2.30434	7.28697	5.81	33.7561	2.41039	7.62234
5.32	28.3024	2.30651	7.29383	5.82	33.8724	2.41247	7.62889
5.33	28.4089	2.30868	7.30068	5.83	33.9889	2.41454	7.63544
5.34	28.5156	2.31084	7.30753	5.84	34.1056	2.41661	7.64199
5.35	28.6225	2.31301	7.31437	5.85	34.2225	2.41868	7.64853
5.36	28.7296	2.31517	7.32120	5.86	34.3396	2.42074	7.65506
5.37	28.8369	2.31733	7.32803	5.87	34.4569	2.42281	7.66159
5.38	28.9444	2.31948	7.33485	5.88	34.5744	2.42487	7.66812
5.39	29.0521	2.32164	7.34166	5.89	34.6921	2.42693	7.67463
5.40	29.1600	2.32379	7.34847	**5.90**	34.8100	2.42899	7.68115
5.41	29.2681	2.32594	7.35527	5.91	34.9281	2.43105	7.68765
5.42	29.3764	2.32809	7.36206	5.92	35.0464	2.43311	7.69415
5.43	29.4849	2.33024	7.36885	5.93	35.1649	2.43516	7.70065
5.44	29.5936	2.33238	7.37564	5.94	35.2836	2.43721	7.70714
5.45	29.7025	2.33452	7.38241	5.95	35.4025	2.43926	7.71362
5.46	29.8116	2.33666	7.38918	5.96	35.5216	2.44131	7.72010
5.47	29.9209	2.33880	7.39594	5.97	35.6409	2.44336	7.72658
5.48	30.0304	2.34094	7.40270	5.98	35.7604	2.44540	7.73305
5.49	30.1401	2.34307	7.40945	5.99	35.8801	2.44745	7.73951
5.50	30.2500	2.34521	7.41620	**6.00**	36.0000	2.44949	7.74597
N	N^2	$\sqrt{N}$	$\sqrt{10N}$	N	N^2	$\sqrt{N}$	$\sqrt{10N}$

TABLE I (Continued)

N	N^2	$\sqrt{N}$	$\sqrt{10N}$
6.00	36.0000	2.44949	7.74597
6.01	36.1201	2.45153	7.75242
6.02	36.2404	2.45357	7.75887
6.03	36.3609	2.45561	7.76531
6.04	36.4816	2.45764	7.77174
6.05	36.6025	2.45967	7.77817
6.06	36.7236	2.46171	7.78460
6.07	36.8449	2.46374	7.79102
6.08	36.9664	2.46577	7.79744
6.09	37.0881	2.46779	7.80385
6.10	37.2100	2.46982	7.81025
6.11	37.3321	2.47184	7.81665
6.12	37.4544	2.47386	7.82304
6.13	37.5769	2.47588	7.82943
6.14	37.6996	2.47790	7.83582
6.15	37.8225	2.47992	7.84219
6.16	37.9456	2.48193	7.84857
6.17	38.0689	2.48395	7.85493
6.18	38.1924	2.48596	7.86130
6.19	38.3161	2.48797	7.86766
6.20	38.4400	2.48998	7.87401
6.21	38.5641	2.49199	7.88036
6.22	38.6884	2.49399	7.88670
6.23	38.8129	2.49600	7.89303
6.24	38.9376	2.49800	7.89937
6.25	39.0625	2.50000	7.90569
6.26	39.1876	2.50200	7.91202
6.27	39.3129	2.50400	7.91833
6.28	39.4384	2.50599	7.92465
6.29	39.5641	2.50799	7.93095
6.30	39.6900	2.50998	7.93725
6.31	39.8161	2.51197	7.94355
6.32	39.9424	2.51396	7.94984
6.33	40.0689	2.51595	7.95613
6.34	40.1956	2.51794	7.96241
6.35	40.3225	2.51992	7.96869
6.36	40.4496	2.52190	7.97496
6.37	40.5769	2.52389	7.98123
6.38	40.7044	2.52587	7.98749
6.39	40.8321	2.52784	7.99375
6.40	40.9600	2.52982	8.00000
6.41	41.0881	2.53180	8.00625
6.42	41.2164	2.53377	8.01249
6.43	41.3449	2.53574	8.01873
6.44	41.4736	2.53772	8.02496
6.45	41.6025	2.53969	8.03119
6.46	41.7316	2.54165	8.03741
6.47	41.8609	2.54362	8.04363
6.48	41.9904	2.54558	8.04984
6.49	42.1201	2.54755	8.05605
6.50	42.2500	2.54951	8.06226
N	N^2	$\sqrt{N}$	$\sqrt{10N}$

N	N^2	$\sqrt{N}$	$\sqrt{10N}$
6.50	42.2500	2.54951	8.06226
6.51	42.3801	2.55147	8.06846
6.52	42.5104	2.55343	8.07465
6.53	42.6409	2.55539	8.08084
6.54	42.7716	2.55734	8.08703
6.55	42.9025	2.55930	8.09321
6.56	43.0336	2.56125	8.09938
6.57	43.1649	2.56320	8.10555
6.58	43.2964	2.56515	8.11172
6.59	43.4281	2.56710	8.11788
6.60	43.5600	2.56905	8.12404
6.61	43.6921	2.57099	8.13019
6.62	43.8244	2.57294	8.13634
6.63	43.9569	2.57488	8.14248
6.64	44.0896	2.57682	8.14862
6.65	44.2225	2.57876	8.15475
6.66	44.3556	2.58070	8.16088
6.67	44.4889	2.58263	8.16701
6.68	44.6224	2.58457	8.17313
6.69	44.7561	2.58650	8.17924
6.70	44.8900	2.58844	8.18535
6.71	45.0241	2.59037	8.19146
6.72	45.1584	2.59230	8.19756
6.73	45.2929	2.59422	8.20366
6.74	45.4276	2.59615	8.20975
6.75	45.5625	2.59808	8.21584
6.76	45.6976	2.60000	8.22192
6.77	45.8329	2.60192	8.22800
6.78	45.9684	2.60384	8.23408
6.79	46.1041	2.60576	8.24015
6.80	46.2400	2.60768	8.24621
6.81	46.3761	2.60960	8.25227
6.82	46.5124	2.61151	8.25833
6.83	46.6489	2.61343	8.26438
6.84	46.7856	2.61534	8.27043
6.85	46.9225	2.61725	8.27647
6.86	47.0596	2.61916	8.28251
6.87	47.1969	2.62107	8.28855
6.88	47.3344	2.62298	8.29458
6.89	47.4721	2.62488	8.30060
6.90	47.6100	2.62679	8.30662
6.91	47.7481	2.62869	8.31264
6.92	47.8864	2.63059	8.31865
6.93	48.0249	2.63249	8.32466
6.94	48.1636	2.63439	8.33067
6.95	48.3025	2.63629	8.33667
6.96	48.4416	2.63818	8.34266
6.97	48.5809	2.64008	8.34865
6.98	48.7204	2.64197	8.35464
6.99	48.8601	2.64386	8.36062
7.00	49.0000	2.64575	8.36660
N	N^2	$\sqrt{N}$	$\sqrt{10N}$

Table I (Continued)

N	N^2	$\sqrt{N}$	$\sqrt{10N}$
7.00	49.0000	2.64575	8.36660
7.01	49.1401	2.64764	8.37257
7.02	49.2804	2.64953	8.37854
7.03	49.4209	2.65141	8.38451
7.04	49.5616	2.65330	8.39047
7.05	49.7025	2.65518	8.39643
7.06	49.8436	2.65707	8.40238
7.07	49.9849	2.65895	8.40833
7.08	50.1264	2.66083	8.41427
7.09	50.2681	2.66271	8.42021
7.10	50.4100	2.66458	8.42615
7.11	50.5521	2.66646	8.43208
7.12	50.6944	2.66833	8.43801
7.13	50.8369	2.67021	8.44393
7.14	50.9796	2.67208	8.44985
7.15	51.1225	2.67395	8.45577
7.16	51.2656	2.67582	8.46168
7.17	51.4089	2.67769	8.46759
7.18	51.5524	2.67955	8.47349
7.19	51.6961	2.68142	8.47939
7.20	51.8400	2.68328	8.48528
7.21	51.9841	2.68514	8.49117
7.22	52.1284	2.68701	8.49706
7.23	52.2729	2.68887	8.50294
7.24	52.4176	2.69072	8.50882
7.25	52.5625	2.69258	8.51469
7.26	52.7076	2.69444	8.52056
7.27	52.8529	2.69629	8.52643
7.28	52.9984	2.69815	8.53229
7.29	53.1441	2.70000	8.53815
7.30	53.2900	2.70185	8.54400
7.31	53.4361	2.70370	8.54985
7.32	53.5824	2.70555	8.55570
7.33	53.7289	2.70740	8.56154
7.34	53.8756	2.70924	8.56738
7.35	54.0225	2.71109	8.57321
7.36	54.1696	2.71293	8.57904
7.37	54.3169	2.71477	8.58487
7.38	54.4644	2.71662	8.59069
7.39	54.6121	2.71846	8.59651
7.40	54.7600	2.72029	8.60233
7.41	54.9081	2.72213	8.60814
7.42	55.0564	2.72397	8.61394
7.43	55.2049	2.72580	8.61974
7.44	55.3536	2.72764	8.62554
7.45	55.5025	2.72947	8.63134
7.46	55.6516	2.73130	8.63713
7.47	55.8009	2.73313	8.64292
7.48	55.9504	2.73496	8.64870
7.49	56.1001	2.73679	8.65448
7.50	56.2500	2.73861	8.66025
N	N^2	$\sqrt{N}$	$\sqrt{10N}$

N	N^2	$\sqrt{N}$	$\sqrt{10N}$
7.50	56.2500	2.73861	8.66025
7.51	56.4001	2.74044	8.66603
7.52	56.5504	2.74226	8.67179
7.53	56.7009	2.74408	8.67756
7.54	56.8516	2.74591	8.68332
7.55	57.0025	2.74773	8.68907
7.56	57.1536	2.74955	8.69483
7.57	57.3049	2.75136	8.70057
7.58	57.4564	2.75318	8.70632
7.59	57.6081	2.75500	8.71206
7.60	57.7600	2.75681	8.71780
7.61	57.9121	2.75862	8.72353
7.62	58.0644	2.76043	8.72926
7.63	58.2169	2.76225	8.73499
7.64	58.3696	2.76405	8.74071
7.65	58.5225	2.76586	8.74643
7.66	58.6756	2.76767	8.75214
7.67	58.8289	2.76948	8.75785
7.68	58.9824	2.77128	8.76356
7.69	59.1361	2.77308	8.76926
7.70	59.2900	2.77489	8.77496
7.71	59.4441	2.77669	8.78066
7.72	59.5984	2.77849	8.78635
7.73	59.7529	2.78029	8.79204
7.74	59.9076	2.78209	8.79773
7.75	60.0625	2.78388	8.80341
7.76	60.2176	2.78568	8.80909
7.77	60.3729	2.78747	8.81476
7.78	60.5284	2.78927	8.82043
7.79	60.6841	2.79106	8.82610
7.80	60.8400	2.79285	8.83176
7.81	60.9961	2.79464	8.83742
7.82	61.1524	2.79643	8.84308
7.83	61.3089	2.79821	8.84873
7.84	61.4656	2.80000	8.85438
7.85	61.6225	2.80179	8.86002
7.86	61.7796	2.80357	8.86566
7.87	61.9369	2.80535	8.87130
7.88	62.0944	2.80713	8.87694
7.89	62.2521	2.80891	3.88257
7.90	62.4100	2.81069	8.88819
7.91	62.5681	2.81247	8.89382
7.92	62.7264	2.81425	8.89944
7.93	62.8849	2.81603	8.90505
7.94	63.0436	2.81780	8.91067
7.95	63.2025	2.81957	8.91628
7.96	63.3616	2.82135	8.92188
7.97	63.5209	2.82312	8.92749
7.98	63.6804	2.82489	8.93308
7.99	63.8401	2.82666	8.93868
8.00	64.0000	2.82843	8.94427
N	N^2	$\sqrt{N}$	$\sqrt{10N}$

TABLE I (Continued)

N	N^2	$\sqrt{N}$	$\sqrt{10N}$
8.00	64.0000	2.82843	8.94427
8.01	64.1601	2.83019	8.94986
8.02	64.3204	2.83196	8.95545
8.03	64.4809	2.83373	8.96103
8.04	64.6416	2.83549	8.96660
8.05	64.8025	2.83725	8.97218
8.06	64.9636	2.83901	8.97775
8.07	65 1249	2.84077	8.98332
8.08	65.2864	2.84253	8.98888
8.09	65.4481	2.84429	8.99444
8.10	65.6100	2.84605	9.00000
8.11	65.7721	2.84781	9.00555
8.12	65.9344	2.84956	9.01110
8.13	66.0969	2.85132	9.01665
8.14	66.2596	2.85307	9.02219
8.15	66.4225	2.85482	9.02774
8.16	66.5856	2.85657	9.03327
8.17	66.7489	2.85832	9.03881
8.18	66.9124	2.86007	9.04434
8.19	67.0761	2.86182	9.04986
8.20	67.2400	2.86356	9.05539
8.21	67.4041	2.86531	9.06091
8.22	67.5684	2.86705	9.06642
8.23	67.7329	2.86880	9.07193
8.24	67.8976	2.87054	9.07744
8.25	68.0625	2.87228	9.08295
8.26	68.2276	2.87402	9.08845
8.27	68.3929	2.87576	9.09395
8.28	68.5584	2.87750	9.09945
8.29	68.7241	2.87924	9.10494
8.30	68.8900	2.88097	9.11043
8.31	69.0561	2.88271	9.11592
8.32	69.2224	2.88444	9.12140
8.33	69.3889	2.88617	9.12688
8.34	69.5556	2.88791	9.13236
8.35	69.7225	2.88964	9.13783
8.36	69.8896	2.89137	9.14330
8.37	70.0569	2.89310	9.14877
8.38	70.2244	2.89482	9.15423
8.39	70.3921	2.89655	9.15969
8.40	70.5600	2.89828	9.16515
8.41	70.7281	2.90000	9.17061
8.42	70.8964	2.90172	9.17606
8.43	71.0649	2.90345	9.18150
8.44	71.2336	2.90517	9.18695
8.45	71.4025	2.90689	9.19239
8.46	71.5716	2.90861	9.19783
8.47	71.7409	2.91033	9.20326
8.48	71.9104	2.91204	9.20869
8.49	72.0801	2.91376	9.21412
8.50	72.2500	2.91548	9.21954
N	N^2	$\sqrt{N}$	$\sqrt{10N}$

N	N^2	$\sqrt{N}$	$\sqrt{10N}$
8.50	72.2500	2.91548	9.21954
8.51	72.4201	2.91719	9.22497
8.52	72.5904	2.91890	9.23038
8.53	72.7609	2.92062	9.23580
8.54	72.9316	2.92233	9.24121
8.55	73.1025	2.92404	9.24662
8.56	73.2736	2.92575	9.25203
8.57	73.4449	2.92746	9.25743
8.58	73.6164	2.92916	9.26283
8.59	73.7881	2.93087	9.26823
8.60	73.9600	2.93258	9.27362
8.61	74.1321	2.93428	9.27901
8.62	74.3044	2.93598	9.28440
8.63	74.4769	2.93769	9.28978
8.64	74.6496	2.93939	9.29516
8.65	74.8225	2.94109	9.30054
8.66	74.9956	2.94279	9.30591
8.67	75.1689	2.94449	9.31128
8.68	75.3424	2.94618	9.31665
8.69	75.5161	2.94788	9.32202
8.70	75.6900	2.94958	9.32738
8.71	75.8641	2.95127	9.33274
8.72	76.0384	2.95296	9.33809
8.73	76.2129	2.95466	9.34345
8.74	76.3876	2.95635	9.34880
8.75	76.5625	2.95804	9.35414
8.76	76.7376	2.95973	9.35949
8.77	76.9129	2.96142	9.36483
8.78	77.0884	2.96311	9.37017
8.79	77.2641	2.96479	9.37550
8.80	77.4400	2.96648	9.38083
8.81	77.6161	2.96816	9.38616
8.82	77.7924	2.96985	9.39149
8.83	77.9689	2.97153	9.39681
8.84	78.1456	2.97321	9.40213
8.85	78.3225	2.97489	9.40744
8.86	78.4996	2.97658	9.41276
8.87	78.6769	2.97825	9.41807
8.88	78.8544	2.97993	9.42338
8.89	79.0321	2.98161	9.42868
8.90	79.2100	2.98329	9.43398
8.91	79.3881	2.98496	9.43928
8.92	79.5664	2.98664	9.44458
8.93	79.7449	2.98831	9.44987
8.94	79.9236	2.98998	9.45516
8.95	80.1025	2.99166	9.46044
8.96	80.2816	2.99333	9.46573
8.97	80.4609	2.99500	9.47101
8.98	80.6404	2.99666	9.47629
8.99	80.8201	2.99833	9.48156
9.00	81.0000	3.00000	9.48683
N	N^2	$\sqrt{N}$	$\sqrt{10N}$

TABLE I (Continued)

N	N²	√N	√10N	N	N²	√N	√10N
9.00	81.0000	3.00000	9.48683	**9.50**	90.2500	3.08221	9.74679
9.01	81.1801	3.00167	9.49210	9.51	90.4401	3.08383	9.75192
9.02	81.3604	3.00333	9.49737	9.52	90.6304	3.08545	9.75705
9.03	81.5409	3.00500	9.50263	9.53	90.8209	3.08707	9.76217
9.04	81.7216	3.00666	9.50789	9.54	91.0116	3.08869	9.76729
9.05	81.9025	3.00832	9.51315	9.55	91.2025	3.09031	9.77241
9.06	82.0836	3.00998	9.51840	9.56	91.3936	3.09192	9.77753
9.07	82.2649	3.01164	9.52365	9.57	91.5849	3.09354	9.78264
9.08	82.4464	3.01330	9.52890	9.58	91.7764	3.09516	9.78775
9.09	82.6281	3.01496	9.53415	9.59	91.9681	3.09677	9.79285
9.10	82.8100	3.01662	9.53939	**9.60**	92.1600	3.09839	9.79796
9.11	82.9921	3.01828	9.54463	9.61	92.3521	3.10000	9.80306
9.12	83.1744	3.01993	9.54987	9.62	92.5444	3.10161	9.80816
9.13	83.3569	3.02159	9.55510	9.63	92.7369	3.10322	9.81326
9.14	83.5396	3.02324	9.56033	9.64	92.9296	3.10483	9.81835
9.15	83.7225	3.02490	9.56556	9.65	93.1225	3.10644	9.82344
9.16	83.9056	3.02655	9.57079	9.66	93.3156	3.10805	9.82853
9.17	84.0889	3.02820	9.57601	9.67	93.5089	3.10966	9.83362
9.18	84.2724	3.02985	9.58123	9.68	93.7024	3.11127	9.83870
9.19	84.4561	3.03150	9.58645	9.69	93.8961	3.11288	9.84378
9.20	84.6400	3.03315	9.59166	**9.70**	94.0900	3.11448	9.84886
9.21	84.8241	3.03480	9.59687	9.71	94.2841	3.11609	9.85393
9.22	85.0084	3.03645	9.60208	9.72	94.4784	3.11769	9.85901
9.23	85.1929	3.03809	9.60729	9.73	94.6729	3.11929	9.86408
9.24	85.3776	3.03974	9.61249	9.74	94.8676	3.12090	9.86914
9.25	85.5625	3.04138	9.61769	9.75	95.0625	3.12250	9.87421
9.26	85.7476	3.04302	9.62289	9.76	95.2576	3.12410	9.87927
9.27	85.9329	3.04467	9.62808	9.77	95.4529	3.12570	9.88433
9.28	86.1184	3.04631	9.63328	9.78	95.6484	3.12730	9.88939
9.29	86.3041	3.04795	9.63846	9.79	95.8441	3.12890	9.89444
9.30	86.4900	3.04959	9.64365	**9.80**	96.0400	3.13050	9.89949
9.31	86.6761	3.05123	9.64883	9.81	96.2361	3.13209	9.90454
9.32	86.8624	3.05287	9.65401	9.82	96.4324	3.13369	9.90959
9.33	87.0489	3.05450	9.65919	9.83	96.6289	3.13528	9.91464
9.34	87.2356	3.05614	9.66437	9.84	96.8256	3.13688	9.91968
9.35	87.4225	3.05778	9.66954	9.85	97.0225	3.13847	9.92472
9.36	87.6096	3.05941	9.67471	9.86	97.2196	3.14006	9.92975
9.37	87.7969	3.06105	9.67988	9.87	97.4169	3.14166	9.93479
9.38	87.9844	3.06268	9.68504	9.88	97.6144	3.14325	9.93982
9.39	88.1721	3.06431	9.69020	9.89	97.8121	3.14484	9.94485
9.40	88.3600	3.06594	9.69536	**9.90**	98.0100	3.14643	9.94987
9.41	88.5481	3.06757	9.70052	9.91	98.2081	3.14802	9.95490
9.42	88.7364	3.06920	9.70567	9.92	98.4064	3.14960	9.95992
9.43	88.9249	3.07083	9.71082	9.93	98.6049	3.15119	9.96494
9.44	89.1136	3.07246	9.71597	9.94	98.8036	3.15278	9.96995
9.45	89.3025	3.07409	9.72111	9.95	99.0025	3.15436	9.97497
9.46	89.4916	3.07571	9.72625	9.96	99.2016	3.15595	9.97998
9.47	89.6809	3.07734	9.73139	9.97	99.4009	3.15753	9.98499
9.48	89.8704	3.07896	9.73653	9.98	99.6004	3.15911	9.98999
9.49	90.0601	3.08058	9.74166	9.99	99.8001	3.16070	9.99500
9.50	90.2500	3.08221	9.74679	**10.00**	100.000	3.16228	10.0000
N	N²	√N	√10N	N	N²	√N	√10N

TABLE II*a* Random Digits

39 65 76 45 45	19 90 69 64 61	20 26 36 31 62	58 24 97 14 97	95 06 70 99 00
73 71 23 70 90	65 97 60 12 11	31 56 34 19 19	47 83 75 51 33	30 62 38 20 46
72 20 47 33 84	51 67 47 97 19	98 40 07 17 66	23 05 09 51 80	59 78 11 52 49
75 17 25 69 17	17 95 21 78 58	24 33 45 77 48	69 81 84 09 29	93 22 70 45 80
37 48 79 88 74	63 52 06 34 30	01 31 60 10 27	35 07 79 71 53	28 99 52 01 41
02 89 08 16 94	85 53 83 29 95	56 27 09 24 43	21 78 55 09 82	72 61 88 73 61
87 18 15 70 07	37 79 49 12 38	48 13 93 55 96	41 92 45 71 51	09 18 25 58 94
98 83 71 70 15	89 09 39 59 24	00 06 41 41 20	14 36 59 25 47	54 45 17 24 89
10 08 58 07 04	76 62 16 48 68	58 76 17 14 86	59 53 11 52 21	66 04 18 72 87
47 90 56 37 31	71 82 13 50 41	27 55 10 24 92	28 04 67 53 44	95 23 00 84 47
93 05 31 03 07	34 18 04 52 35	74 13 39 35 22	68 95 23 92 35	36 63 70 35 33
21 89 11 47 99	11 20 99 45 18	76 51 94 84 86	13 79 93 37 55	98 16 04 41 67
95 18 94 06 97	27 37 83 28 71	79 57 95 13 91	09 61 87 25 21	56 20 11 32 44
97 08 31 55 73	10 65 81 92 59	77 31 61 95 46	20 44 90 32 64	26 99 76 75 63
69 26 88 86 13	59 71 74 17 32	48 38 75 93 29	73 37 32 04 05	60 82 29 20 25
41 47 10 25 03	87 63 93 95 17	81 83 83 04 49	77 45 85 50 51	79 88 01 97 30
91 94 14 63 62	08 61 74 51 69	92 79 43 89 79	29 18 94 51 23	14 85 11 47 23
80 06 54 18 47	08 52 85 08 40	48 40 35 94 22	72 65 71 08 86	50 03 42 99 36
67 72 77 63 99	89 85 84 46 06	64 71 06 21 66	89 37 20 70 01	61 65 70 22 12
59 40 24 13 75	42 29 72 23 19	06 94 76 10 08	81 30 15 39 14	81 83 17 16 33
63 62 06 34 41	79 53 36 02 95	94 61 09 43 62	20 21 14 68 86	84 95 48 46 45
78 47 23 53 90	79 93 96 38 63	34 85 52 05 09	85 43 01 72 73	14 93 87 81 40
87 68 62 15 43	97 48 72 66 48	53 16 71 13 81	59 97 50 99 52	24 62 20 42 31
47 60 92 10 77	26 97 05 73 51	88 46 38 03 58	72 68 49 29 31	75 70 16 08 24
56 88 87 59 41	06 87 37 78 48	65 88 69 58 39	88 02 84 27 83	85 81 56 39 38
22 17 68 65 84	87 02 22 57 51	68 69 80 95 44	11 29 01 95 80	49 34 35 86 47
19 36 27 59 46	39 77 32 77 09	79 57 92 36 59	89 74 39 82 15	08 58 94 34 74
16 77 23 02 77	28 06 24 25 93	22 45 44 84 11	87 80 61 65 31	09 71 91 74 25
78 43 76 71 61	97 67 63 99 61	80 45 67 93 82	59 73 19 85 23	53 33 65 97 21
03 28 28 26 08	69 30 16 09 05	53 58 47 70 93	66 56 45 65 79	45 56 20 19 47
04 31 17 21 56	33 73 99 19 87	26 72 39 27 67	53 77 57 68 93	60 61 97 22 61
61 06 98 03 91	87 14 77 43 96	43 00 65 98 50	45 60 33 01 07	98 99 46 50 47
23 68 35 26 00	99 53 93 61 28	52 70 05 48 34	56 65 05 61 86	90 92 10 70 80
15 39 25 70 99	93 86 52 77 65	15 33 59 05 28	22 87 26 07 47	86 96 98 29 06
58 71 96 30 24	18 46 23 34 27	85 13 99 24 44	49 18 09 79 49	74 16 32 23 02
93 22 53 64 39	07 10 63 76 35	87 03 04 79 88	08 13 13 85 51	55 34 57 72 69
78 76 58 54 74	92 38 70 96 92	52 06 79 79 45	82 63 18 27 44	69 66 92 19 09
61 81 31 96 82	00 57 25 60 59	46 72 60 18 77	55 66 12 62 11	08 99 55 64 57
42 88 07 10 05	24 98 65 63 21	47 21 61 88 32	27 80 30 21 60	10 92 35 36 12
77 94 30 05 39	28 10 99 00 27	12 73 73 99 12	49 99 57 94 82	96 88 57 17 91

TABLE IIb Random Normal Numbers, $\mu = 0$, $\sigma = 1$

0.464	0.137	2.455	−0.323	−0.068	0.296	−0.288	1.298	0.241	−0.957
0.060	−2.526	−0.531	−0.194	0.543	−1.558	0.187	−1.190	0.022	0.525
1.486	−0.354	−0.634	0.697	0.926	1.375	0.785	−0.963	−0.853	−1.865
1.022	−0.472	1.279	3.521	0.571	−1.851	0.194	1.192	−0.501	−0.273
1.394	−0.555	0.046	0.321	2.945	1.974	−0.258	0.412	0.439	−0.035
0.906	−0.513	−0.525	0.595	0.881	−0.934	1.579	0.161	−1.885	0.371
1.179	−1.055	0.007	0.769	0.971	0.712	1.090	−0.631	−0.255	−0.702
−1.501	−0.488	−0.162	−0.136	1.033	0.203	0.448	0.748	−0.423	−0.432
−0.690	0.756	−1.618	−0.345	−0.511	−2.051	−0.457	−0.218	0.857	−0.465
1.372	0.225	0.378	0.761	0.181	−0.736	0.960	−1.530	−0.260	0.120
−0.482	1.678	−0.057	−1.229	−0.486	0.856	−0.491	−1.983	−2.830	−0.238
−1.376	−0.150	1.356	−0.561	−0.256	−0.212	0.219	0.779	0.953	−0.869
−1.010	0.598	−0.918	1.598	0.065	0.415	−0.169	0.313	−0.973	−1.016
−0.005	−0.899	0.012	−0.725	1.147	−0.121	1.096	0.481	−1.691	0.417
1.393	−1.163	−0.911	1.231	−0.199	−0.246	1.239	−2.574	−0.558	0.056
−1.787	−0.261	1.237	1.046	−0.508	−1.630	−0.146	−0.392	−0.627	0.561
−0.105	−0.375	−1.384	0.360	−0.992	−0.116	−1.698	−2.832	−1.108	−2.357
−1.339	1.827	−0.959	0.424	0.969	−1.141	−1.041	0.362	−1.726	1.956
1.041	0.535	0.731	1.377	0.983	−1.330	1.620	−1.040	0.524	−0.281
0.279	−2.056	0.717	−0.873	−1.096	−1.396	1.047	0.089	−0.573	0.932
−1.805	−2.008	−1.633	0.542	0.250	−0.166	0.032	0.079	0.471	−1.029
−1.186	1.180	1.114	0.882	1.265	−0.202	0.151	−0.376	−0.310	0.479
0.658	−1.141	1.151	−1.210	−0.927	0.425	0.290	−0.902	0.610	1.709
−0.439	0.358	−1.939	0.891	−0.227	0.602	0.873	−0.437	−0.220	−0.057
−1.399	−0.230	0.385	−0.649	−0.577	0.237	−0.289	0.513	0.738	−0.300
0.199	0.208	−1.083	−0.219	−0.291	1.221	1.119	0.004	−2.015	−0.594
0.159	0·272	−0.313	0.084	−2.828	−0.439	−0.792	−1.275	−0.623	−1.047
2.273	0.606	0.606	−0.747	0.247	1.291	0.063	−1.793	−0.699	−1.347
0.041	−0.307	0.121	0.790	−0.584	0.541	0.484	−0.986	0.481	0.996
−1.132	−2.098	0.921	0.145	0.446	−1.661	1.045	−1.363	−0.586	−1.023
0.768	0.079	−1.473	0.034	−2.127	0.665	0.084	−0.880	−0.579	0.551
0.375	−1.658	−0.851	0.234	−0.656	0.340	−0.086	−0.158	−0.120	0.418
−0.513	−0.344	0.210	−0.736	1.041	0.008	0.427	−0.831	0.191	0.074
0.292	−0.521	1.266	−1.206	−0.899	0.110	−0.528	−0.813	0.071	0.524
1.026	2.990	−0.574	−0.491	−1.114	1.297	−1.433	−1.345	−3.001	0.479
−1.334	1.278	−0.568	−0.109	−0.515	−0.566	2.923	0.500	0.359	0.326
−0.287	−0.144	−0.254	0.574	−0.451	−1.181	−1.190	−0.318	−0.094	1.114
0.161	−0.886	−0.921	−0.509	1.410	−0.518	0.192	−0.432	1.501	1.068
−1.346	0.193	−1.202	0.394	−1.045	0.843	0.942	1.045	0.031	0.772
−1.250	−0.199	−0.288	1.810	1.378	0.584	1.216	0.733	0.402	0.226
0.630	−0.537	0.782	0.060	0.499	−0.431	1.705	1.164	0.884	−0.298
0.375	−1.941	0.247	−0.491	0.665	−0.135	−0.145	−0.498	0.457	1.064
−1.420	0.489	−1.711	−1.186	0.754	−0.732	−0.066	1.006	−0.798	0.162
−0.151	−0.243	−0.430	−0.762	0.298	1.049	1.810	2.885	−0.768	−0.129
−0.309	0.531	0.416	−1.541	1.456	2.040	−0.124	0.196	0.023	−1.204
0.424	−0.444	0.593	0.993	−0.106	0.116	0.484	−1.272	1.066	1.097
0.593	0.658	−1.127	−1.407	−1.579	−1.616	1.458	1.262	0.736	−0.916
0.862	−0.885	−0.142	−0.504	0.532	1.381	0.022	−0.281	−0.342	1.222
0.235	−0.628	−0.023	−0.463	−0.899	−0.394	−0.538	1.707	−0.188	−1.153
−0.853	0.402	0.777	0.833	0.410	−0.349	−1.094	0.580	1.395	1.298

TABLE IIIa Binomial Coefficients

n	$\binom{n}{0}$	$\binom{n}{1}$	$\binom{n}{2}$	$\binom{n}{3}$	$\binom{n}{4}$	$\binom{n}{5}$	$\binom{n}{6}$	$\binom{n}{7}$	$\binom{n}{8}$	$\binom{n}{9}$	$\binom{n}{10}$
0	1										
1	1	1									
2	1	2	1								
3	1	3	3	1							
4	1	4	6	4	1						
5	1	5	10	10	5	1					
6	1	6	15	20	15	6	1				
7	1	7	21	35	35	21	7	1			
8	1	8	28	56	70	56	28	8	1		
9	1	9	36	84	126	126	84	36	9	1	
10	1	10	45	120	210	252	210	120	45	10	1
11	1	11	55	165	330	462	462	330	165	55	11
12	1	12	66	220	495	792	924	792	495	220	66
13	1	13	78	286	715	1287	1716	1716	1287	715	286
14	1	14	91	364	1001	2002	3003	3432	3003	2002	1001
15	1	15	105	455	1365	3003	5005	6435	6435	5005	3003
16	1	16	120	560	1820	4368	8008	11440	12870	11440	8008
17	1	17	136	680	2380	6188	12376	19448	24310	24310	19448
18	1	18	153	816	3060	8568	18564	31824	43758	48620	43758
19	1	19	171	969	3876	11628	27132	50388	75582	92378	92378
20	1	20	190	1140	4845	15504	38760	77520	125970	167960	184756

Note. $\binom{n}{x} = \dfrac{n(n-1)(n-2)\cdots(n-m+1)}{x(x-1)(x-2)\cdots 3.2.1}$; $\binom{n}{0} = 1$; $\binom{n}{1} = n$. For coefficients missing from the above table, use the relation

$$\binom{n}{x} = \binom{n}{n-x}, \qquad \text{e.g.,} \qquad \binom{20}{11} = \binom{20}{9} = 167960.$$

TABLE III*b* Individual Binomial Probabilities $p(x)$

n	x	π .05	.10	.15	.20	.25	.30	.35	.40	.45	.50
1	0	.9500	.9000	.8500	.8000	.7500	.7000	.6500	.6000	.5500	.5000
	1	.0500	.1000	.1500	.2000	.2500	.3000	.3500	.4000	.4500	.5000
2	0	.9025	.8100	.7225	.6400	.5625	.4900	.4225	.3600	.3025	.2500
	1	.0950	.1800	.2550	.3200	.3750	.4200	.4550	.4800	.4950	.5000
	2	.0025	.0100	.0225	.0400	.0625	.0900	.1225	.1600	.2025	.2500
3	0	.8574	.7290	.6141	.5120	.4219	.3430	.2746	.2160	.1664	.1250
	1	.1354	.2430	.3251	.3840	.4219	.4410	.4436	.4320	.4084	.3750
	2	.0071	.0270	.0574	.0960	.1406	.1890	.2389	.2880	.3341	.3750
	3	.0001	.0010	.0034	.0080	.0156	.0270	.0429	.0640	.0911	.1250
4	0	.8145	.6561	.5220	.4096	.3164	.2401	.1785	.1296	.0915	.0625
	1	.1715	.2916	.3685	.4096	.4219	.4116	.3845	.3456	.2995	.2500
	2	.0135	.0486	.0975	.1536	.2109	.2646	.3105	.3456	.3675	.3750
	3	.0005	.0036	.0115	.0256	.0469	.0756	.1115	.1536	.2005	.2500
	4	.0000	.0001	.0005	.0016	.0039	.0081	.0150	.0256	.0410	.0625
5	0	.7738	.5905	.4437	.3277	.2373	.1681	.1160	.0778	.0503	.0312
	1	.2036	.3280	.3915	.4096	.3955	.3602	.3124	.2592	.2059	.1562
	2	.0214	.0729	.1382	.2048	.2637	.3087	.3364	.3456	.3369	.3125
	3	.0011	.0081	.0244	.0512	.0879	.1323	.1811	.2304	.2757	.3125
	4	.0000	.0004	.0022	.0064	.0146	.0284	.0488	.0768	.1128	.1562
	5	.0000	.0000	.0001	.0003	.0010	.0024	.0053	.0102	.0185	.0312
6	0	.7351	.5314	.3771	.2621	.1780	.1176	.0754	.0467	.0277	.0156
	1	.2321	.3543	.3993	.3932	.3560	.3025	.2437	.1866	.1359	.0938
	2	.0305	.0984	.1762	.2458	.2966	.3241	.3280	.3110	.2780	.2344
	3	.0021	.0146	.0415	.0819	.1318	.1852	.2355	.2765	.3032	.3125
	4	.0001	.0012	.0055	.0154	.0330	.0595	.0951	.1382	.1861	.2344
	5	.0000	.0001	.0004	.0015	.0044	.0102	.0205	.0369	.0609	.0938
	6	.0000	.0000	.0000	.0001	.0002	.0007	.0018	.0041	.0083	.0156
7	0	.6983	.4783	.3206	.2097	.1335	.0824	.0490	.0280	.0152	.0078
	1	.2573	.3720	.3960	.3670	.3115	.2471	.1848	.1306	.0872	.0547
	2	.0406	.1240	.2097	.2753	.3115	.3177	.2985	.2613	.2140	.1641
	3	.0036	.0230	.0617	.1147	.1730	.2269	.2679	.2903	.2918	.2734
	4	.0002	.0026	.0109	.0287	.0577	.0972	.1442	.1935	.2388	.2734
	5	.0000	.0002	.0012	.0043	.0115	.0250	.0466	.0774	.1172	.1641
	6	.0000	.0000	.0001	.0004	.0013	.0036	.0084	.0172	.0320	.0547
	7	.0000	.0000	.0000	.0000	.0001	.0002	.0006	.0016	.0037	.0078

If $\pi > .50$, interchange π and $(1 - \pi)$.

TABLE IIIb (Continued)

n	x	.05	.10	.15	.20	.25	π .30	.35	.40	.45	.50
8	0	.6634	.4305	.2725	.1678	.1001	.0576	.0319	.0168	.0084	.0039
	1	.2793	.3826	.3847	.3355	.2670	.1977	.1373	.0896	.0548	.0312
	2	.0515	.1488	.2376	.2936	.3115	.2965	.2587	.2090	.1569	.1094
	3	.0054	.0331	.0839	.1468	.2076	.2541	.2786	.2787	.2568	.2188
	4	.0004	.0046	.0185	.0459	.0865	.1361	.1875	.2322	.2627	.2734
	5	.0000	.0004	.0026	.0092	.0231	.0467	.0808	.1239	.1719	.2188
	6	.0000	.0000	.0002	.0011	.0038	.0100	.0217	.0413	.0703	.1094
	7	.0000	.0000	.0000	.0001	.0004	.0012	.0033	.0079	.0164	.0312
	8	.0000	.0000	.0000	.0000	.0000	.0001	.0002	.0007	.0017	.0039
9	0	.6302	.3874	.2316	.1342	.0751	.0404	.0207	.0101	.0046	.0020
	1	.2985	.3874	.3679	.3020	.2253	.1556	.1004	.0605	.0339	.0176
	2	.0629	.1722	.2597	.3020	.3003	.2668	.2162	.1612	.1110	.0703
	3	.0077	.0446	.1069	.1762	.2336	.2668	.2716	.2508	.2119	.1641
	4	.0006	.0074	.0283	.0661	.1168	.1715	.2194	.2508	.2600	.2461
	5	.0000	.0008	.0050	.0165	.0389	.0735	.1181	.1672	.2128	.2461
	6	.0000	.0001	.0006	.0028	.0087	.0210	.0424	.0743	.1160	.1641
	7	.0000	.0000	.0000	.0003	.0012	.0039	.0098	.0212	.0407	.0703
	8	.0000	.0000	.0000	.0000	.0001	.0004	.0013	.0035	.0083	.0176
	9	.0000	.0000	.0000	.0000	.0000	.0000	.0001	.0003	.0008	.0020
10	0	.5987	.3487	.1969	.1074	.0563	.0282	.0135	.0060	.0025	.0010
	1	.3151	.3874	.3474	.2684	.1877	.1211	.0725	.0403	.0207	.0098
	2	.0746	.1937	.2759	.3020	.2816	.2335	.1757	.1209	.0763	.0439
	3	.0105	.0574	.1298	.2013	.2503	.2668	.2522	.2150	.1665	.1172
	4	.0010	.0112	.0401	.0881	.1460	.2001	.2377	.2508	.2384	.2051
	5	.0001	.0015	.0085	.0264	.0584	.1029	.1536	.2007	.2340	.2461
	6	.0000	.0001	.0012	.0055	.0162	.0368	.0689	.1115	.1596	.2051
	7	.0000	.0000	.0001	.0008	.0031	.0090	.0212	.0425	.0746	.1172
	8	.0000	.0000	.0000	.0001	.0004	.0014	.0043	.0106	.0229	.0439
	9	.0000	.0000	.0000	.0000	.0000	.0001	.0005	.0016	.0042	.0098
	10	.0000	.0000	.0000	.0000	.0000	.0000	.0000	.0001	.0003	.0010

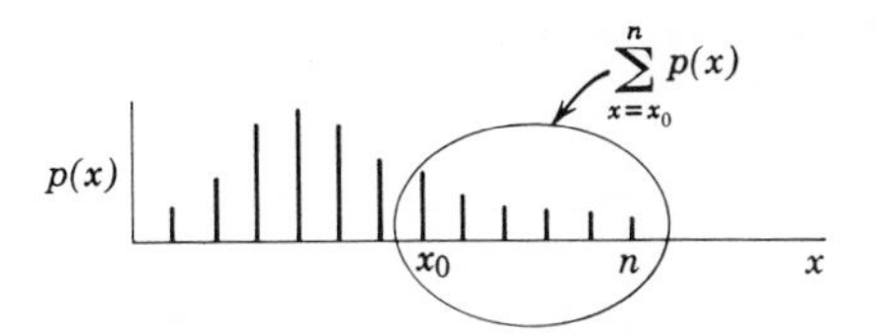

TABLE IIIc Cumulative Binomial Probabilities in Right-hand Tail

n	x_0	π .05	.10	.15	.20	.25	.30	.35	.40	.45	.50
2	1	.0975	.1900	.2775	.3600	.4375	.5100	.5775	.6400	.6975	.7500
	2	.0025	.0100	.0225	.0400	.0625	.0900	.1225	.1600	.2025	.2500
3	1	.1426	.2710	.3859	.4880	.5781	.6570	.7254	.7840	.8336	.8750
	2	.0072	.0280	.0608	.1040	.1562	.2160	.2818	.3520	.4252	.5000
	3	.0001	.0010	.0034	.0080	.0156	.0270	.0429	.0640	.0911	.1250
4	1	.1855	.3439	.4780	.5904	.6836	.7599	.8215	.8704	.9085	.9375
	2	.0140	.0523	.1095	.1808	.2617	.3483	.4370	.5248	.6090	.6875
	3	.0005	.0037	.0120	.0272	.0508	.0837	.1265	.1792	.2415	.3125
	4	.0000	.0001	.0005	.0016	.0039	.0081	.0150	.0256	.0410	.0625
5	1	.2262	.4095	.5563	.6723	.7627	.8319	.8840	.9222	.9497	.9688
	2	.0226	.0815	.1648	.2627	.3672	.4718	.5716	.6630	.7438	.8125
	3	.0012	.0086	.0266	.0579	.1035	.1631	.2352	.3174	.4069	.5000
	4	.0000	.0005	.0022	.0067	.0156	.0308	.0540	.0870	.1312	.1875
	5	.0000	.0000	.0001	.0003	.0010	.0024	.0053	.0102	.0185	.0312
6	1	.2649	.4686	.6229	.7379	.8220	.8824	.9246	.9533	.9723	.9844
	2	.0328	.1143	.2235	.3447	.4661	.5798	.6809	.7667	.8364	.8906
	3	.0022	.0158	.0473	.0989	.1694	.2557	.3529	.4557	.5585	.6562
	4	.0001	.0013	.0059	.0170	.0376	.0705	.1174	.1792	.2553	.3438
	5	.0000	.0001	.0004	.0016	.0046	.0109	.0223	.0410	.0692	.1094
	6	.0000	.0000	.0000	.0001	.0002	.0007	.0018	.0041	.0083	.0156
7	1	.3017	.5217	.6794	.7903	.8665	.9176	.9510	.9720	.9848	.9922
	2	.0444	.1497	.2834	.4233	.5551	.6706	.7662	.8414	.8976	.9375
	3	.0038	.0257	.0738	.1480	.2436	.3529	.4677	.5801	.6836	.7734
	4	.0002	.0027	.0121	.0333	.0706	.1260	.1998	.2898	.3917	.5000
	5	.0000	.0002	.0012	.0047	.0129	.0288	.0556	.0963	.1529	.2266
	6	.0000	.0000	.0001	.0004	.0013	.0038	.0090	.0188	.0357	.0625
	7	.0000	.0000	.0000	.0000	.0001	.0002	.0006	.0016	.0037	.0078

TABLE IIIc (Continued)

n	x_0	.05	.10	.15	.20	.25	π .30	.35	.40	.45	.50
8	1	.3366	.5695	.7275	.8322	.8999	.9424	.9681	.9832	.9916	.9961
	2	.0572	.1869	.3428	.4967	.6329	.7447	.8309	.8936	.9368	.9648
	3	.0058	.0381	.1052	.2031	.3215	.4482	.5722	.6846	.7799	.8555
	4	.0004	.0050	.0214	.0563	.1138	.1941	.2936	.4059	.5230	.6367
	5	.0000	.0004	.0029	.0104	.0273	.0580	.1061	.1737	.2604	.3633
	6	.0000	.0000	.0002	.0012	.0042	.0113	.0253	.0498	.0885	.1445
	7	.0000	.0000	.0000	.0001	.0004	.0013	.0036	.0085	.0181	.0352
	8	.0000	.0000	.0000	.0000	.0000	.0001	.0002	.0007	.0017	.0039
9	1	.3698	.6126	.7684	.8658	.9249	.9596	.9793	.9899	.9954	.9980
	2	.0712	.2252	.4005	.5638	.6997	.8040	.8789	.9295	.9615	.9805
	3	.0084	.0530	.1409	.2618	.3993	.5372	.6627	.7682	.8505	.9102
	4	.0006	.0083	.0339	.0856	.1657	.2703	.3911	.5174	.6386	.7461
	5	.0000	.0009	.0056	.0196	.0489	.0988	.1717	.2666	.3786	.5000
	6	.0000	.0001	.0006	.0031	.0100	.0253	.0536	.0994	.1658	.2539
	7	.0000	.0000	.0000	.0003	.0013	.0043	.0112	.0250	.0498	.0898
	8	.0000	.0000	.0000	.0000	.0001	.0004	.0014	.0038	.0091	.0195
	9	.0000	.0000	.0000	.0000	.0000	.0000	.0001	.0003	.0008	.0020
10	1	.4013	.6513	.8031	.8926	.9437	.9718	.9865	.9940	.9975	.9990
	2	.0861	.2639	.4557	.6242	.7560	.8507	.9140	.9536	.9767	.9893
	3	.0115	.0702	.1798	.3222	.4744	.6172	.7384	.8327	.9004	.9453
	4	.0010	.0128	.0500	.1209	.2241	.3504	.4862	.6177	.7340	.8281
	5	.0001	.0016	.0099	.0328	.0781	.1503	.2485	.3669	.4956	.6230
	6	.0000	.0001	.0014	.0064	.0197	.0473	.0949	.1662	.2616	.3770
	7	.0000	.0000	.0001	.0009	.0035	.0106	.0260	.0548	.1020	.1719
	8	.0000	.0000	.0000	.0001	.0004	.0016	.0048	.0123	.0274	.0547
	9	.0000	.0000	.0000	.0000	.0000	.0001	.0005	.0017	.0045	.0107
	10	.0000	.0000	.0000	.0000	.0000	.0000	.0000	.0001	.0003	.0010

TABLE IV Areas for a Standard Normal Distribution

An entry in the table is the area under the curve, between $z = 0$ and a positive value of z. Areas for negative values of z are obtained by symmetry.

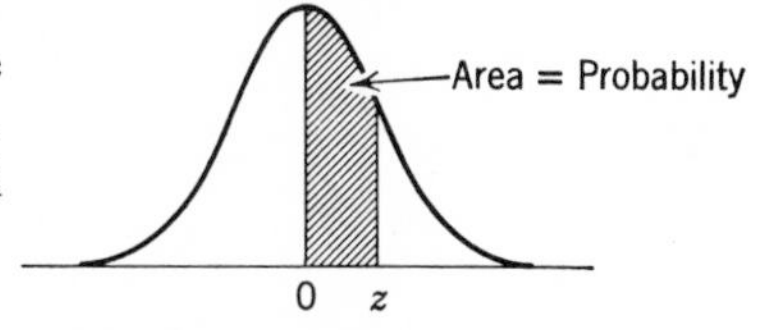

Second Decimal Place of z

$\downarrow z$ $\rightarrow$	.00	.01	.02	.03	.04	.05	.06	.07	.08	.09
0.0	.0000	.0040	.0080	.0120	.0160	.0199	.0239	.0279	.0319	.0359
0.1	.0398	.0438	.0478	.0517	.0557	.0596	.0636	.0675	.0714	.0753
0.2	.0793	.0832	.0871	.0910	.0948	.0987	.1026	.1064	.1103	.1141
0.3	.1179	.1217	.1255	.1293	.1331	.1368	.1406	.1443	.1480	.1517
0.4	.1554	.1591	.1628	.1664	.1700	.1736	.1772	.1808	.1844	.1879
0.5	.1915	.1950	.1985	.2019	.2054	.2088	.2123	.2157	.2190	.2224
0.6	.2257	.2291	.2324	.2357	.2389	.2422	.2454	.2486	.2517	.2549
0.7	.2580	.2611	.2642	.2673	.2703	.2734	.2764	.2794	.2823	.2852
0.8	.2881	.2910	.2939	.2967	.2995	.3023	.3051	.3078	.3106	.3133
0.9	.3159	.3186	.3212	.3238	.3264	.3289	.3315	.3340	.3365	.3389
1.0	.3413	.3438	.3461	.3485	.3508	.3531	.3554	.3577	.3599	.3621
1.1	.3643	.3665	.3686	.3708	.3729	.3749	.3770	.3790	.3810	.3830
1.2	.3849	.3869	.3888	.3907	.3925	.3944	.3962	.3980	.3997	.4015
1.3	.4032	.4049	.4066	.4082	.4099	.4115	.4131	.4147	.4162	.4177
1.4	.4192	.4207	.4222	.4236	.4251	.4265	.4279	.4292	.4306	.4319
1.5	.4332	.4345	.4357	.4370	.4382	.4394	.4406	.4418	.4429	.4441
1.6	.4452	.4463	.4474	.4484	.4495	.4505	.4515	.4525	.4535	.4545
1.7	.4554	.4564	.4573	.4582	.4591	.4599	.4608	.4616	.4625	.4633
1.8	.4641	.4649	.4656	.4664	.4671	.4678	.4686	.4693	.4699	.4706
1.9	.4713	.4719	.4726	.4732	.4738	.4744	.4750	.4756	.4761	.4767
2.0	.4772	.4778	.4783	.4788	.4793	.4798	.4803	.4808	.4812	.4817
2.1	.4821	.4826	.4830	.4834	.4838	.4842	.4846	.4850	.4854	.4857
2.2	.4861	.4864	.4868	.4871	.4875	.4878	.4881	.4884	.4887	.4890
2.3	.4893	.4896	.4898	.4901	.4904	.4906	.4909	.4911	.4913	.4916
2.4	.4918	.4920	.4922	.4925	.4927	.4929	.4931	.4932	.4934	.4936
2.5	.4938	.4940	.4941	.4943	.4945	.4946	.4948	.4949	.4951	.4952
2.6	.4953	.4955	.4956	.4957	.4959	.4960	.4961	.4962	.4963	.4964
2.7	.4965	.4966	.4967	.4968	.4969	.4970	.4971	.4972	.4973	.4974
2.8	.4974	.4975	.4976	.4977	.4977	.4978	.4979	.4979	.4980	.4981
2.9	.4981	.4982	.4982	.4983	.4984	.4984	.4985	.4985	.4986	.4986
3.0	.4987	.4987	.4987	.4988	.4988	.4989	.4989	.4989	.4990	.4990

TABLE V Student's t Critical Points

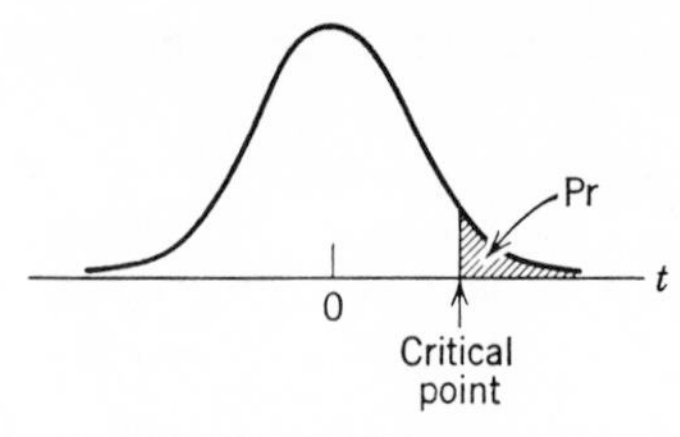

d.f. \ Pr	.10	.05	.025	.01	.005
1	3.078	6.314	12.706	31.821	63.657
2	1.886	2.920	4.303	6.965	9.925
3	1.638	2.353	3.182	4.541	5.841
4	1.533	2.132	2.776	3.747	4.604
5	1.476	2.015	2.571	3.365	4.032
6	1.440	1.943	2.447	3.143	3.707
7	1.415	1.895	2.365	2.998	3.499
8	1.397	1.860	2.306	2.896	3.355
9	1.383	1.833	2.262	2.821	3.250
10	1.372	1.812	2.228	2.764	3.169
11	1.363	1.796	2.201	2.718	3.106
12	1.356	1.782	2.179	2.681	3.055
13	1.350	1.771	2.160	2.650	3.012
14	1.345	1.761	2.145	2.624	2.977
15	1.341	1.753	2.131	2.602	2.947
16	1.337	1.746	2.120	2.583	2.921
17	1.333	1.740	2.110	2.567	2.898
18	1.330	1.734	2.101	2.552	2.878
19	1.328	1.729	2.093	2.539	2.861
20	1.325	1.725	2.086	2.528	2.845
21	1.323	1.721	2.080	2.518	2.831
22	1.321	1.717	2.074	2.508	2.819
23	1.319	1.714	2.069	2.500	2.807
24	1.318	1.711	2.064	2.492	2.797
25	1.316	1.708	2.060	2.485	2.787
26	1.315	1.706	2.056	2.479	2.779
27	1.314	1.703	2.052	2.473	2.771
28	1.313	1.701	2.048	2.467	2.763
29	1.311	1.699	2.045	2.462	2.756
30	1.310	1.697	2.042	2.457	2.750
40	1.303	1.684	2.021	2.423	2.704
60	1.296	1.671	2.000	2.390	2.660
120	1.289	1.658	1.980	2.358	2.617
∞	1.282	1.645	1.960	2.326	2.576

TABLE VI C^2 Critical Points* ($C^2 = \chi^2/\text{d.f.}$)

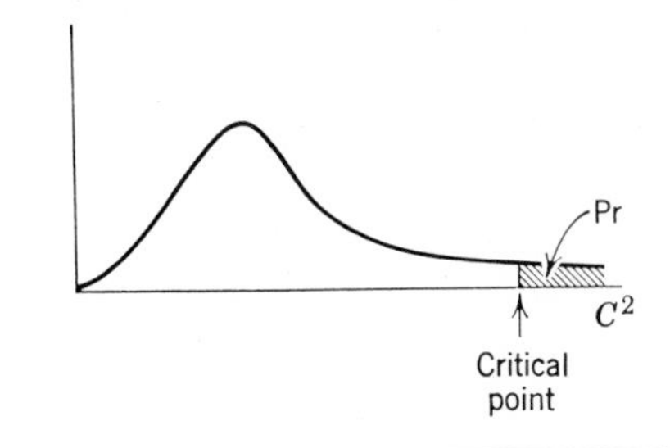

df \ Pr	.995	.99	.975	.95	.90	.10	.05	.025	.01	.005
1	.000039	.00016	.00098	.0039	.0158	2.71	3.84	5.02	6.63	7.88
2	.00501	.0101	.0253	.0513	.1054	2.30	3.00	3.69	4.61	5.30
3	.0239	.0383	.0719	.117	.195	2.08	2.60	3.12	3.78	4.28
4	.0517	.0743	.121	.178	.266	1.94	2.37	2.79	3.32	3.72
5	.0823	.111	.166	.229	.322	1.85	2.21	2.57	3.02	3.35
6	.113	.145	.206	.273	.367	1.77	2.10	2.41	2.80	3.09
7	.141	.177	.241	.310	.405	1.72	2.01	2.29	2.64	2.90
8	.168	.206	.272	.342	.436	1.67	1.94	2.19	2.51	2.74
9	.193	.232	.300	.369	.463	1.63	1.88	2.11	2.41	2.62
10	.216	.256	.325	.394	.487	1.60	1.83	2.05	2.32	2.52
11	.237	.278	.347	.416	.507	1.57	1.79	1.99	2.25	2.43
12	.256	.298	.367	.435	.525	1.55	1.75	1.94	2.18	2.36
13	.274	.316	.385	.453	.542	1.52	1.72	1.90	2.13	2.29
14	.291	.333	.402	.469	.556	1.50	1.69	1.87	2.08	2.24
15	.307	.349	.417	.484	.570	1.49	1.67	1.83	2.04	2.19
16	.321	.363	.432	.498	.582	1.47	1.64	1.80	2.00	2.14
18	.348	.390	.457	.522	.604	1.44	1.60	1.75	1.93	2.06
20	.372	.413	.480	.543	.622	1.42	1.57	1.71	1.88	2.00
24	.412	.452	.517	.577	.652	1.38	1.52	1.64	1.79	1.90
30	.460	.498	.560	.616	.687	1.34	1.46	1.57	1.70	1.79
40	.518	.554	.611	.663	.726	1.30	1.39	1.48	1.59	1.67
60	.592	.625	.675	.720	.774	1.24	1.32	1.39	1.47	1.53
120	.699	.724	.763	.798	.839	1.17	1.22	1.27	1.32	1.36
∞	1.000	1.000	1.000	1.000	1.000	1.00	1.00	1.00	1.00	1.00

Interpolation should be performed using reciprocals of the degrees of freedom.

* To obtain critical values of χ^2, multiply the critical value of C^2 by (d.f.)

TABLE VII F Distribution Critical Points 5% (Roman Type) and 1% (Boldface Type) Points

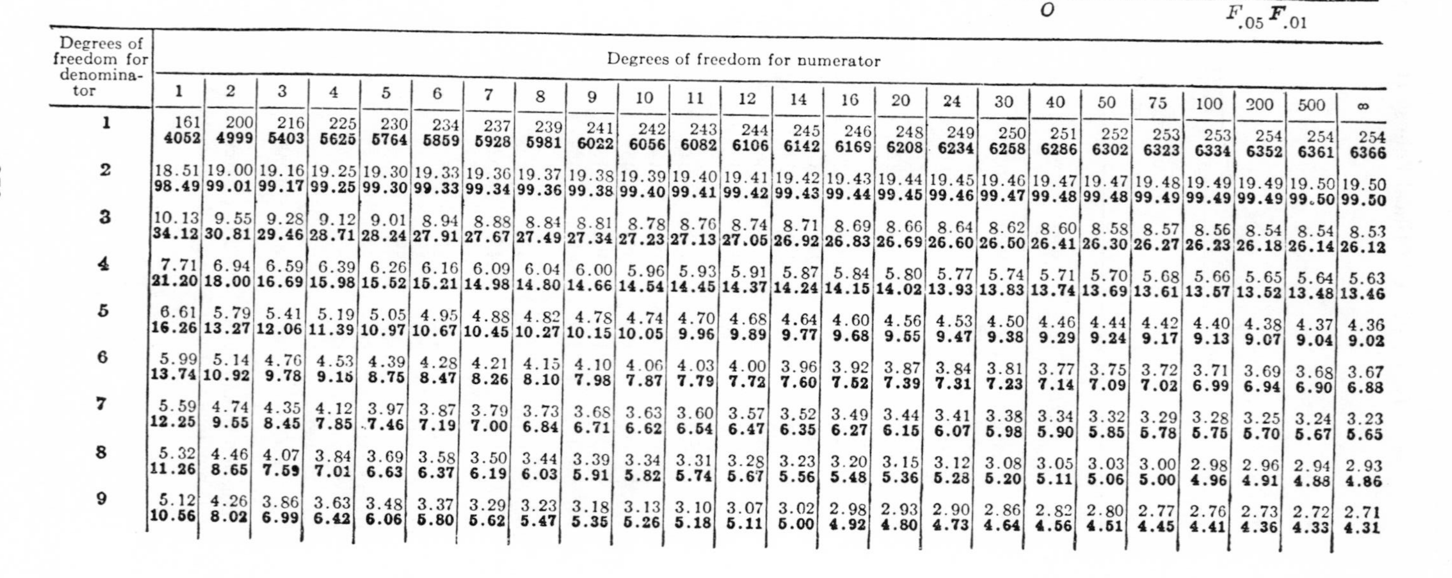

Degrees of freedom for denominator	Degrees of freedom for numerator																						
	1	2	3	4	5	6	7	8	9	10	11	12	14	16	20	24	30	40	50	75	100	200	500	∞
1	161	200	216	225	230	234	237	239	241	242	243	244	245	246	248	249	250	251	252	253	253	254	254	254
	4052	**4999**	**5403**	**5625**	**5764**	**5859**	**5928**	**5981**	**6022**	**6056**	**6082**	**6106**	**6142**	**6169**	**6208**	**6234**	**6258**	**6286**	**6302**	**6323**	**6334**	**6352**	**6361**	**6366**
2	18.51	19.00	19.16	19.25	19.30	19.33	19.36	19.37	19.38	19.39	19.40	19.41	19.42	19.43	19.44	19.45	19.46	19.47	19.47	19.48	19.49	19.49	19.50	19.50
	98.49	**99.01**	**99.17**	**99.25**	**99.30**	**99.33**	**99.34**	**99.36**	**99.38**	**99.40**	**99.41**	**99.42**	**99.43**	**99.44**	**99.45**	**99.46**	**99.47**	**99.48**	**99.48**	**99.49**	**99.49**	**99.49**	**99.50**	**99.50**
3	10.13	9.55	9.28	9.12	9.01	8.94	8.88	8.84	8.81	8.78	8.76	8.74	8.71	8.69	8.66	8.64	8.62	8.60	8.58	8.57	8.56	8.54	8.54	8.53
	34.12	**30.81**	**29.46**	**28.71**	**28.24**	**27.91**	**27.67**	**27.49**	**27.34**	**27.23**	**27.13**	**27.05**	**26.92**	**26.83**	**26.69**	**26.60**	**26.50**	**26.41**	**26.30**	**26.27**	**26.23**	**26.18**	**26.14**	**26.12**
4	7.71	6.94	6.59	6.39	6.26	6.16	6.09	6.04	6.00	5.96	5.93	5.91	5.87	5.84	5.80	5.77	5.74	5.71	5.70	5.68	5.66	5.65	5.64	5.63
	21.20	**18.00**	**16.69**	**15.98**	**15.52**	**15.21**	**14.98**	**14.80**	**14.66**	**14.54**	**14.45**	**14.37**	**14.24**	**14.15**	**14.02**	**13.93**	**13.83**	**13.74**	**13.69**	**13.61**	**13.57**	**13.52**	**13.48**	**13.46**
5	6.61	5.79	5.41	5.19	5.05	4.95	4.88	4.82	4.78	4.74	4.70	4.68	4.64	4.60	4.56	4.53	4.50	4.46	4.44	4.42	4.40	4.38	4.37	4.36
	16.26	**13.27**	**12.06**	**11.39**	**10.97**	**10.67**	**10.45**	**10.27**	**10.15**	**10.05**	**9.96**	**9.89**	**9.77**	**9.68**	**9.55**	**9.47**	**9.38**	**9.29**	**9.24**	**9.17**	**9.13**	**9.07**	**9.04**	**9.02**
6	5.99	5.14	4.76	4.53	4.39	4.28	4.21	4.15	4.10	4.06	4.03	4.00	3.96	3.92	3.87	3.84	3.81	3.77	3.75	3.72	3.71	3.69	3.68	3.67
	13.74	**10.92**	**9.78**	**9.15**	**8.75**	**8.47**	**8.26**	**8.10**	**7.98**	**7.87**	**7.79**	**7.72**	**7.60**	**7.52**	**7.39**	**7.31**	**7.23**	**7.14**	**7.09**	**7.02**	**6.99**	**6.94**	**6.90**	**6.88**
7	5.59	4.74	4.35	4.12	3.97	3.87	3.79	3.73	3.68	3.63	3.60	3.57	3.52	3.49	3.44	3.41	3.38	3.34	3.32	3.29	3.28	3.25	3.24	3.23
	12.25	**9.55**	**8.45**	**7.85**	**7.46**	**7.19**	**7.00**	**6.84**	**6.71**	**6.62**	**6.54**	**6.47**	**6.35**	**6.27**	**6.15**	**6.07**	**5.98**	**5.90**	**5.85**	**5.78**	**5.75**	**5.70**	**5.67**	**5.65**
8	5.32	4.46	4.07	3.84	3.69	3.58	3.50	3.44	3.39	3.34	3.31	3.28	3.23	3.20	3.15	3.12	3.08	3.05	3.03	3.00	2.98	2.96	2.94	2.93
	11.26	**8.65**	**7.59**	**7.01**	**6.63**	**6.37**	**6.19**	**6.03**	**5.91**	**5.82**	**5.74**	**5.67**	**5.56**	**5.48**	**5.36**	**5.28**	**5.20**	**5.11**	**5.06**	**5.00**	**4.96**	**4.91**	**4.88**	**4.86**
9	5.12	4.26	3.86	3.63	3.48	3.37	3.29	3.23	3.18	3.13	3.10	3.07	3.02	2.98	2.93	2.90	2.86	2.82	2.80	2.77	2.76	2.73	2.72	2.71
	10.56	**8.02**	**6.99**	**6.42**	**6.06**	**5.80**	**5.62**	**5.47**	**5.35**	**5.26**	**5.18**	**5.11**	**5.00**	**4.92**	**4.80**	**4.73**	**4.64**	**4.56**	**4.51**	**4.45**	**4.41**	**4.36**	**4.33**	**4.31**

TABLE VIII (Continued)

N	0	1	2	3	4	5	6	7	8	9
55	7404	7412	7419	7427	7435	7443	7451	7459	7466	7474
56	7482	7490	7497	7505	7513	7520	7528	7536	7543	7551
57	7559	7566	7574	7582	7589	7597	7604	7612	7619	7627
58	7634	7642	7649	7657	7664	7672	7679	7686	7694	7701
59	7709	7716	7723	7731	7738	7745	7752	7760	7767	7774
60	7782	7789	7796	7803	7810	7818	7825	7832	7839	7846
61	7853	7860	7868	7875	7882	7889	7896	7903	7910	7917
62	7924	7931	7938	7945	7952	7959	7966	7973	7980	7987
63	7993	8000	8007	8014	8021	8028	8035	8041	8048	8055
64	8062	8069	8075	8082	8089	8096	8102	8109	8116	8122
65	8129	8136	8142	8149	8156	8162	8169	8176	8182	8189
66	8195	8202	8209	8215	8222	8228	8235	8241	8248	8254
67	8261	8267	8274	8280	8287	8293	8299	8306	8312	8319
68	8325	8331	8338	8344	8351	8357	8363	8370	8376	8382
69	8388	8395	8401	8407	8414	8420	8426	8432	8439	8445
70	8451	8457	8463	8470	8476	8482	8488	8494	8500	8506
71	8513	8519	8525	8531	8537	8543	8549	8555	8561	8567
72	8573	8579	8585	8591	8597	8603	8609	8615	8621	8627
73	8633	8639	8645	8651	8657	8663	8669	8675	8681	8686
74	8692	8698	8704	8710	8716	8722	8727	8733	8739	8745
75	8751	8756	8762	8768	8774	8779	8785	8791	8797	8802
76	8808	8814	8820	8825	8831	8837	8842	8848	8854	8859
77	8865	8871	8876	8882	8887	8893	8899	8904	8910	8915
78	8921	8927	8932	8938	8943	8949	8954	8960	8965	9971
79	8976	8982	8987	8993	8998	9004	9009	9015	9020	9025
80	9031	9036	9042	9047	9053	9058	9063	9069	9074	9079
81	9085	9090	9096	9101	9106	9112	9117	9122	9128	9133
82	9138	9143	9149	9154	9159	9165	9170	9175	9180	9186
83	9191	9196	9201	9206	9212	9217	9222	9227	9232	9238
84	9243	9248	9253	9258	9263	9269	9274	9279	9284	9289
85	9294	9299	9304	9309	9315	9320	9325	9330	9335	9340
86	9345	9350	9355	9360	9365	9370	9375	9380	9385	9390
87	9395	9400	9405	9410	9415	9420	9425	9430	9435	9440
88	9445	9450	9455	9460	9465	9469	9474	9479	9484	9489
89	9494	9499	9504	9509	9513	9518	9523	9528	9533	9538
90	9542	9547	9552	9557	9562	9566	9571	9576	9581	9586
91	9590	9595	9600	9605	9609	9614	9619	9624	9628	9633
92	9638	9643	9647	9652	9657	9661	9666	9671	9675	9680
93	9685	9689	9694	9699	9703	9708	9713	9717	9722	9727
94	9731	9736	9741	9745	9750	9754	9759	9763	9768	9773
95	9777	9782	9786	9791	9795	9800	9805	9809	9814	9818
96	9823	9827	9832	9836	9841	9845	9850	9854	9859	9863
97	9868	9872	9877	9881	9886	9890	9894	9899	9903	9908
98	9912	9917	9921	9926	9930	9934	9939	9943	9948	9952
99	9956	9961	9965	9969	9974	9978	9983	9987	9991	9996

* The log of N is "the power to which 10 must be raised to yield N." Thus log 100 = 2, because $10^2 = 100$. In this table, only the "mantissa" (the digits to the right of the decimal) is given for each log. The characteristic (the integer to the left of the decimal) is 1; for example log 19.1 = 1.281. Log $\frac{1}{10} N$ requires the characteristic 0, log 10 N the characteristic 2, log 100 N the characteristic 3, and so on. Thus log 537 = 2.73.

CITATIONS FOR TABLES

I. Reproduced, by permission, from the *Wiley Trigonometric Tables*, John Wiley and Sons, 1945.

II. (a) Reproduced, by permission, from R. C. Clelland et al., *Basic Statistics with Business Applications*, John Wiley and Sons, 1966.

(b) Reproduced, by permission, from the RAND Corporation.

III. Reproduced, by permission, from the *Chemical Rubber Company Standard Mathematical Tables*, 16th Student Edition.

IV. Reproduced, by permission, from P. Hoel, *Elementary Statistics*, 2nd Edition, John Wiley and Sons, 1966.

V. Reproduced, by permission, from R. Fisher and F. Yates, *Statistical Tables*, Oliver and Boyd, Edinburgh, 1938.

VI. Reproduced, by permission, from W. J. Dixon and F. J. Massey, *Introduction to Statistical Analysis*, 2nd Edition, McGraw-Hill, 1957.

VII. Reproduced, by permission, from *Statistical Methods*, 6th Edition, by George W. Snedecor and William G. Cochrane, 1967, by the Iowa State University Press, Ames, Iowa.

VIII. Reproduced from John E. Freund, *Modern Elementary Statistics*, 3rd Edition, © 1967, by permission of Prentice-Hall Inc., Englewood Cliffs, New Jersey.

Answers to Odd-Numbered Problems

The student is *not* expected always to calculate the answer as precisely as the given answers below. These answers are given to a fairly high degree of precision merely for the benefit of those who want it; even so, the last digit may be slightly in error because of slide rule inaccuracy.

2-1 Mode < median < mean. The mode is not a bad central measure in this case, which is not very asymmetrical.

2-3 77.4, 81.25, 85

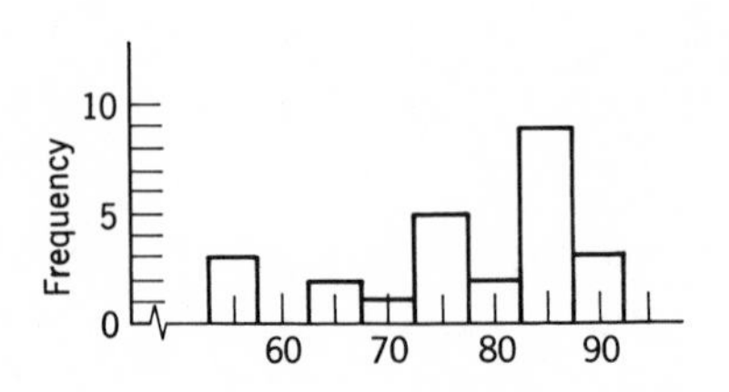

2-5

	Mean	Median	Mode
raw	77.78	81.47	hardly defined
fine	77.4	81.25	85
coarse	78.4	80.00	80

(a) Mode depends too much on the degree of grouping.

(b) Usually, but not always, does the coarse grouping give worse approximations.

2-7 range = 30 or 40

MAD = 8.58

$$s = \sqrt{\frac{2736}{24}} = \sqrt{114.0} = 10.67$$

2-9 Text coding is preferred because it has integral and small values of y, which are easy to compute.

2-11 239, 483

2-13 (a) 77.4, $5\sqrt{4.84} = 11.0$

(b) 8.60, $\sqrt{14.4} = 3.8$

2-15 121.50

2-17 27.6% (NOT .952/4)

3-1 (c) $\frac{n_1}{n}$ (authors' answer $= \frac{23}{50} = .46$)

$$\frac{n - n_1}{n} = 1 - \frac{n_1}{n} \left(= \frac{27}{50} = .54 \right)$$

3-5 (b) not equally likely; 1/4, 1/4, 1/2

(c) 3/4

3-7 (a) .50, .30, .65, .15

(c) .50, .70, .85, .35

3-9 (a) $\frac{6}{16} = .375$

(b) $\frac{6}{16} = .375$

(c) No

3-11 (a) .40

(b) .60

(c) .55

(d) .78

(e) .17

(f) $\left.\begin{matrix} \frac{17}{40} = .42 \\ \frac{23}{40} = .58 \end{matrix}\right\}$ sum = 1

3-15 (a) $\frac{.06}{.21} = .29$

(b) Yes. $\Pr(A/A \cup B) =$

$$\frac{\Pr(A)}{\Pr(A \cup B)} = \frac{\Pr(A)}{\Pr(A) + \Pr(B)}$$

3-17 (a) $\frac{4}{52} \times \frac{3}{51} = \frac{1}{221} = .0046$

(b) $\frac{2}{52} \times \frac{1}{51} = \frac{1}{1326} = .00076$

(c) $\frac{20}{52} \times \frac{19}{51} = .143$

3-19 (a) $\frac{8}{10} \times \frac{7}{9} = \frac{28}{45} = .62$

(b) $\frac{8}{10} \times \frac{7}{9} \times \frac{6}{8} \times \frac{5}{7} \times \frac{4}{6} \times \frac{2}{5} = \frac{4}{45} = .089$

(c) $\frac{8}{10} \times \frac{7}{9} \times \cdots \times \frac{1}{3} \times \frac{2}{2} = \frac{1}{45} = .022$

3-21 0

3-25 (a) Yes. $\Pr(E_1 \cap E_2) = \Pr(E_1)\Pr(E_2)$, i.e., $\frac{1}{8} = \frac{1}{4} \cdot \frac{1}{2}$

(b) No. $\Pr(E_1 \cap E_3) \neq \Pr(E_1)\Pr(E_3)$, i.e., $\frac{1}{8} \neq \frac{1}{4} \cdot \frac{1}{2}$

3-27 (a) Yes

(b) Yes

3-29 (a) .3

(b) Impossible conditions—there must be an error of specification.

(c) $0 \leq \Pr(e_4) \leq .2$

(d) Impossible conditions.

3-31 (a) $\frac{2}{6}$

(b) $\frac{3}{6}$

(c) 0

(d) $\frac{1}{6}$

sum = 1

3-33 (a) $\dfrac{.001}{.126} = .0079$

For n tosses, $\dfrac{.001}{.001 + \dfrac{1}{2^n}(.999)}$

(b) .506

(c) .999

See how the probabilities grow toward certainty as $n \to \infty$.

4-1 (a)

x	$p(x)$
0	1/16
1	4/16
2	6/16
3	4/16
4	1/16
	16/16

(b)

y	$p(y)$
0	2/16
1	6/16
2	6/16
3	2/16
	16/16

4-3

x	$p(x)$
0	6/36
1	10/36
2	8/36
3	6/36
4	4/36
5	2/36
	36/36

4-5 (a) $\mu = 2$ $\sigma^2 = 1$

(b) $\mu = 1.5$ $\sigma^2 = .75$

4-7 (a) $\mu_X = 3.5$ $\sigma_X = \sqrt{\frac{35}{12}} = 1.7$

(b), (c) $\mu_Y = 11$ $\sigma_Y = 3.4$

4-9

x	$p(x)$
0	16/81 = .198
1	32/81 = .395
2	24/81 = .296
3	8/81 = .099
4	1/81 = .012
	81/81

$\mu = \frac{4}{3} = 1.33$

$\sigma^2 = \frac{8}{9} = .89$

4-11 (a) $\mu_X = 1.36$ $\sigma_X = \sqrt{2.43} = 1.56$

(b) $\mu_Y = 3.15$ $\sigma_Y = \sqrt{2.16} = 1.47$

(c) μ_X

4-13 (a) $p(x) = \binom{3}{x} .2^x .8^{3-x}$

x	$p(x)$
0	64/125 = .512
1	48/125 = .384
2	12/125 = .096
3	1/125 = .008
	1.00

$\mu = .60$
$\sigma^2 = .48$

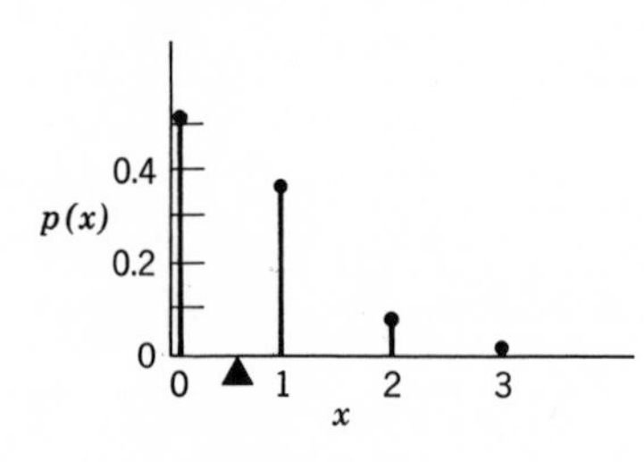

(b) $p(x) = \binom{3}{x} .1^x .9^{3-x}$

x	$p(x)$
0	.729
1	.243
2	.027
3	.001
	1.00

$\mu = .30$
$\sigma^2 = .27$

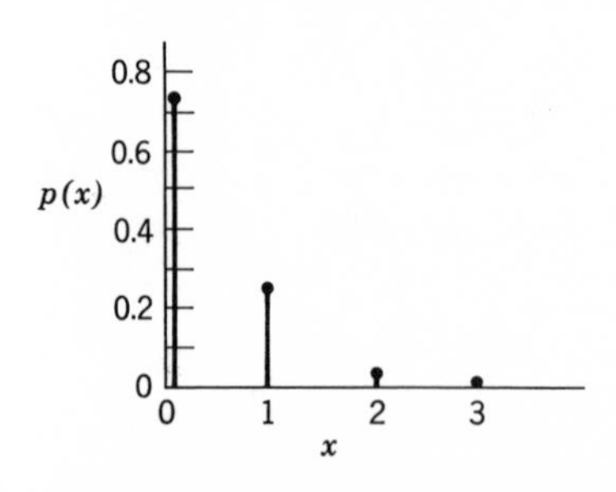

4-15 $p(x) = \binom{3}{x}\left(\frac{1}{6}\right)^x\left(\frac{5}{6}\right)^{3-x}$

x	$p(x)$
0	125/216 = .579
1	75/216 = .347
2	15/216 = .070
3	1/216 = .004
	216/216 = 1.00

$\mu = \frac{1}{2} = .50$
$\sigma^2 = \frac{15}{36} = .416$

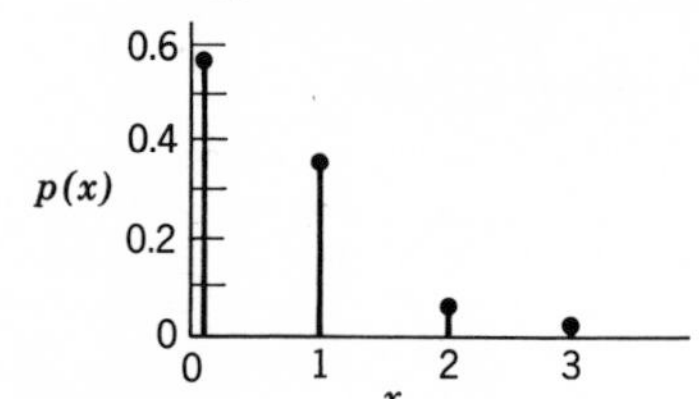

4-17 $\mu = n\pi$
$\sigma^2 = n\pi(1 - \pi)$

4-19 (a) .9544
(b) $.9495 \simeq .95$
(c) $.9901 \simeq .99$
(d) .9772
(e) .9772

4-21

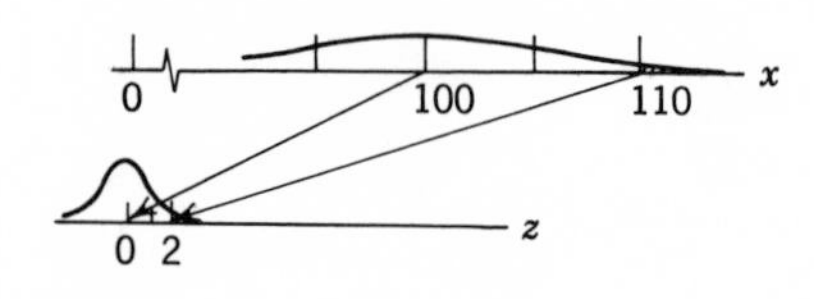

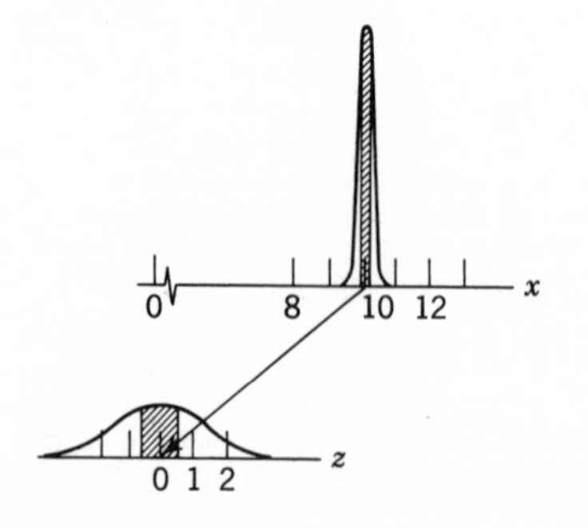

4-23 (a) .092
(b) .251 } sum = 1
(c) .657

4-25 (a)

x	$p(x)$
0	2/16
1	6/16
2	6/16
3	2/16
	16/16

r	$p(r)$
−2	12/16
0	4/16
	16/16

$\mu_R = -1.5$
$\sigma_R^2 = 12/16 = .75$

(b) (i) let $Y = |X - 2|$

y	$p(y)$	$yp(y)$
0	6/16	0
1	8/16	8/16
2	2/16	4/16
	16/16	$\mu = 12/16$

(ii)

x	$p(x)$	$\|x - 2\|$	$\|x - 2\| p(x)$
0	2/16	2	4/16
1	6/16	1	6/16
2	6/16	0	0/16
3	2/16	1	2/16
			$\mu = 12/16$

(c) $E(X^2) = 3$
(d) $E(X - \mu)^2 = 3/4 = \sigma_X^2$ of course

4-27 (a) $.6^5$

(b) $\sum_{x=3}^{5} \binom{5}{x} .6^x .4^{5-x} = .683$

(c) A sample of 5 has a 68% chance of correctly predicting, whereas a single observation has only a 60% chance.

4-29 (a) .0062
(b) 10.124

4-31 (a)

e	Pr (e)
.SSS	15/48
.SSF	3/48
.SFS	5/48
.SFF	1/48
.FSS	15/48
.FSF	3/48
.FFS	5/48
.FFF	1/48
	48/48

x	$p(x)$
0	15/48
1	23/48
2	9/48
3	1/48
	48/48

$\mu = \frac{11}{12} = .92$
(note $= \frac{1}{2} + \frac{1}{4} + \frac{1}{6}$)
(b) 10/48 = .21

5-1 (a) (b)

x \ y	0	1	2	3	$p(x)$
0	1/16				1/16
1		2/16	2/16		4/16
2		2/16	2/16	2/16	6/16
3		2/16	2/16		4/16
4	1/16				1/16
					1

or

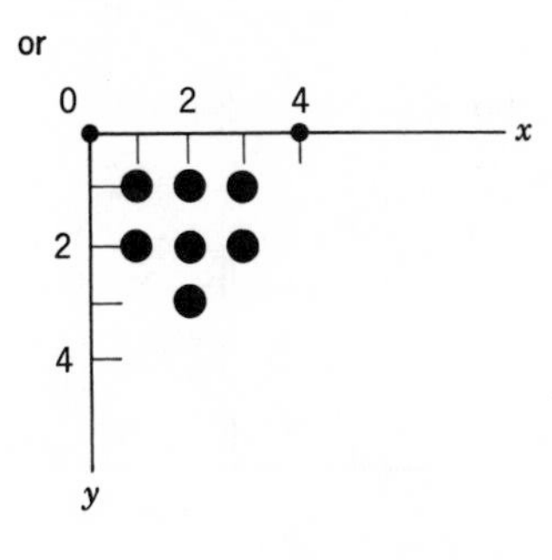

(c) 2, 1

(d)

x	$p(x/Y=2)$
1	1/3
2	1/3
3	1/3

(e) 2, 2/3

(f) No. For example, $p(0, 1) \neq p_X(0)p_Y(1)$.

5-3 (a)

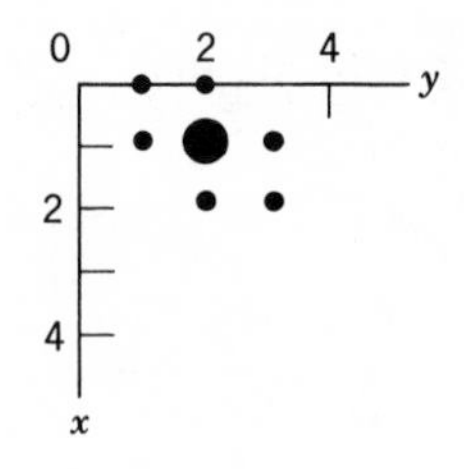

(b)

x	$p(x)$
0	.2
1	.6
2	.2

(c) 1, .4

(d)

x	$p(x)$
0	1/6
1	4/6
2	1/6

(e) 1, 1/3

(f) No, because, for example, $p(0, 3) \neq p_X(0)p_Y(3)$

5-5 (a)

u	$p(u)$	$up(u)$
−1	.2	−.2
0	.1	0
1	.4	.4
2	.1	.2
3	.1	.3
5	.1	.5

$\mu = 1.2$

(b) 1.2

(c) .2

5-7 −.6

5-9 (a) Yes

(b) (i) 0, because of symmetry

(ii) $E(X) + E(Y) = 3\frac{1}{2}$

5-11 (a)

s	$p(s)$
2	1/9
3	2/9
4	3/9
5	2/9
6	1/9

$\mu_S = 4$

$\sigma_S^2 = 4/3$

(b) For X_1 and X_2

$\mu = 2$, $\sigma^2 = 2/3$

(c) $E(X_1 + X_2) = E(X_1) + E(X_2)$

$\text{var}\,(X_1 + X_2) = \text{Var}\,X_1 + \text{Var}\,X_2$.

5-13 (a)

s	$p(s)$
2	.1
3	.2
4	.3
5	.3
6	.1

$\mu_S = 4.10$, $\sigma_S^2 = 1.29$

(b) $\mu_1 = 2.00$, $\sigma_S^2 = .60$

$\mu_2 = 2.10$, $\sigma_S^2 = .69$

(c) $\text{cov}\,(X_1, X_2) = 0$ by symmetry

5-15 (a)

$x \backslash y$	0	1
0	.15	.10
1	.25	.25
2	.10	.15

X and Y are dependent

$\sigma_{XY} = .05$

(b) $\mu_X = 1.0 \quad \sigma_X^2 = .50$

$\mu_Y = .5 \quad \sigma_Y^2 = .25$

$\mu_Z = 1.5 \quad \sigma_Z^2 = .85$

5-17

		400 400
65	10	100
70	11.5	1200/9 = 133.3

Negative covariance means that a high first grade X_1 tends to be followed by a low second grade X_2. This may be because a student who does well on the first exam becomes overconfident and fails to study for the second exam. Similarly, a student who does poorly on the first exam may study very hard for the second.

The negative covariance makes the average grade less fluctuating ($\sigma = 10$ instead of 15).

5-19 (a)

x	$p(x)$
5	.5
6	.5

(b)

y	$p(y)$
5	.2
6	.4
7	.4

$\mu_X = 5.5, \mu_Y = 6.2$

(c)

s	$p(s)$
10	.1
11	.4
12	.2
13	.3

$\mu_S = 11.7 = \mu_X + \mu_Y$

(d)

y	$p(y/X = 5)$
5	.2
6	.6
7	.2

$\mu_{Y/X=5} = 6$

(e) No, because (*b*) and (*d*) are different

(f) .7

5-21 (a)

$h \backslash w$	140	150	160
65	.2	.1	
70	.1	.2	.1
75		.1	.2

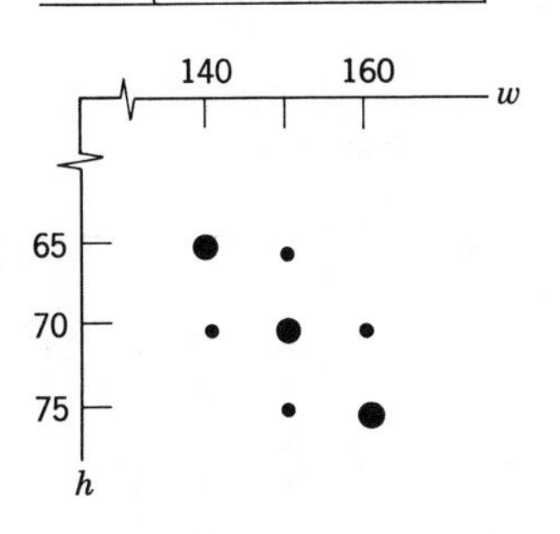

(b)

h	$p(h)$
65	.3
70	.4
75	.3

$\mu_H = 70 \quad \sigma_H^2 = 15$

(c)

w	$p(w)$
140	.3
150	.4
160	.3

$\mu_W = 150 \quad \sigma_W^2 = 60$

(d) 20

(e) 143.3, 156.7

(f) No, because $\sigma_{HW} \neq 0$

(g) $\mu_I = 590 \; (=2\mu_H + 3\mu_{H'})$
$\sigma_I^2 = 840 \; (=4\sigma_H^2 + 9\sigma_{H'}^2 + 12\sigma_{HH'})$

$\sigma_I = 29.0 \; (=\sqrt{840})$

i	$p(i)$	coded $i' = (i - 600)/10$	$i'p(i')$
550	.2	−5	−1.0
560	.1	−4	− .4
570	0	−3	0
580	.1	−2	− .2
590	.2	−1	− .2
600	.1	0	0
610	0	1	0
620	.1	2	.2
630	.2	3	.6
			−1.0

$\mu_I = 600 + 10(-1.0) = 590$
Similarly, $\sigma_I^2 = 840$

5-23 (a)

5¢	10¢	$25¢_1$	$25¢_2$
↓	↓	↓	↓
·(H	H	H	H)
·(H	H	H	T)
·(H	H	T	H)
·(H	H	T	T)
·(H	T	H	H)

etc.

(b)

r	$p(r)$
0	1/16
5	1/16
10	1/16
15	1/16
20	0
25	2/16
30	
·	
·	
· ·	
85	1/16

$\mu_R = 32.5$
$\sigma_R^2 = 343.75$

(c)

x_1	$p(x_1)$
0	1/2
5	1/2

$\mu_1 = 5/2 \qquad \sigma_1^2 = 25/4$

(d) $\mu_2 = 10/2 \qquad \sigma_2^2 = 100/4$
$\mu_3 = 25/2 \qquad \sigma_3^2 = 625/4$
$\mu_4 = 25/2 \qquad \sigma_4^2 = 625/4$

(e) $\sum \mu_i = 65/2 = 32.5 \qquad \sum \sigma_i^2 = 1375/4 = 343.75$

5-25 (a) (b)

$x_1 \backslash x_2$	1	2	···	6	$p(x_1)$
1	0	1/30	···	1/30	1/6
2	1/30	0		·	1/6
·	·	·		·	·
·	·		·	·	·
·	·		·	·	·
6	1/30		···	0	1/6
$p(x_2)$	1/6	1/6	···	1/6	1

(c) No, because, for example,
$p(1, 1) \neq p_X(1)\, p_Y(1)$
i.e., $0 \neq \frac{1}{6} \cdot \frac{1}{6}$

(d) $-7/12 = -.58$

(e) 3.5, $35/12 = 2.92$

(f) $\mu_S = 7.0 \; (=\mu_1 + \mu_2)$
$\sigma_S^2 = 28/6 \; (=\sigma_1^2 + \sigma_2^2 + 2\sigma_{12}) = 4.67$

Or compute from $p(s)$ directly (the hard way).

5-27 (a) 35, $350/12 = 29.2$
(b) $60 - 10 = 50$

6-1 Correction: In the last sentence, interchange "standard deviation" with "range."

6-5 (a) 4, $\sqrt{8/3} = 1.63$

(b)

$\bar{x}$	$p(\bar{x})$
2	1/9
3	2/9
4	3/9
5	2/9
6	1/9

$\mu_{\bar{X}} = 4$

$\sigma_{\bar{X}} = \sqrt{8/6} = 1.154$

(c)

$\bar{x}$	$p(\bar{x})$
2	1/27
8/3	3/27
10/3	6/27
4	7/27
14/3	6/27
16/3	3/27
6	1/27

$\mu_{\bar{X}} = 4$

$\sigma_{\bar{X}} = \sqrt{8/9} = .943$

(d) See Fig. 6-3(a).

6-7 $\Pr(-3.67 < Z < 3.00) = .9987$

6-9 .0154

6-11 .023

6-13 (a) 9000 and 900,000
(b) .147

6-15 (a) .014
(b) .008

6-17 .24

6-19 .018 (.023 without continuity correction)

6-21 (a) $(.309)^5 = .0028$
(b) .131
(c) .131
(d) Since $850 = 170 \times 5$, (*b*) and (*c*) are asking exactly the same event. On the other hand, event (*b*) occurs whenever (*a*) occurs, and some other times as well.

6-23 (a) Equally
(b) $2n$, $3.92\sqrt{n}$
200, 39.2

6-25 (a) $\Pr\left(|Z| > \frac{100}{\sqrt{(200)(8.5)}}\right) = .016$

6-27 (a)

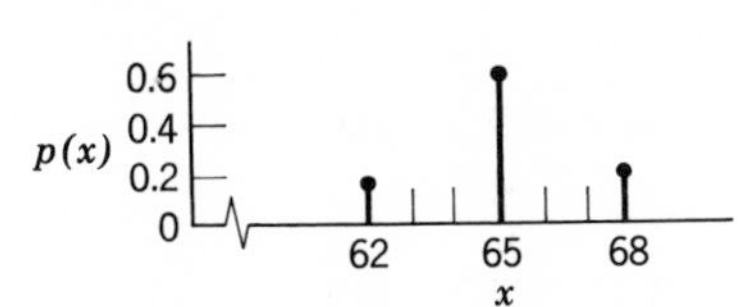

$\mu = 65$, $\sigma^2 = 18/5 = 3.6$

(b)

$\bar{x}$	$p(\bar{x})$
63.5	.3
65	.4
66.5	.3

(c)

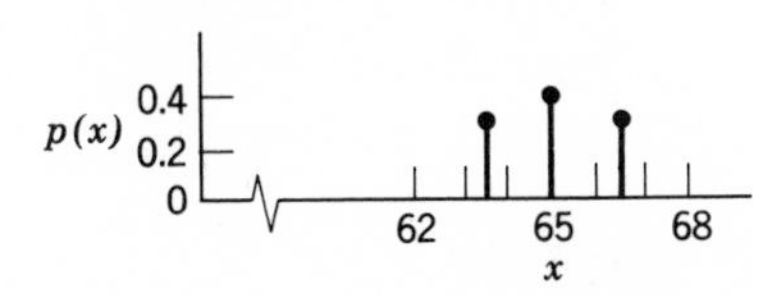

(d) $\mu_{\bar{X}} = 65 = \mu$

(e)

$$\sigma^2_{\bar{X}} = 1.35 = \frac{\sigma^2}{2}\left(\frac{3}{4}\right)$$

(f) $\sigma^2_{\bar{X}} = \frac{\sigma^2}{n}\left(\frac{N-n}{N-1}\right) < \sigma^2$

7-1 (a) $71 \pm 1.96 \frac{(3)}{\sqrt{100}} = 71 \pm .59$

7-3 $.83 \pm .032$

7-5 (a) $20(\frac{19}{20})^{19}(\frac{1}{20})$

(b) $(\frac{19}{20})^{20}$

(c) 1 − (sum of answers above)

Answers (*a*) and (*b*) can be roughly approximated by the normal, as .39 and .30 (or .243, if you like). The correct values are .36 and .38, respectively.

7-7 (b) 9/10. Hence $\bar{X}$ is preferable.

7-9

$\bar{x}$	$p(\bar{x})$	$\bar{x}p(\bar{x})$
2	1/9	2/9
3	2/9	6/9
4	3/9	12/9
5	2/9	10/9
6	1/9	6/9

(a) $E(\bar{X}) = 36/9 = \mu$

(b)

$\bar{x}$	$2\bar{x}+1$	$(2\bar{x}+1)p(\bar{x})$
2	5	5/9
3	7	14/9
4	9	27/9
5	11	22/9
6	13	13/9

$E(2\bar{X}+1) = 81/9 = 2\mu + 1$
unbiased

(c)

$\bar{x}$	$(\bar{x})^2$	$(\bar{x})^2p(\bar{x})$
2	4	4/9
3	9	18/9
4	16	48/9
5	25	50/9
6	36	36/9

$E(\bar{X}^2) = 156/9 \neq \mu^2$
Bias $= E(\bar{X}^2) - \mu^2$
$= 156/9 - 4^2$
$= 4/3$

(d) Similarly,

$E(1/\bar{X}) \neq 1/\mu$
Bias $= .274 - .250$
$= .024$

Theoretically,

(a) unbiased, by (6-10)

(b) unbiased, by Table 4-2

(c) biased; by (4-5), for any random variable:

$E(X^2) - \mu^2 = \sigma^2$

In particular, for $\bar{X}$:

$E(\bar{X}^2) - \mu^2 = \sigma^2_{\bar{X}}$

i.e., bias $= \sigma^2_{\bar{X}}$
$= 4/3$

7-11

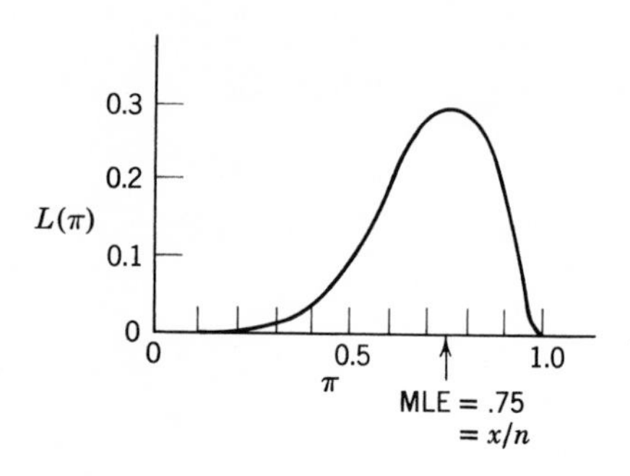

7-13 (a) $\frac{1}{n}\sum (x_i - \mu)^2$

(b) Yes. Proof:

$$E\left\{\frac{1}{n}\sum (X_i - \mu)^2\right\}$$

$$= \frac{1}{n}\sum E(X_i - \mu)^2$$

$$= \frac{1}{n}\sum \sigma^2$$

$$= \sigma^2$$

8-1 $1 \pm 1.96\sqrt{.22} = 1 \pm .92$

8-3 (a) $9 \pm 1.96\sqrt{136/60} = 9 \pm 2.96$

(b) Factor of 4, so that $n = 240$.

(c) $n = 522$

8-5 $69 \pm 1.98(4/\sqrt{100}) = 69 \pm 7.9$

8-7 Supposing σ and μ are both unknown, use $\bar{x} \pm 2.776s/\sqrt{5}$.

8-9. $2 \pm 2.048\sqrt{750/28}\sqrt{1/10 + 1/20}$
$= 2 \pm 4.1$

8-13 $.4820 \pm 1.96\sqrt{(.482)(.518)/10{,}000}$
$= .4820 \pm .0098$

8-15 $.04 < \pi < .49$

8-17 (a) $.1992 \pm 1.96\sqrt{(.199)(.801)/2500}$
$= .1992 \pm .0157$

$.1992 \pm .98/\sqrt{2500}$
$= .1992 \pm .0196$

(b) No. The closer P is to .5, the closer approximation will (8-21) be.

8-19 $.06 \pm 1.96\sqrt{\frac{(.72)(.28)}{100} + \frac{(.66)(.34)}{100}}$
$= .060 \pm .128$

8-21 $.39 \pm 2.58\sqrt{\frac{(.79)(.21)}{1000} + \frac{(.40)(.60)}{300}}$
$= .390 \pm .080$

8-23 $\frac{47.3}{3.12} < \sigma^2 < \frac{47.3}{.072}$ i.e., $15.2 < \sigma^2 < 658$

8-25 $4 \pm 2.776\sqrt{52/4}\sqrt{1/3 + 1/3}$
$= 4 \pm 8.17$

8-27 $(\pi_1 - \pi_2) = (.142 - .114)$

$$\pm 1.96\sqrt{\frac{(.142)(.858)}{2500} + \frac{(.114)(.886)}{2500}}$$

i.e., $\pi_1 - \pi_2 = .028 \pm .018$

Thus $\frac{\Delta\pi}{\pi} = \frac{.028 \pm .018}{.142}$

i.e., relative decline = 20% ± 13%

Although the best guess for the decline is one-fifth, when sampling fluctuation is allowed for, with 95% confidence we can only say that the relative decline was between 7% and 33%.

9-1 I, II

9-3 (a) Reject H_0 iff $\frac{P - .50}{\sqrt{\frac{(.5)(.5)}{100}}} > .67$

i.e., $P > .533$

(b) About 25% of the students will make erroneous rejections of H_0.

(c) .092

9-5 (a) Reject H_0 iff

$$\frac{\bar{X} - 8.5}{\sqrt{1/49}} > 1.645 \text{ i.e., } \bar{X} > 8.74$$

Since $\bar{X} = 8.8$, reject H_0. (It would be more accurate to use the t critical value of 1.68.)

(b)

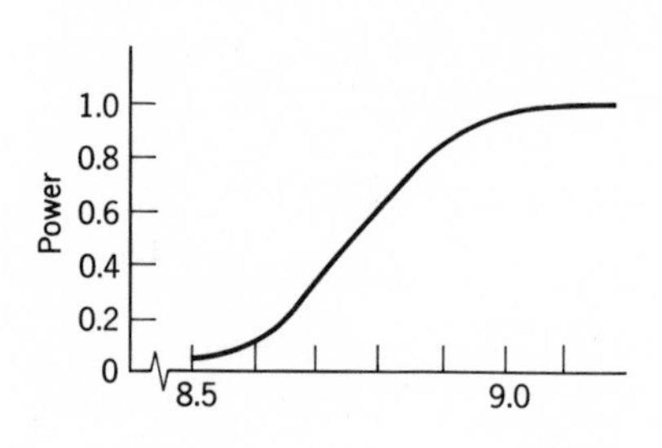

9-7 (a) Prob-value = .04 ($z = 1.77$) i.e., if the claim ($6600) is true, the chance of getting a sample as extreme as this ($6730) is only 4%.

(b) Yes.

(c) I would not reject. However, if possible, I would avoid accepting H_0, in order to avoid the risk of a type II error.

9-9 .007 ($z = 2.48$)

9-11 Reject H_0 ($z = 2.34$ and prob-value $= .010$).

9-13 (a) Three answers:

(i) Using the normal approximation, (which is very rough), reject H_0 if

$P < .19$ or $P > .81$

(ii) Using Fig. 8-4, (which is also rough), reject H_0 if

$P < .14$ or $P > .86$

(iii) Since P is very discrete (tenths), it is better to use the binomial Table II. It is seen that a 5% test is not possible. The best that can be done is a 2.16% test:

Reject H_0 if $P = 0, .1, .9$, or 1.0, i.e., if $P \leq .1$ or $P \geq .9$

(b) Reject H_0 if

$P < .402$ or $P > .598$

Again, because of the discrete nature of P, it would be better to state the answer:

Reject H_0 if

$P \leq .40$ or $P \geq .60$

Then α is found by continuity correction to be 5.7% ($z = 1.90$).

9-15 (a) reject ($t = 8.2$)
accept ($t = .21$)
reject ($t = 2.60$)

(b) $5726 < \mu < 6334$
Therefore reject, accept, reject.

9-17 (a) $12,100 < \mu$

(b) 18.4%

(c) You cannot reject H_0, for either reason (*a*) or (*b*).

9-19 (a) $.080 \pm .043$

(b) prob-value $< .001$ ($z = 3.7$)

(c) The sample difference is statistically significant at the 5% level.

(d) The sociological significance of the difference in populations is a relative matter.

9-21 (a) $1 \pm .64$

(b) .002 ($z = 3.08$)

(c) Yes

10-1 (a) $t = \dfrac{7}{\sqrt{\left(\dfrac{44}{6}\right)\left(\dfrac{4}{4}\right)}} = 3.68$

exceeds $t_{.025} = 2.45$. Therefore reject H_0.

$$F = \frac{4 \times 24.5}{44/6} = 13.36$$

exceeds $F_{.05} = 5.99$. Therefore reject H_0.

(b) Since $t^2 = F$ and $t^2_{.025} = F_{.05}$ we see in this particular case that the test using t gives exactly the same conclusion as the test using F. Mathematicians have proved in general that whenever F has 1 and k df, and t has k df, then t^2 and F have exactly the same distribution.

(c) $7 \pm 2.45\sqrt{44/6}\sqrt{2/4} = 7 \pm 4.7$

10-5 $F = \dfrac{50(114/18)}{(329/3)} = 2.89$

which falls short of $F_{.05} = 3.06$. Therefore do not reject H_0.

10-7 (a) *hour factor:*

$$F = \frac{27}{10/12} = 32.4$$

exceeds $F_{.05} = 6.94$. Therefore reject H_0.

Man factor:

$$F = \frac{16/3}{10/12} = 6.4$$

falls short of $F_{.05} = 6.94$. Therefore do not reject H_0.

(b) For hour factor, the confidence allowance is ± 2.77 for the following differences in μ:

i \ I	1	2	3
1		−3*	3*
2			6*
3			

10-9 95% confidence interval:

$$\mu_Y - \mu_X = 6 \pm 2.77\sqrt{8/5} = 6 \pm 3.5$$

$\therefore$ reject H_0 at 5% level.

11-1 (a) $S = 760 + \dfrac{2.6}{18} y$

$= 760 + .144\, y$

or $= -396 + .144\, Y$

(b) $a = \$760$ = estimate of savings of the average person.

$a_0 = \$-396$ = estimate of savings of a person with zero income. However, this is extrapolating recklessly.

11-3 (a) .068 bushel (all units are "per acre").

(b) Not economical (net return = 13.6¢ − 25¢).

(c) 13.6¢

11-5 (a) $S(a_0, b) = \sum (Y_i - a_0 - bX_i)^2$

(b) $\dfrac{\partial S}{\partial a_0} = -2 \sum (Y_i - a_0 - bX_i) = 0$

$\dfrac{\partial S}{\partial b} = -2 \sum X_i(Y_i - a_0 - bX_i) = 0$

(c) $a_0 = -396$, $b = .144$, as before

(d) The method in the text is easier than the method in this problem.

(c) 878

(d) $s^2 = 7863/2 = 3931$.

(e) Two degrees of freedom for s^2 are almost too few. It would be better to collect more data. This scarcity of data is even more acute in Problem 13-3.

12-1 (a) $\beta = \frac{2.6}{18} \pm 3.18\sqrt{\frac{.0388}{18}}$

$= .144 \pm .148$

(b) $\beta^* = \frac{15.4}{18} \pm 3.18\sqrt{\frac{.0388}{18}}$

$= .856 \pm .148$

Note that $\beta^* = 1 - \beta$, and the error allowances for β^* and β are the same.

12-3 $t = \frac{2.6/18}{\sqrt{.0388/18}} = 3.11$

which falls short of $t_{.01} = 4.54$. Therefore do not reject H_0.

12-5 It is preferable to observe i in a period of wide fluctuation.

13-1 (a) $S = 760 + .115y - .029w$

(b) Coefficient of y is .115, which is less than the former value, .144. The multiple correlation coefficient is the proper measure of "the relation of S to Y, other things being equal." The simple correlation coefficient measures the relation of S to Y, taking no account of W. In fact, W is (negatively) correlated with both S and Y, and thereby produces a misleadingly high correlation between S and Y themselves.

13-3 (a) (1) $760 = a$

(2) $2.6 = 18b - 18c - .007d$

(3) $-6.3 = -18b + 144c + .024d$

(4) $-.0017 = -.007b + .024c + .000010d$

(b) $a = 760$

$b = .1054$

$c = -.0242$

$d = -38.1$

13-5 (a) 36.7

(b) 25.5, which is much better.

13-7 (a) $S = \frac{269}{8} - \frac{52.5}{42}(T - 7.5)$

$= 33.6 - 1.25(T - 7.5)$

(b) There is serious bias caused by the fact that we started at a seasonal high (Christmas), so that of course the time trend is downwards.

13-9 $\beta = -8 \pm -2.45\sqrt{48/2}$

$= -8 \pm 12.0$

13-11 Make β is better by .38 mpg.

14-1 (a) $24/\sqrt{(44)(34)} = .62$
(b) $-.49 < \rho < .95$
(c) No

14-3 (a) $35^2/(100)(20) = .62$
(b) .38
(c) $F = 4.7$ while $F_{.05} = 10.13$
$t = 2.2$ while $t_{.025} = 3.18$
$\beta = .35 \pm .51$

For any of these 3 reasons, do not reject H_0.

14-5 (e) alone is false, and should be:
"If $b < 1$, no strict conclusion can be drawn about b_*."

14-7 (a) .874
(b) .982
(c) .992
(d) .76, .984
(e) $R^2 \geq r^2$ necessarily
(f) $r_{SY \cdot W}$

14-9 (a) .22
(b) .016
(c) 1, 2
(d) 28
(e) -5.3

15-1 (a) .1, .4, .5
(b) .28, .44, .28

15-3 (a) a_1
(b) a_1
(c) a_3
(d) False. It should be:
"Action a_1 is best when 'rain' is predicted, and also when no prediction is possible. Action a_3 is best when 'shine' is predicted. Action a_2 is never best."

15-5 (a) Midrange, median, mean, mode.
(b) Correct.

15-7 (a) Mode, 73 or 74
(b) Median, 73 to 74
(c) Mean, 73.5

15-9 Closer to 20, because the data are twice as reliable.

15-11 (a) is less believable, because it puts complete faith in a very small and unreliable sample.

15-13 (a) 103.54, $\alpha = .33$, $\beta = .020$
(b) 113.12, $\alpha = .05$, $\beta = .195$
Average loss increases by a factor of 3.16.
(c) $r_0/r_1 = 4/1$, which is unreasonable.

15-15 (a) You do not want to play, because the value of the game is 10/8 to me, which I could win by using the strategy mix: H played 5/8 of the time, T played 3/8.
(b) Each play H and T equally often, which results in a zero payoff. I would secretly choose my penny only if my opponent was also secretly choosing *and* seemed easy to outwit.

Glossary of Important Symbols

Symbol	*Meaning*	*Definition or Other Important Reference*
(a) ENGLISH LETTERS		
a	estimated regression intercept	(11-7), (11-13)
ANOVA	analysis of variance	Table 10-6
b	estimated regression slope	(11-7), (11-16), (12-13)
B	bias	(7-12)
c	number of columns in analysis of variance, or	(10-27)
	estimated regression coefficient	(13-3)
C	constant coefficient in a contrast	(10-22)
C^2	modified chi-square variable	(8-23)
d.f.	degrees of freedom	(8-11)
e	regression error	(12-3), (12-4)
E	(also F, G, etc.) = event	(3-6)
$\bar{E}$	not E	(3-17)
$E(X)$	expected value of $X = \mu_X$	(4-17b)
F	variance ratio	(10-7), (10-17), (10-28), (14-24)
H_0	null hypothesis	(9-1)
H_1	alternate hypothesis	(9-2)
iff	if and only if	
$L(\ )$	likelihood function	(7-24), (12-48)
MLE	maximum likelihood estimate(tion)	Table 7-2
MSD	mean squared deviation	(2-5), (7-13)

Symbol	*Meaning*	*Definition or Other Important Reference*
MSS	mean sum of squares	Table 10-6
n	sample size	(6-17)
N	population size	(6-17)
$N(,)$	normal distribution, with specified mean and variance	(6-31)
P	sample proportion	(1-2), (6-28)
Pr (E)	probability of event E	(3-1)
Pr (E/F)	conditional probability of E, given F	(3-22)
$p(x)$	probability function of X	(5-5)
$p(x, y)$	joint probability function of X and Y	(5-2b)
$p(x/y)$	conditioned probability function of X, given $Y = y$	(5-10)
r	simple correlation, or	(14-4), (14-16)
	number of rows in analysis of variance	(10-27)
r^2	coefficient of determination	(14-29)
$r_{XY.Z}$	partial correlation of X and Y, if Z were held constant	(14-39), (14-40)
R	multiple correlation, or decision rule	(14-43), (14-44) (9-3)
s^2	variance of sample, or residual variance in regression	(2-6) (12-24)
s_p^2	pooled variance of samples	(8-16), (10-26)
S	sample sum	(5-16), (6-2)
SS	sum of squares, or variation	Table (10-6)
t	student's t variable	(8-10), (12-26)
T	time	(13-24)
var	variance $= \sigma^2$	(5-32)
W	weighted sum	(5-30)
X	(also Y, V, W, etc.) = random variable, or	(4-1), (4-2)
	regressor in original form	(11-4)
x	(also y, v, etc.) = value of X, or regressor in terms of deviations from the mean	(Fig. 4-2) (11-5)
$\bar{X}$	sample mean of X (note this is a different usage than $\bar{E}$)	(2-1), (6-9)

Symbol	*Meaning*	*Definition or Other Important Reference*
$\bar{x}$	(realized) value of $\bar{X}$. After Chapter 8 this distinction between capital and little letters is forgotten	(7-9)
Z	standard normal variable, or a second regressor	(4-13), (8-9) Table 13-1

(b) GREEK LETTERS are generally reserved for population parameters as follows:

α	probability of type I error, or population regression intercept	(9-18) (12-3)
β	probability of type II error, or population regression slope	(9-9) (12-3)
γ	population regression coefficient	(13-1)
θ	any population parameter	(7-11)
$\hat{\theta}$	sample estimator of θ	(7-11)
μ	population mean	(4-3), (4-10), (4-17a)
π	population proportion	(1-2), (4-7), (6-20)
$\prod$	product of	(7-30)
ρ_{xy}	population correlation of X and Y	(14-3)
σ	population standard deviation	(4-4)
σ^2	population variance	(4-4), (4-5), (4-19)
σ_{xy}	population covariance of X and Y	(5-21), (5-22), (5-23)
$\sum$	sum of	Table 2-2

(c) OTHER MATHEMATICAL SYMBOLS

$E \cup F$	E or F, or both	(3-10)
$E \cap F$	E and F	(3-11)

$\triangleq$	equals, by definition	(2-1a)
$\simeq$	approximately equals	(2-1b)
$\sim$	is distributed as	(6-31)

(d) GREEK ALPHABET

Letters	*Names*	*English Equivalent*	*Letters*	*Names*	*English Equivalent*
$\mathrm{A}\alpha$	Alpha	a	$\mathrm{N}\nu$	Nu	n
$\mathrm{B}\beta$	Beta	b	$\Xi\xi$	Xi	x
$\Gamma\gamma$	Gamma	g	$\mathrm{O}o$	Omicron	o
$\Delta\delta$	Delta	d	$\Pi\pi$	Pi	p
$\mathrm{E}\epsilon$	Epsilon	e	$\mathrm{P}\rho$	Rho	r
$\mathrm{Z}\zeta$	Zeta	z	$\Sigma\sigma$	Sigma	s
$\mathrm{H}\eta$	Eta	—	$\mathrm{T}\tau$	Tau	t
$\Theta\theta$	Theta	—	$\Upsilon\upsilon$	Upsilon	u or y
$\mathrm{I}\iota$	Iota	i	$\Phi\phi$	Phi	—
$\mathrm{K}\kappa$	Kappa	k	$\mathrm{X}\chi$	Chi	—
$\Lambda\lambda$	Lambda	l	$\Psi\psi$	Psi	—
$\mathrm{M}\mu$	Mu	m	$\Omega\omega$	Omega	—

Index